Zünkler

Electrical Power Systems for Lunar and Space Exploration

Jan-Henrik Zünkler

Electrical Power Systems for Lunar and Space Exploration

Fundamentals and Design

HANSER

About the author:
Jan-Henrik Zünkler, Nieheim

Print-ISBN: 978-1-56990-413-8
E-Book-ISBN: 978-1-56990-478-7

Bibliographic information of the German National Library:
The German National Library lists this publication in the German National Bibliography; detailed bibliographic data are available on the Internet at http://dnb.d-nb.de.

Vilshofener Straße 10 | 81679 Munich | info@hanser.de
www.hanserpublications.com
www.hanser-fachbuch.de
Editor: Julia Stepp
Production Management: Melanie Zinsler
Cover concept: Marc Müller-Bremer, www.rebranding.de, München
Cover design: Max Kostopoulos
Cover picture: © shutterstock.com/Dabarti CGI
Typesetting: Eberl & Koesel Studio, Kempten

Contents

Preface

In 2019, I joined the think tank Spaceship EAC at the European Astronaut Centre (EAC) for an internship as part of my master's studies. My project was the development of a new sensor system utilizing technologies from the Internet of Things (IoT). The EAC is located at the German Aerospace Center (DLR) in Cologne, and for me, this environment, as well as our focus on the human exploration of the Moon, was highly inspiring. When my internship ended, I went back to university to finish my last courses. As my project with EAC went very well, I was offered to work on my master's thesis with Spaceship EAC as a visiting researcher. Because I studied electrical engineering and industrial engineering with a focus on Electrical Power Systems (EPSs), Spaceship EAC asked me to work on a software-based simulation system for lunar Electrical Power Systems. For Spaceship EAC, the intention was to move from hardware-based research to software-based research. However, the scope of software projects often expands during development. This was also true for my project, and in the end, it took me from the beginning of 2020 to mid-2023 to develop the first complete version of my simulation software. During that time, the need for such a software solution was also discussed in the space community. So I was invited by the German Aerospace Center and by the Technical University of Berlin to present my project.

As one can imagine, the scope of this development project was much too big for what was initially intended to be a master's thesis. After completing the development, I wrote down everything I did and ended up with around 1000+ pages. I discussed the situation with my professor, and we both agreed on handing in Chapter 2 of the work as my master's thesis. This chapter was about the simulation of environmental parameters influencing the operation of photovoltaic generators on the lunar surface, as well as about a global study of different parameters. In addition to this, my professor offered me to hand in the last chapter, which was about the simulation of Electrical Power Systems on the Moon, as a PhD thesis. I followed this path and was enrolled as a PhD student immediately after completing my master's degree.

At that time, however, there was still one more chapter left in the original work I had written. It was the introductory chapter to the Artemis program, to the lunar environment and to Electrical Power Systems (EPSs). Since the chapters on Electrical Power Systems and their subsystems are typically not very detailed in standard literature for space engineers, and since there was no book available at that time focusing on Electrical Power Systems for space and lunar exploration, I reached out to my favorite publisher Hanser, who was highly interested in my work and decided to publish it.

This book focuses on the design of possible EPSs for crewed lunar missions and a lunar base within the Artemis program to enable an optimal system development. It provides the necessary mission-specific background, the required engineering background and related design methods. It is aimed at engineers and engineering students who need information on Electrical Power Systems in space. Although the focus is on the Artemis program, the book provides general information on space Electrical Power Systems. Chapter 1 gives a general introduction to the book. Chapter 2 focuses on lunar missions within the Artemis program, related vehicles and space systems. The lunar environment is discussed in Chapter 3. Chapter 4 covers possible EPS technologies, as well as subsystems for lunar EPSs. The Electrical Power System of the International Space Station (ISS), as well as possible systematic options for lunar EPSs, are discussed in Chapter 5. In Chapter 6, related design methods for lunar and space EPSs and EPS subsystems can be found. A list of formula symbols (Chapter 7) and a bibliography (Chapter 8) round off the contents.

However, I think at this point it is important to note that the Earth is currently the only known planet where humans can live. When giving tours at EAC, I often asked visitors what their favorite planet was. Mars, Saturn and Jupiter were among the top answers, but no one ever answered Earth. Working on the exploration of the solar system has made me even more aware of the luxury we have living on this planet, where, from a space engineering perspective, a livable environment, clean breathing air, and water are free and available to everyone. If we were looking for other planets outside the solar system to colonize in the future, we would be looking for an exact copy of the Earth. We should not forget what we have here and what we are about to lose. While building bases on other celestial bodies or the terraforming of Mars could be a cool vision for the future, this won't protect us from a dying planet Earth.

June 2025

Jan-Henrik Zünkler

1 Introduction

Over the years, human space exploration has had different focuses and has taken place in different locations. The focus of missions included technology verification, in-orbit construction, as well as science activities in fields like planetary science, remote observation, space science, and zero-gravity science. Besides the lunar missions within the Apollo program and the Hubble missions, human space exploration mainly took place in Low Earth Orbit (LEO). The construction of space stations, such as the Mir or the International Space Station (ISS), implemented laboratories in LEO, which enabled constant and ongoing research activities. Today, human space missions are focused on the ISS and experiments in the space and zero-gravity environment. A picture of the ISS taken from a spacecraft in front of it is shown in Figure 1.1.

Figure 1.1 Picture of the International Space Station (ISS) taken from a spacecraft in front of it (image: © ESA/NASA/T. Pesquet [1])

The ISS has been in operation for a long period of time. Because of this, degradation in different types of systems can be observed. Degradation in some systems is uncritical, as long as these systems can be replaced in a relatively simple way. There are also other parts of the ISS, though, that cannot be replaced in a simple way. Examples of such systems are the load-bearing structures that have critical parts of oxidation and the protection system against impacts that is slowly degrading over time. As the ISS is approaching the end of its operational period, the international partner agencies are currently discussing the future of human spaceflight. It is intended to move the location, on which human space missions are focused, from LEO to the near lunar environment. NASA's Artemis program will focus on human exploration in lunar orbit and on the lunar surface, together with international partner agencies like the European Space Agency (ESA), the Canadian Space Agency (CSA) and the Japanese Aerospace Exploration Agency (JAXA). Current plans include crewed landings on the lunar surface, as well as the implementation of the space station Lunar Gateway in lunar orbit. Figure 1.2 shows a visualization of the future Lunar Gateway.

Figure 1.2 Visualization of the future Lunar Gateway (image: © NASA [5])

The Artemis 1 Mission was an uncrewed technology verification mission employing the Orion spacecraft on a trajectory around the Moon. Currently, the focus lies on the missions Artemis 2 to 5. The first crewed lunar mission within the Artemis program will be Artemis 2, which will be a flyby mission around the Moon featuring a crew of 4 astronauts. The first mission within the Artemis program including a landing on the lunar surface will be Artemis 3. Possible landing sites on the lunar south pole for the Artemis 3 mission have already been identified. Later missions will focus on the construction of the Lunar Gateway, which will from then on be employed in lunar surface missions as an interim step, before the descent to the surface. With the Artemis 4 mission, some of the initial Lunar Gateway modules will be delivered to lunar orbit. With reference to industry contracts, up to 13 Artemis missions and the implementation of the Lunar Gateway are currently scheduled.

The Lunar Gateway will be developed and implemented in collaboration of the international partner agencies. Resulting from its location in lunar orbit, it will be less accessible than the ISS. The altitude of the ISS in LEO has a value of around 400 km. Contrary to that, the distance between the Earth and the Moon varies between 363,300 km and 405,500 km and the Lunar Gateway will be implemented in a highly eccentric lunar polar orbit featuring an altitude between 3000 km and 70,000 km above the surface [10] [11]. The Lunar Gateway will be much smaller than the ISS and it will also not be crewed permanently. It is intended to utilize the Lunar Gateway as an orbital science laboratory and as an interim step in crewed lunar surface missions. Science experiments on board of the Lunar Gateway will be mainly teleoperated. The Orion spacecraft will be employed within the Artemis program for the transfer of astronauts from the Earth to the Lunar Gateway and back. During a mission, two astronauts will stay on board of the Lunar Gateway and carry out science, construction and maintenance activities. The remaining two astronauts will transfer from the Lunar Gateway to a lunar landing vehicle, docked to the Lunar Gateway, and decent to the lunar surface on board of this vehicle. Currently, the construction of a lunar base remains a vision for later Artemis missions, and initial lunar surface missions are designed without a need for a base or other infrastructure on the lunar surface. A visualization of a possible lunar base is shown in Figure 1.3.

Figure 1.3 Visualization of a Moon base concept (image: © ESA [13])

Research activities on the lunar surface will focus on the lunar environment, lunar geo-physics, lunar geology, heliophysics, or radio astronomy. Another interesting topic could be the study of lunar resources and their utilization in In-Situ Resource Utilization (ISRU) processes. Lunar surface missions are also interesting for engineering in order to test systems for human deep space exploration in an environment that offers a fast return to the Earth. Engineering challenges on the Moon will be the extreme temperature range from around –178.2 °C to around +123.9 °C, the extreme durations of the lunar day and night in a magnitude of around 328 hours in most locations, radiation, dust mitigation, as well as the protection against impacts [11] [14] [18] [19] [134]. Regions on the lunar south pole are of particular interest for the upcoming Artemis missions. These regions offer extended lunar day durations in special locations. Especially lunar craters are of potential interest, since these can contain resources like frozen water.

Space systems are usually developed by matching the mission goals with the environmental conditions during a mission, the technical system boundaries and the mission specific boundaries. The process of matching mission goals, requirements, and limitations takes a number of design iterations and is usually carried out by engineers and mission specialists. Power systems play an important role in all space missions because these systems are required in order to supply other systems with electrical energy, without interruptions and during all mission phases. However, power systems also have a relatively great amount of mass and volume, which is limited by the capacity of a certain launch vehicle. Energy is generated during the mission employing solar power systems in the majority of space missions, but some missions also rely on other systems like nuclear-based generators. Besides the generative subsystems, electrical power systems consist of a complex arrangement of different subsystems, providing different functionalities like power conditioning, power distribution, the transformation between different voltage levels or energy storage functionalities.

2 Space and Lunar Exploration

The crewed exploration of the Moon started with NASA's Apollo program in the 1960s and 1970s. Figure 2.1 shows an image taken during the Apollo 11 mission and another one taken during the Apollo 17 mission. While the duration of stay on the lunar surface was limited to up to 3 days within the Apollo program, longer durations of stay on the surface will be scheduled for the upcoming lunar missions within the Artemis program and the ultimate vision for the exploration of the lunar surface is a permanently crewed base on the lunar surface [4] [14] [20]. This chapter focuses on future lunar missions. General lunar mission topologies, currently scheduled lunar missions, employed technologies, scientific goals, and mission goals as well as basic parameters for the design of lunar power systems, are pointed out.

Figure 2.1 Astronaut E. B. Aldrin (Apollo 11, left) and scientist astronaut H. H. Schmitt (Apollo 17, right) during Extravehicular Activities (EVAs) on the lunar surface (image on the left: © NASA, JSC, E. B. Aldrin; image on the right: © NASA, JSC, H. H. Schmitt [21] [22])

2.1 Design Reference Missions

Mission design studies the combination of different mission types, mission topologies and technologies with focus on maximizing mission return and minimizing the required input factors. The Exploration Systems Architecture Study (ESAS) was carried out by NASA in 2005. A number of different Design Reference Missions (DRMs) were developed for human exploration missions to deep space during this study. DRMs can be defined as basic mission concepts fulfilling different explorational requirements and goals. Three DRMs were developed for crewed missions to the lunar surface within the ESAS study. These did not take into account the utilization of a space station in lunar orbit. The DRMs developed for the ESAS study utilized the vehicles developed for the Constellation program. These vehicles were developed with focus on the exploration of the Moon and Mars. The vehicles were utilized as the basis for the design of the space systems that will be utilized in the Artemis program.

The Design Reference Mission 1 (DRM 1) has a relatively short duration of stay in order to enable a global access to the lunar surface. DRM 1 fulfills the mission goal of an exploration of areas that are of high scientific interest, independently of lunar surface infrastructure, as well as the scouting for possible lunar base construction sites. Missions on the surface include up to four astronauts who stay on the surface for periods of up to 7 days within a DRM 1 [15]. DRM 1 can have global destinations. Subsequently, DRM 1 need to take everything required for the mission to the lunar surface, including a Crew Habitation Module (CHM) and the required crew supplies for the stay on the surface. The Crew Habitation Module for the lunar surface is in this case referred to as the Surface Exploration Module (SEM). The complete topology of a DRM 1 is visualized in Figure 2.2.

This mission type requires two launches. The Earth Departure Stage (EDS) and the Lunar Surface Access Module (LSAM) are launched together on top of a heavy-lift Launch Vehicle (LV) to Earth orbit, while the Crew Exploration Vehicle (CEV), consisting of the Crew Exploration Module (CEM) and the Service Module (SM), is launched on top of a smaller Launch Vehicle to Earth orbit [15]. CEM and LSAM rendezvous and dock in Earth orbit. The Trans Lunar Injection (TLI) maneuver is carried out by the Earth Departure Stage (EDS) that is docked to the LSAM. After that, the EDS will be expended, and the crew will transit to the Moon. On arrival in lunar orbit, the LSAM will perform the lunar orbit insertion maneuver [15]. Following this, the crew will transfer to the LSAM and undock from the CEV. The CEV will stay in lunar orbit, while the crew on board of the LSAM will descent to the lunar surface. After landing on the lunar surface, the crew will carry out operational and scientific activities for up to 7 days. During this period, up to daily Extravehicular Activities (EVAs) will be scheduled in order to maximize the mission return. After the stay on the lunar surface, the crew will ascent to lunar orbit utilizing the Ascent Vehicle (AV), which is usually implemented as a part of the LSAM. The Ascent Vehicle is sometimes also referred to as

Ascent Stage (AS). In lunar orbit, Ascent Module (AM) and CEV will rendezvous and dock. After that, the crew will transfer to the CEV, and the Ascent Vehicle will be expended, so that no fuel for the transit of the AV to Earth needs to be reserved. Following this, the Service Module of the CEV will perform the lunar orbit departure maneuver and the crew will transit back to the Earth for several days [15]. Approaching the Earth, the Service Module will be expended and the Crew Exploration Module will directly enter the Earth's atmosphere, followed by a splashdown landing.

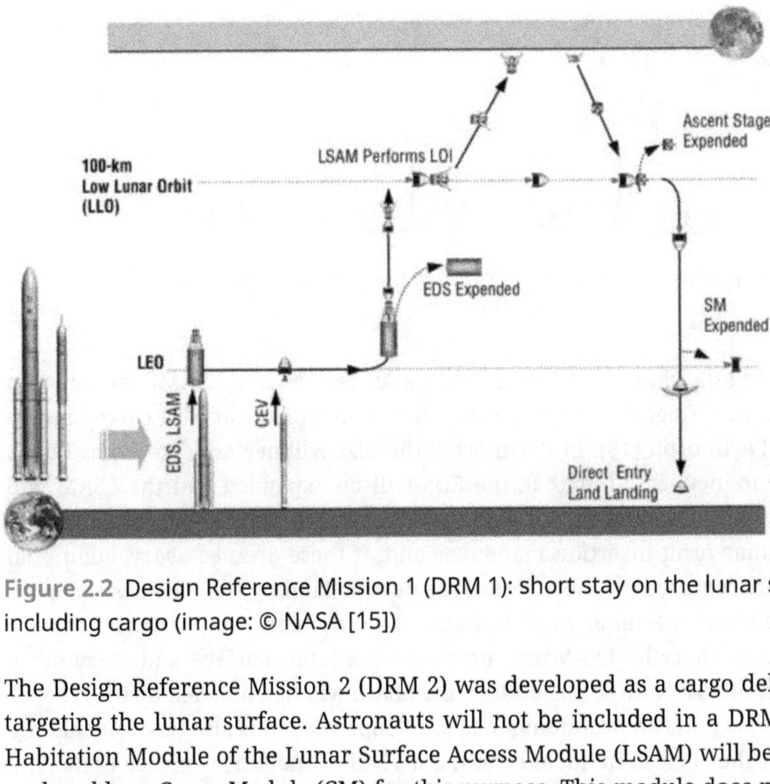

Figure 2.2 Design Reference Mission 1 (DRM 1): short stay on the lunar surface including cargo (image: © NASA [15])

The Design Reference Mission 2 (DRM 2) was developed as a cargo delivery mission targeting the lunar surface. Astronauts will not be included in a DRM 2. The Crew Habitation Module of the Lunar Surface Access Module (LSAM) will be utilized as or replaced by a Cargo Module (CM) for this purpose. This module does not necessarily need to be pressurized. Cargo missions can be employed to deliver construction material and infrastructure to the lunar surface during lunar base construction, such as lunar base modules, inflatable modules, robotic exploration systems, or other equipment. This type of mission can also be employed as a periodic logistic resupply mission in order to deliver crew supplies during the operational phase of a lunar base. In order to maximize the capacity of a Cargo Vehicle (CV), these vehicles will stay on the lunar surface and they will not return to the Earth or lunar orbit, so that the fuel mass for the ascent to lunar orbit and the return to the Earth does not need be reserved. Instead, the additional mass can be utilized as additional cargo capacity. The topology of a DRM 2 is visualized in Figure 2.3.

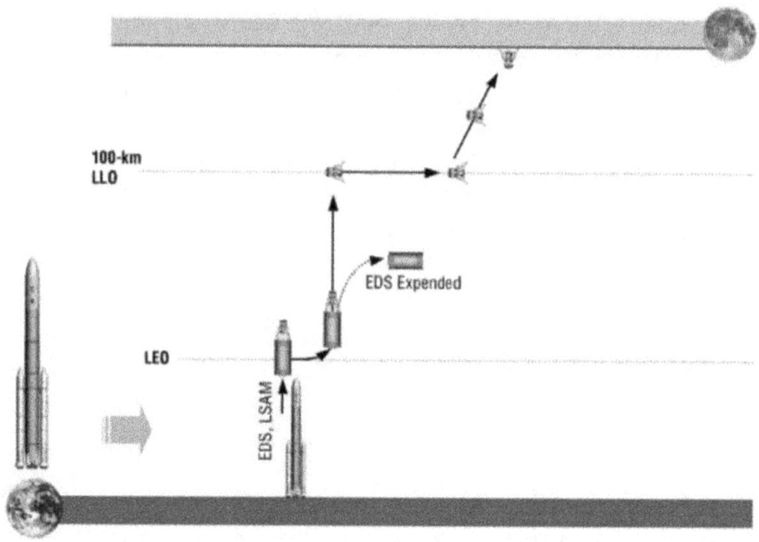

Figure 2.3 Design Reference Mission 2 (DRM 2): lunar surface cargo delivery (image: © NASA [15])

In the beginning of a DRM 2, the Lunar Surface Access Module (LSAM), mounted on the Earth Departure Stage (EDS), will be launched on top of a heavy-lift cargo Launch Vehicle to the Earth orbit [15]. In Earth orbit, the EDS will perform the Trans Lunar Injection (TLI) maneuver. After that, the EDS will be expended and the LSAM will transit to lunar orbit for a number of days. At the arrival in lunar orbit, the LSAM will perform the lunar orbit insertion maneuver and, if these are necessary, additional orbit maneuvers required for the access to the desired landing site. The desired landing site can be close to a lunar base, but also cargo missions to other locations are conceivable. After that, the LSAM will decent to the lunar surface and carry out a lunar landing maneuver. After the landing, the LSAM will be unloaded by the crew or by robotic systems that can be autonomous or teleoperated. It is also possible that the LSAM lands on the surface in advance to the arrival of the lunar surface crew [15]. Only a minor fuel reserve will be present in the LSAM tanks after landing. Because of this, possible future use cases for a cargo LSAM are limited.

The Design Reference Mission 3 (DRM 3) has a similar topology as the DRM 1, but it also offers the utilization of a previously constructed lunar base and longer durations of stay. Missions on the lunar surface featuring a duration of up to 6 months shall be possible within a DRM 3. Following this, a DRM 3 requires at least the integration of a minimum lunar infrastructure that allows crew habitation for the duration of a DRM 3. The construction could be carried out by robotic systems, but for some scheduled or unscheduled tasks, the employment of astronauts might also be required. In the beginning, the scope of the lunar facilities will be limited and astronauts will actively work on the construction and integration of the lunar base. Once the infrastructure and the lunar base are deployed, the main mission focus will shift to scientific activities. Because of this, in advance of the arrival of the first crew, at least one DRM 2 cargo delivery mission to the desired lunar base construction site will be required, which is delivering the infrastructure for the following DRM 3. Two possible return vehicles have to stay on the lunar surface during a DRM 3 for redundancy reasons. Because of this, another Lunar Surface Access Module (LSAM) has to be deployed on the lunar surface in an uncrewed mission, in advance of the arrival of the first crewed mission. This LSAM will serve as Ascent Vehicle for the following crewed mission. Since every crewed mission will also include a LSAM, the LSAM of a previous DRM 3 will serve as an Ascent Vehicle for the following DRM 3. The topology of a DRM 3 is visualized in Figure 2.4.

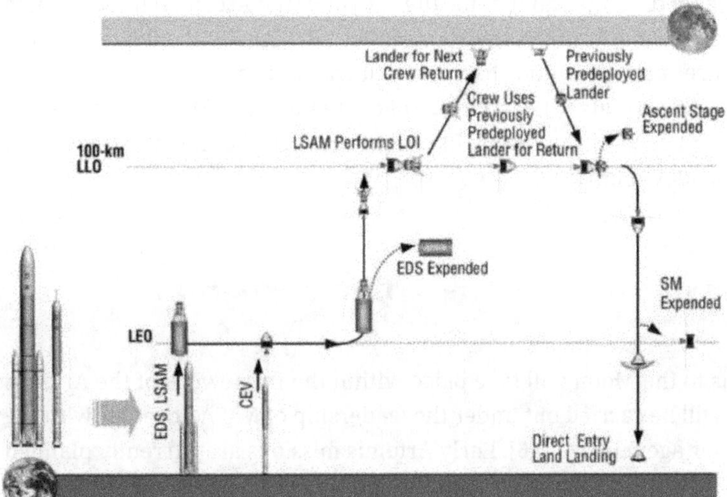

Figure 2.4 Design Reference Mission 3 (DRM 3): delivery of astronauts and cargo to the lunar surface (image: © NASA [15])

In the beginning of a DRM 3, the Lunar Surface Access Module (LSAM) will be launched together with the Earth Departure Stage (EDS) on top of a heavy-lift Launch Vehicle to Earth orbit. Briefly after that, the Crew Exploration Vehicle (CEV), crewed with up to 4 astronauts, will be launched on top of smaller LV to Earth orbit. In Earth orbit, the CEV will rendezvous and dock with the LSAM and the EDS. The EDS will perform the Trans Lunar Injection maneuver after system verification activities have been completed. This stage will be expended after that and the crew on board of the docked CEW and LSAM will transit to lunar orbit for the following days. On the arrival in lunar orbit, the LSAM will carry out the lunar orbit insertion maneuver. In case that additional maneuvers would be required in order to get access to the desired landing site, the LSAM will also carry out additional orbit maneuvers. Before the descent to the lunar surface, the crew will transfer to the LSAM and undock from the CEV, which will stay in lunar orbit for the duration of the surface mission [15]. The LSAM will then carry out the descent and landing maneuvers after that. After the landing on the lunar surface, the crew will depart from the LSAM and unload cargo. For the upcoming mission duration, the crew will inhabit the lunar surface infrastructure modules and the systems of the LSAM will shut down into a deep sleep state featuring a minimum of system operation. During the mission, the crew will carry out construction, maintenance and scientific activities. At the end of their mission on the lunar surface, the crew will transfer to the previously deployed LSAM. The ascent into lunar orbit will be carried out by the Ascent Vehicle, which can be implemented as LSAM or as an Ascent Stage (AS) submodule of it. In lunar orbit, the Ascent Vehicle will rendezvous and dock with the CEV. After that, the crew will transfer to the CEV, and the Ascent Vehicle will be expended, so that no fuel for the transit of the Ascent Vehicle to Earth orbit needs to be reserved. After that, the Service Module of the CEV will perform the lunar orbit departure maneuver and the crew will transit back to the Earth for the following days. At the arrival in near Earth space, the Service Module will be expended, the crew on board of the CEM will decent through the Earth's atmosphere and perform a splashdown landing.

2.2 Lunar Missions within the Artemis Program

Upcoming missions to the Moon will take place within the framework of the Artemis program and they will be carried out under the leadership of NASA, together with the international partner agencies [4] [16]. Early Artemis missions are currently planned and prepared. Artemis 1, an uncrewed mission to lunar orbit with focus on system verification, took place between November and December 2022. The Artemis missions 2 to 5 have already been scheduled. These future missions include astronaut crews as well as the implementation of the Lunar Gateway space station and they will be focused on the exploration of the lunar surface with astronauts. The construction of a permanent crewed lunar base still remains a vision for later phase missions.

2.2.1 Artemis Mission Topologies

Initial phase lunar missions and later phase lunar missions can be differentiated, whereby the later phase missions are sometimes also referred to as sustaining missions. Both mission types can be differentiated by the utilization of pre-deployed lunar infrastructure. Initial phase missions do not rely on lunar infrastructure and sustaining phase missions rely on the utilization of lunar infrastructure. Subsequently, everything required for initial phase missions needs to be brought on board of the utilized vehicle. Besides others, a mission goal of initial lunar missions will be technology demonstration. Later phase (sustaining) missions have the mission goal of the exploration of the lunar surface for longer durations building up on already existing lunar infrastructure.

In an Artemis mission, a crew of 4 astronauts will launch on top of the Space Launch System (SLS) Launch Vehicle (LV) on board of the Orion spacecraft. When a landing on the lunar surface is also part of the mission, a Human Landing System (HLS) has to be launched and transferred to lunar orbit in advance of the mission. In lunar orbit, the Orion spacecraft and the HLS will rendezvous and dock. After this, a part of the crew will descent to the lunar surface on board of the HLS. The Lunar Gateway space station will be utilized in later Artemis missions. During such a mission, the Orion spacecraft will rendezvous and dock with the Lunar Gateway in lunar Near Rectilinear Halo Orbit (NRHO). Half of the Crew will transfer to the Human Landing System, docked to the Lunar Gateway, and the other half of the crew will stay on board of the Lunar Gateway. Crew supplies, as well as required equipment for the exploration of the lunar surface, have to be delivered to the Gateway with additional cargo missions in advance of crewed missions.

An initial Artemis mission requires between 2 and 4 launches. One launch of the Space Launch System (SLS) will be required in order to deliver a crew of 4 astronauts on board of an Orion spacecraft to lunar orbit. Additional 1 to 3 launches will be required in order to deliver the Human Landing System (HLS) to lunar Near Rectilinear Halo Orbit (NRHO) in advance of the crew arrival. The HLS shall be equipped with Extravehicular Mobility Units (EMUs), crew supplies and science equipment. In lunar NRHO, the HLS will rendezvous and dock with the Orion spacecraft. Two astronauts will stay on board of Orion in lunar NRHO for the duration of the lunar surface mission, while the other two astronauts will descent on board of the HLS to the lunar south pole. Durations of stay of 6.5 days will be scheduled during initial phase Artemis missions. During the surface missions, between 3 and 5 Extravehicular Activities (EVA) will be scheduled in a radius of up to 2 km around the landing site. After the surface mission, the surface exploration crew will ascent back to lunar orbit with collected samples utilizing the HLS. In lunar orbit, the HLS will again rendezvous and dock with Orion in lunar NRHO. After that, the surface exploration crew will transfer back from the HLS to Orion, and Orion will undock from the HLS in NRHO and depart from lunar orbit. At the end of the mission, the Crew Exploration Module of Orion will perform a splashdown landing in the Pacific Ocean. The topology of the initial phase of the Artemis mission is visualized in Figure 2.5.

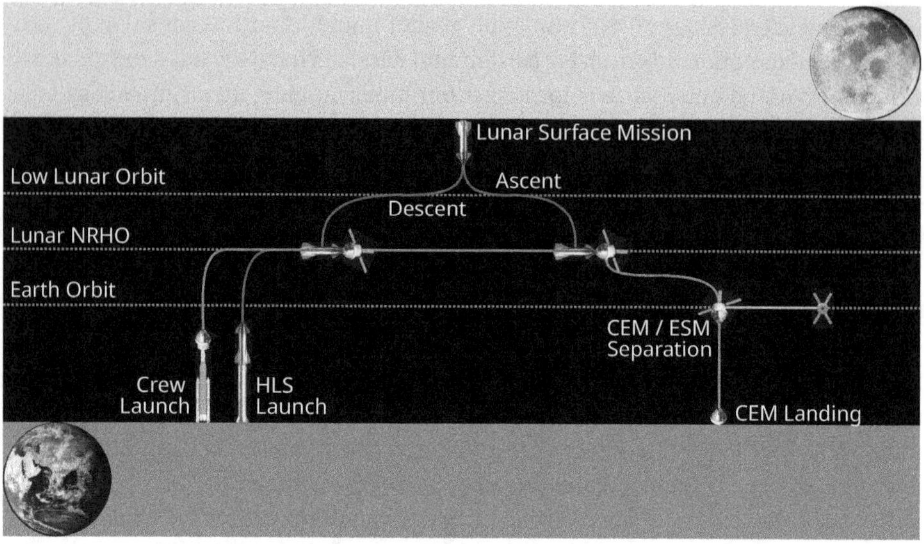

Figure 2.5 Topology of the initial phase of the Artemis mission (used images: © NASA, ESA, ESA – P. Carril, SpaceX [42] [43] [64] [80] [90] [91] [92])

Three different sustaining Artemis mission topologies have been developed. These include the utilization of the Lunar Gateway space station. The topology of the sustaining phase of the Artemis missions is visualized in Figure 2.6.

A "short stay on the lunar Poles" mission will focus on the exploration of the lunar poles. Possible mission goals are the collection of lunar samples, the deployment and construction of surface infrastructure, as well as maintenance tasks on already implemented infrastructure. Periods of up to 6.5 days on the lunar surface will be scheduled during this type of missions.

A "short stay in a non-polar location" mission will focus on the exploration of the lunar surface in non-lunar locations. Resulting from the longer transit duration from Near Rectilinear Halo Orbit to a non-polar location, only 1 to 2 extravehicular activities will be possible during this type of missions.

An "extended stay on the lunar poles" mission will focus on the exploration of the lunar surface during a period of up to 30 days. This mission type will be relying on lunar surface infrastructure systems that have to be constructed incrementally during previous lunar missions. After landing, a walking transfer using Extravehicular Mobility Units (EMUs) to the lunar habitat will be scheduled. The habitat could be a lunar base or a pressurized rover vehicle. Subsequently, one Extravehicular Mobility Unit (EMU) needs to be present in the HLS for each astronaut involved in the surface mission. The collection of lunar samples in the magnitude between 530 kg and 1.1 T will be planned during sustaining Artemis missions.

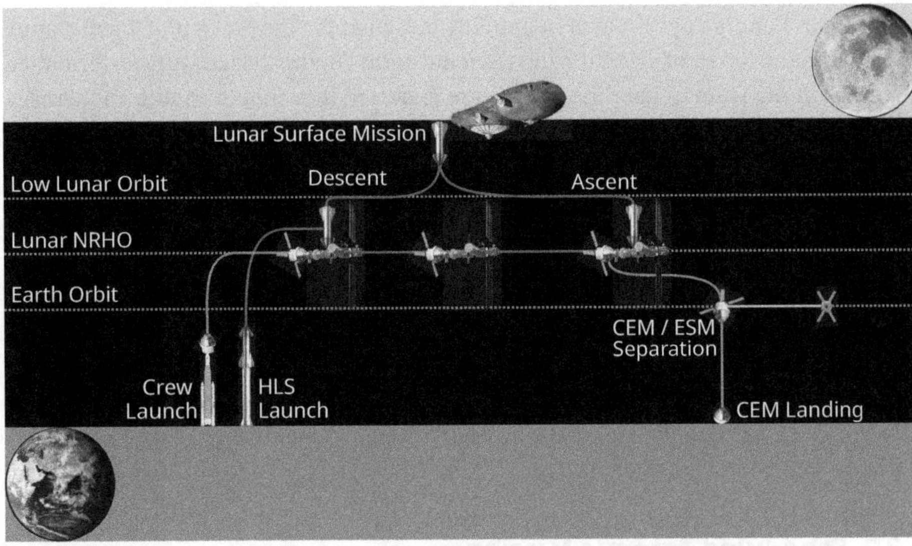

Figure 2.6 Topology of the sustaining phase of the Artemis mission (used images: © NASA, ESA, ESA – P. Carril, SpaceX [42] [43] [64] [80] [90] [91] [92]

Depending on the utilized approach for the Human Landing System, between 2 and 4 launches are required for a sustaining type of the Artemis mission. In advance of an Artemis mission, the Human Landing System has to be launched to lunar NRHO utilizing between 1 and 3 launches. In lunar NRHO, the HLS will rendezvous and dock with the Lunar Gateway. The crew of a sustaining type of the Artemis mission will launch on top of a Space Launch System Launch Vehicle inside the Orion Spacecraft to lunar NRHO, where the crew and Orion will rendezvous and dock with the Lunar Gateway. There, up to 4 astronauts will transfer to the Human Landing System and descent to the lunar surface as a surface exploration crew. The remaining crew on board of the Lunar Gateway will carry out science, construction and maintenance tasks during their mission in lunar orbit. After their mission, the surface exploration crew will ascent again to lunar NRHO utilizing the Human Landing System, where they will rendezvous and dock with the Lunar Gateway. On board of the Lunar Gateway, the crew will transfer back to the Orion spacecraft and depart from the Lunar Gateway. Approaching near Earth space, the European Service Module (ESM) will be separated from the Orion spacecraft to expose the CEM's heatshield to the Earth's atmosphere and Orion will perform a splashdown landing in the Pacific Ocean.

The Human Landing System will be initially designed for the descent of 2 astronauts and additional 100 kg of scientific instruments to the lunar surface [16]. Performance increases in the Human Landing System are expected that should enable an increase in its capacity to a crew of up to 4 astronauts. The crew cabin of the Human Landing System will be utilized as a Crew Habitation Module during the first lunar surface missions, when no lunar infrastructure is available to house the astronauts. During later lunar surface missions, the crew will inhabit a lunar surface habitat and the Human Landing System will be operated in a low power mode. This habitat will consist out of pressurized crew modules and unpressurized modules for systems and autonomous robotic units. It should enable missions featuring a longer duration of stay on the lunar surface. During Extravehicular Activities (EVAs), up to 1.5 h will be accounted for construction, configuration and set up tasks and 4 h will be accounted for scientific activities.

2.2.2 Scheduled Artemis Missions

One Artemis mission has already taken place and a number of four additional lunar missions have already been scheduled and prepared within the Artemis program. Artemis 1 was the first mission within the Artemis program and took place between November and December 2022. During the mission, the Orion spacecraft was uncrewed. The objective of the mission was the systems verification of the Orion spacecraft, the Space Launch System (SLS) Launch Vehicle (LV) and the exploration ground systems. Engineering data regarding system performance was collected and new technologies were demonstrated. In addition to this, 10 CubeSats were deployed. For the transit phases between the Earth and the Moon, around 10 days were accounted for one way, and in between the transit phases, the Orion spacecraft spent around 6 days in lunar Distant Retrograde Orbit (DRO). The topology of the Artemis 1 mission is visualized in Figure 2.7.

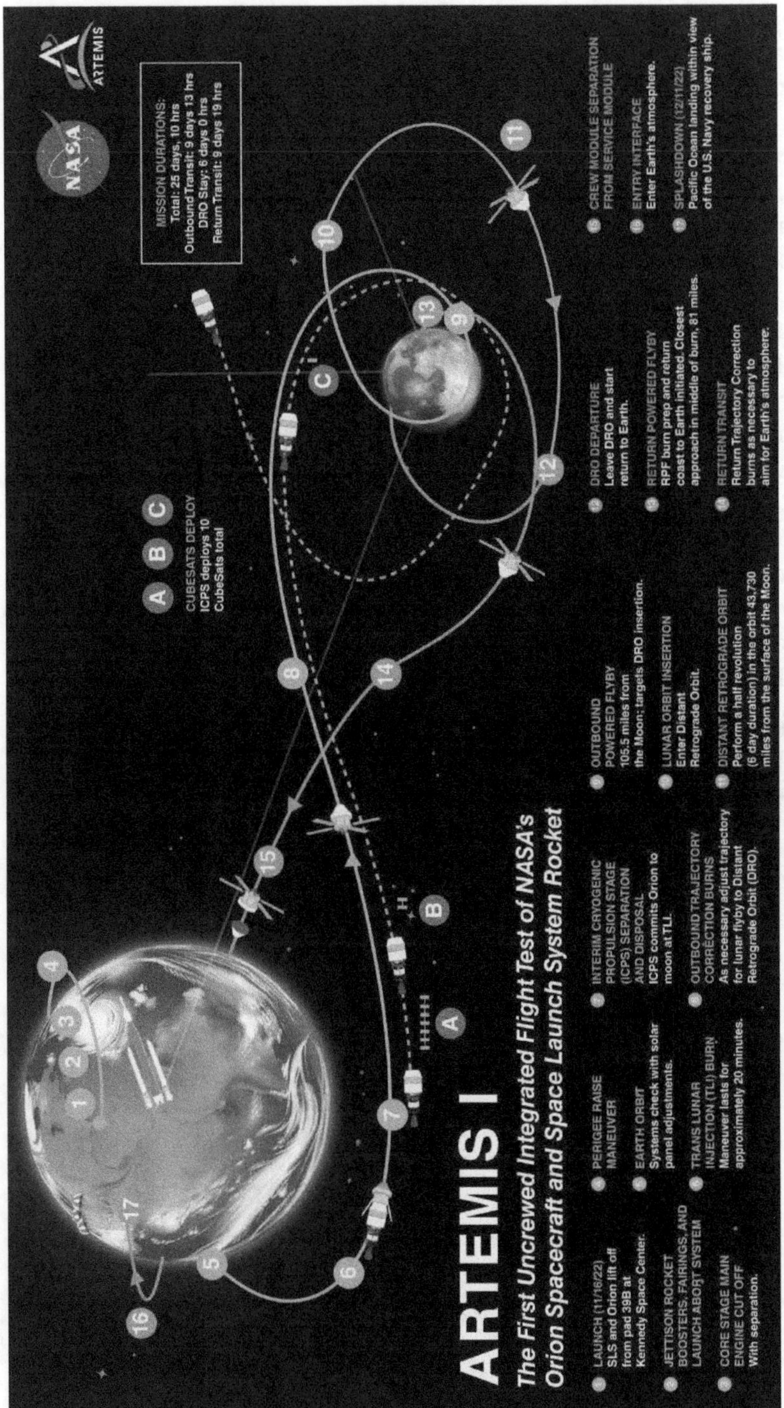

Figure 2.7 Artemis 1 mission map (image: © NASA [8])

The Orion spacecraft was launched on top of a Space Launch System (SLS) Block 1 Launch Vehicle (LV) from Kennedy Space Center (KSC). SLS Block 1 LV employed the Interim Cryogenic Propulsion Stage (ICPS) as the second stage. The Orion spacecraft, together with the ICPS, was delivered into Earth orbit by the SLS core stage. Following this, solar arrays were deployed, system checks were carried out, and perigee raising maneuvers were performed. The Trans Lunar Injection maneuver was carried out with a 20-minute-long engine burn by the ICPS. After that, the ICPS was separated from the Orion spacecraft and expended. In the following, the Orion spacecraft spent around 10 days on the transit trajectory to lunar Distant Retrograde Orbit (DRO). The closest approach to the lunar surface took place in an altitude of 97 km over the lunar surface. The lunar orbit injection maneuver was carried out by the engines of the European Service Module (ESM), which is part of the Orion spacecraft. After that, the Orion spacecraft spent around 6 days in lunar DRO. In DRO, the maximum altitude over the lunar surface was around 65,000 km. After the stay in lunar orbit, the engines of the ESM carried out the DRO departure maneuver, followed by a close flyby over the lunar surface in an altitude between 97 km and 130 km, and the transit of the Orion spacecraft back to the Earth. After around 10 days on the transit trajectory, the European Service Module (ESM) was separated from the Crew Exploration Module (CEM) in order to expose the CEM's heat shield. The CEM entered the Earth's atmosphere directly, followed by a splashdown landing in the Pacific Ocean.

Artemis 2 will be a lunar flyby mission. The Orion spacecraft will be crewed with a crew consisting of the four astronauts Reid Wiseman, Victor Glover, Christina Koch, and Jeremy Hansen. The duration of the mission will be approximately 10 days. The Orion spacecraft will fly on a direct return trajectory and only execute one lunar flyby before returning back to the Earth. The topology of the Artemis 2 mission is visualized in Figure 2.8.

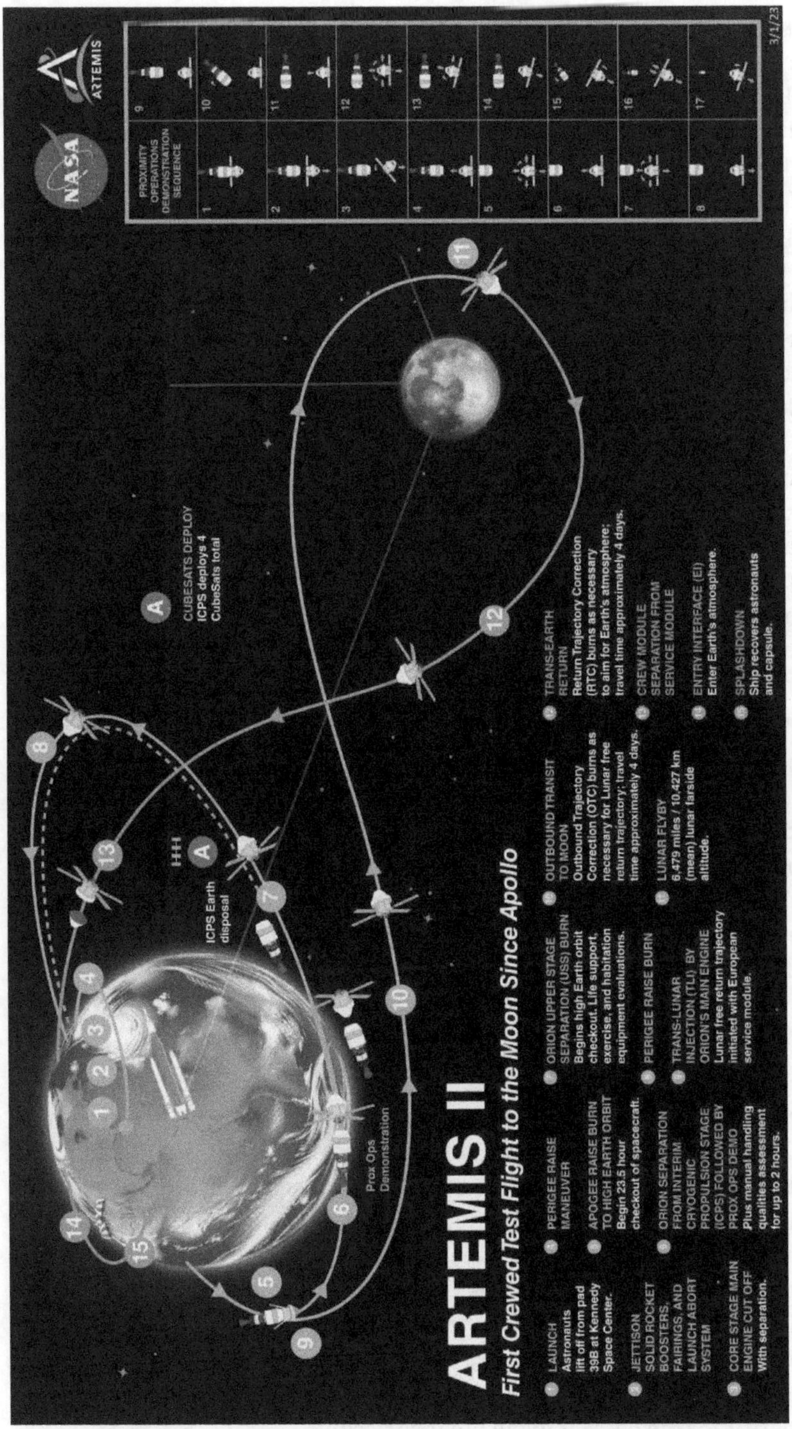

Figure 2.8 Artemis 2 mission map (image: © NASA [8])

At the beginning of the mission, the Orion spacecraft will be launched from Kennedy Space Center (KSC) into Earth orbit utilizing the SLS Block 1 Launch Vehicle (LV). The SLS Block 1 LV will deliver Orion, the crew, as well as the Interim Cryogenic Propulsion Stage (ICPS) to a low elliptical Earth orbit in an altitude between 185 km and 2800 km, featuring an orbital period of around 90 minutes. During the following 2 orbits, the ICPS will carry out apogee and perigee raising maneuvers to bring the Orion spacecraft in an elliptical high Earth orbit. This orbit has an orbital period of around 24 hours in an altitude of between 380 km and 110,000 km. After the orbit raising maneuvers have been completed, the Orion spacecraft will separate from the ICPS and the Crew will perform a manual maneuvering demonstration around the ICPS for up to 2 hours. The ICPS will be expended after the flight demonstration and final system checks of the Orion spacecraft will be carried out. Following this, the Orion main engine will perform the Trans Lunar Injection maneuver that places the Orion spacecraft on a direct return trajectory around the Moon. The crew on board of the Orion spacecraft will then transit to lunar orbit for a period of 4 days, complete 1 lunar flyby in an altitude of 10,427 km and transit back to the Earth for a period of additional 4 days. When the Orion spacecraft is approaching the Earth, the European Service Module will be separated from Orion in order to expose Orion's heat shield to the Earth's atmosphere. Followed by this, the Orion spacecraft will directly transit through the Earth's atmosphere and perform a splashdown landing in the Pacific Ocean.

Artemis 3 will be the first mission within the Artemis program including a landing on the lunar surface. Orion will be crewed with 4 astronauts, from which 2 will descent to the lunar surface during the mission. A Human Landing System (HLS) will be utilized for landing on the lunar surface. The HLS will be placed in lunar Near Rectilinear Halo Orbit (NRHO) in advance of the Artemis 3 mission. 2 astronauts will stay for approximately 1 week on the lunar surface, while the other 2 astronauts will stay on board of the Orion spacecraft. During this period, a minimum number of 3 Extravehicular Activities (EVAs) will be scheduled. The topology of the Artemis 3 mission is visualized in Figure 2.9.

At the beginning of Artemis 3, the crew will launch on board of the Orion spacecraft on top of a Space Launch System (SLS) Block 1 Launch Vehicle (LV) from Kennedy Space Center. After separation from the SLS core stage, the Interim Cryogenic Propulsion Stage (ICPS) will perform perigee raising maneuvers, the solar panels will be deployed and final system checks will be carried out. In the next step, the ICPS will perform the Trans Lunar Injection maneuver. Following this, the Orion spacecraft will separate from the ICPS, the ICPS will be expended and the crew on board of the Orion spacecraft will transit to lunar orbit. After an outbound lunar transit and a lunar flyby, the Orion main engine will perform the maneuver for the insertion into lunar Near Rectilinear Halo Orbit (NRHO), where the Orion spacecraft will rendezvous and dock with the Starship HLS from SpaceX. In lunar NRHO, the crew will activate the HLS and prepare it for the descent to the lunar surface.

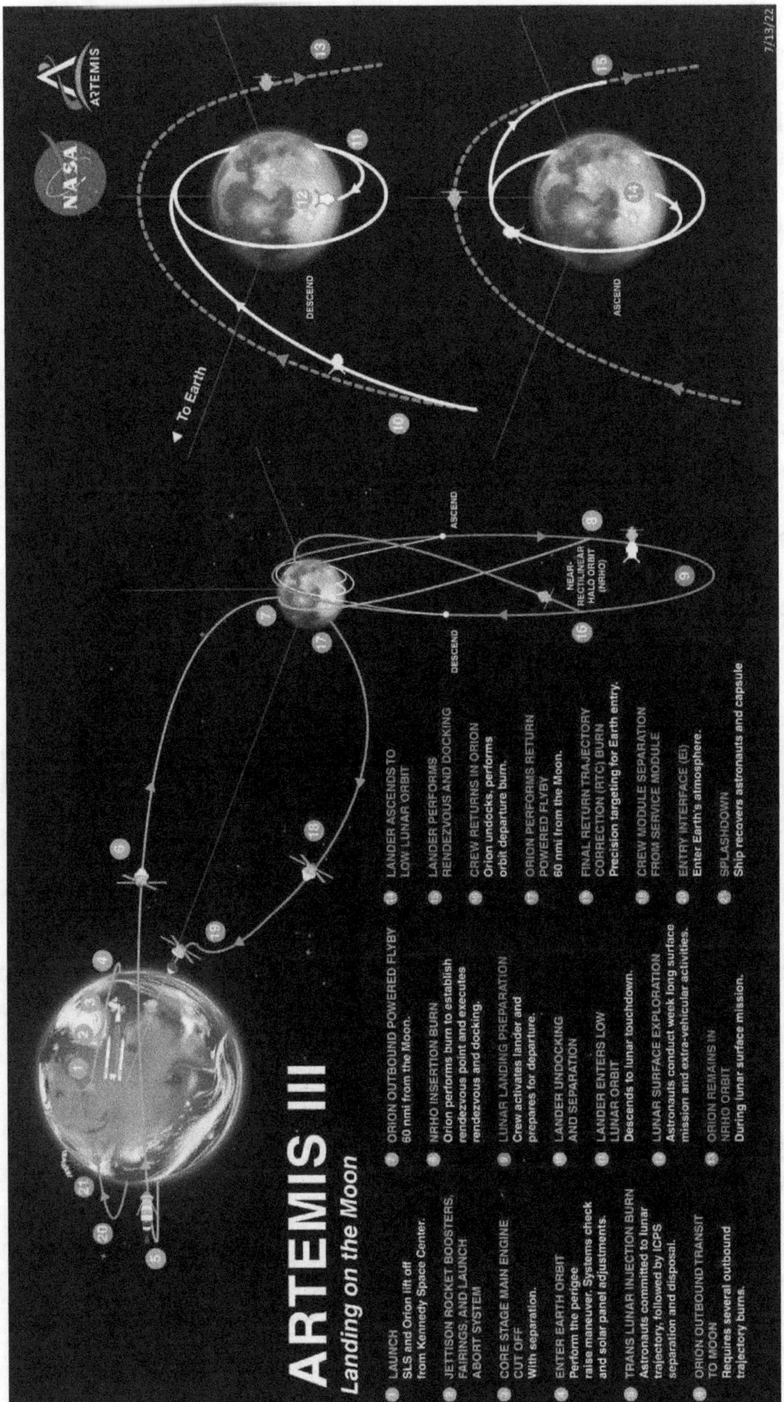

Figure 2.9 Artemis 3 mission map (image: © NASA [8])

Finally, half of the crew will transfer to the Human Landing System, undock and separate from the Orion spacecraft, while the other half of the crew will stay on board of the Orion spacecraft. The HLS will then perform the maneuver to enter low lunar orbit. From there, it will descent to the lunar surface and land in a location close to the lunar south pole. The crew will stay on the lunar surface for around 1 week and during this period, the astronauts will carry out 3 or even more Extravehicular Activities. After the stay on the lunar surface, the HLS will be utilized as an Ascent Vehicle into low lunar orbit, where the rendezvous and docking maneuver with the Orion spacecraft will take place. After that, the crew will spend additional 5 days on the docked spacecraft in lunar orbit. In the end of the mission in lunar orbit, the crew will transfer back to the Orion spacecraft, undock and separate from the HLS. The HLS will be expended in a heliocentric orbit after this maneuver. The Orion spacecraft will perform the lunar orbit departure maneuver and leave the NRHO with a final lunar flyby into the transit trajectory. When the Orion spacecraft approaches the Earth, the European Service Module (ESM) will be separated from the Crew Exploration Module (CEM) in order to expose its heat shield to the Earth's atmosphere. The crew on board of the CEM will directly enter the Earth's atmosphere, followed by a splashdown landing in the Pacific Ocean.

Artemis 4 will be the first crewed mission within the Artemis program delivering infrastructure to lunar orbit and it will also be the first mission utilizing the Lunar Gateway. During the mission, the International Habitation Module (I-Hab) will be delivered to the Lunar Gateway in lunar Near Rectilinear Halo Orbit (NRHO). Artemis 4 will follow up on cargo missions, delivering the Human Landing System, the Power and Propulsion Element (PPE) of the modules and the Habitation and Logistics Outpost (HALO) module to lunar NRHO. In addition to this, Artemis 4 will also be the first crewed mission utilizing the Exploration Upper Stage (EUS) instead of the Interim Cryogenic Propulsion Stage (ICPS) as the upper state of the Space Launch System (SLS), which will increase the cargo capacity of the Launch Vehicle (LV). The topology of the Artemis 4 mission is visualized in Figure 2.10.

The Lunar International Habitation module (Lunar I-Hab) will be mounted together with the Orion spacecraft on top of the Space Launch System (SLS) Block 1-B Launch Vehicle and launch from Kennedy Space Center (KSC). After separation from the SLS core stage, the Exploration Upper Stage (EUS) will carry out maneuvers for the circularization of the orbit in Low Earth Orbit (LEO). After additional system checks have been carried out and the solar arrays of the Orion spacecraft have been deployed, the EUS will perform the Trans Lunar Injection maneuver to place Orion and Lunar I-Hab on the transit trajectory to lunar orbit. During the transit to lunar orbit, Orion and Lunar I-Hab will separate from the EUS, Orion will dock to the docking port of Lunar I-Hab and the EUS will be expended. A visualization of Orion and Lunar I-Hab docked in near Earth space is shown in Figure 2.11.

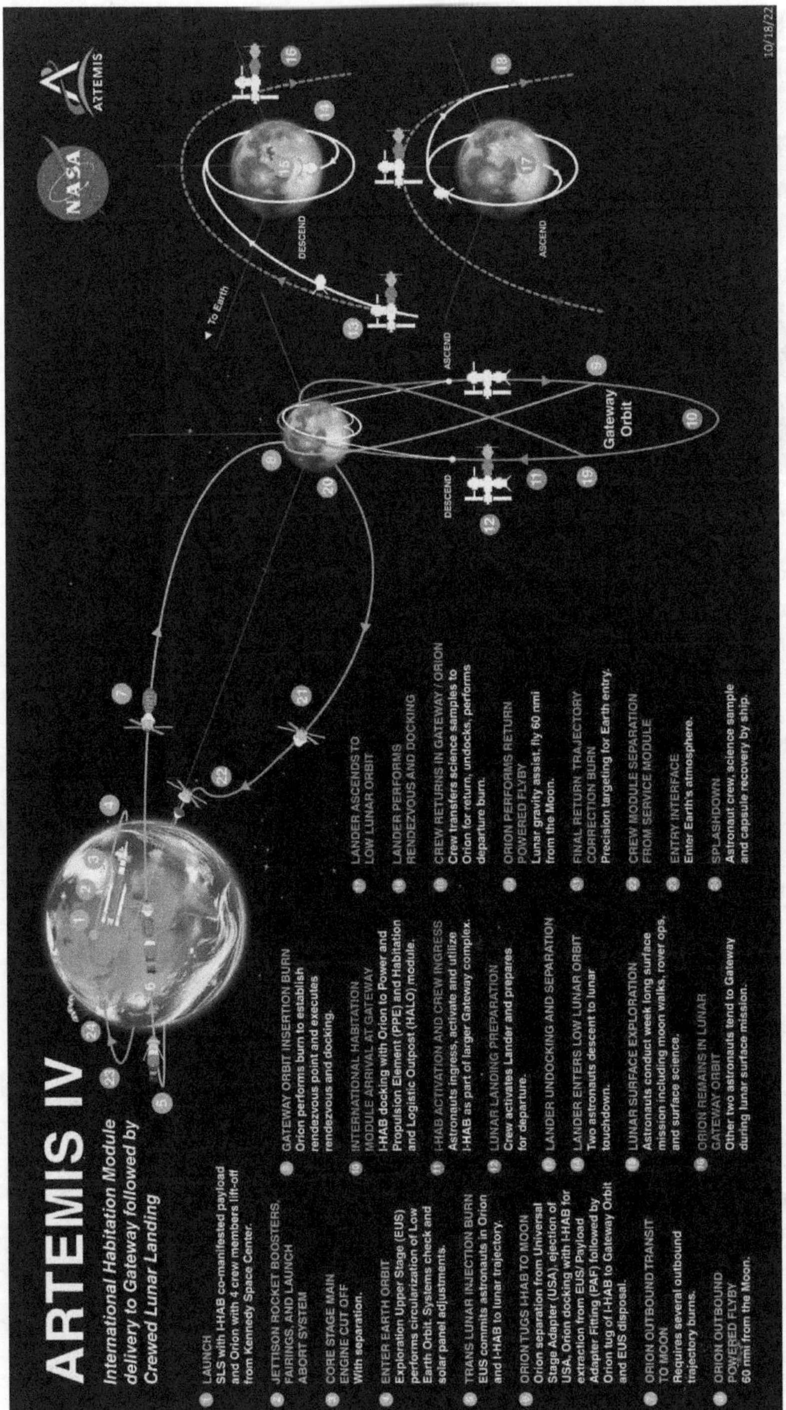

Figure 2.10 Artemis 4 mission map (image: © NASA [8])

Figure 2.11 Visualization of Orion and Lunar I-Hab near Earth space (image: © ESA [32])

After approaching lunar orbit, the Orion spacecraft and Lunar I-Hab will perform an outbound lunar transit, followed by an outbound powered flyby, while the Orion spacecraft will carry out several maneuvers to insert both into the Lunar Gateway orbit NRHO. In NRHO, Orion and Lunar I-Hab will rendezvous and dock with the initial Lunar Gateway modules. After the arrival, the crew will activate the Gateway and the Human Landing System (HLS). Half of the crew will transfer to the HLS, undock from the Gateway, descent and land on the lunar surface. The surface exploration crew will spend around one week on the lunar surface, where they will carry out Extravehicular Activities (EVAs), which will also include surface science and rover operation. The remaining crew of 2 astronauts in lunar orbit will carry out initial setup activities on the Gateway and Lunar I-Hab during this period. After their one-week-long exploration mission on the surface, the surface exploration crew will ascent to lunar orbit again on board of the Human Landing System (HLS), where the HLS will rendezvous and dock with the Gateway. The crew will spend additional 5 days in lunar orbit after the arrival of the HLS. After that, the crew will again transfer back to the Orion spacecraft. The crew on board of Orion will undock and separate from the Gateway after that. Orion will carry out the lunar orbit departure maneuver, followed by a return powered flyby around the Moon to insert Orion into its return trajectory. When approaching the Earth, the European Service Module (ESM) will be separated from the Crew Exploration Module (CEM) and will be expended in order to expose the CEM's heat shield to the Earth's atmosphere. The crew will directly enter and transit through the Earth's atmosphere, followed by a splashdown landing into the Pacific Ocean.

The Artemis 5 mission will deliver the Lunar View module to the Lunar Gateway and it will also have a lunar surface mission. The topology of the Artemis 5 mission is visualized in Figure 2.12 and a visualization of Orion and Lunar View docking with the Gateway is shown in Figure 2.13.

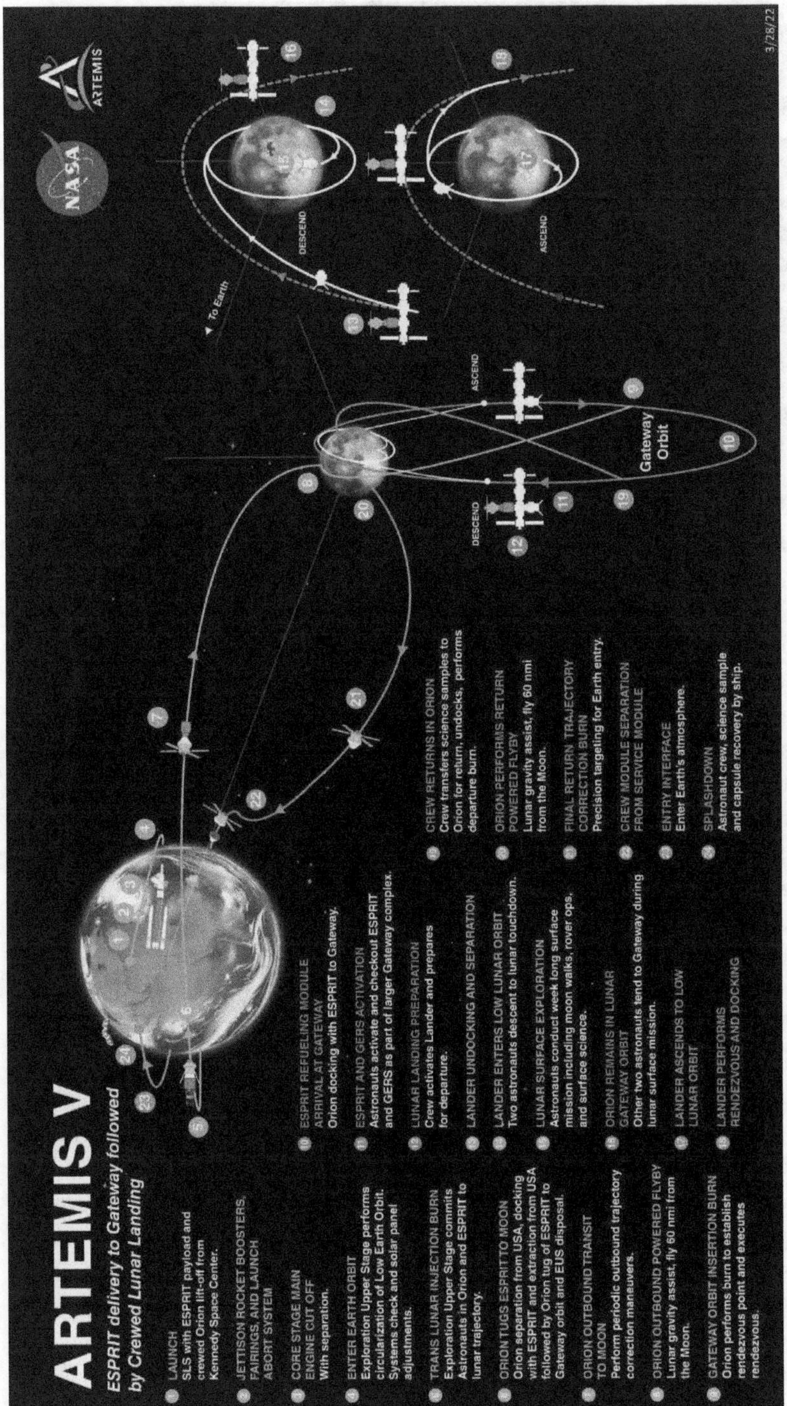

Figure 2.12 Artemis 5 mission map (image: © NASA [8])

Figure 2.13 Visualization of Orion and Lunar View before the initial docking with the Lunar Gateway during the Artemis 5 mission (image: © ESA [37])

At the beginning of the Artemis 5 mission, the Lunar View module including propellant for the Gateway, the Orion spacecraft and a crew of 4 astronauts will launch with a Space Launch System (SLS) Block 1-B Launch Vehicle from Kennedy Space Center (KSC). After separation from the SLS core stage, the Exploration Upper Stage (EUS) will carry out maneuvers to circularize the orbit of the EUS, Lunar View and Orion in Low Earth Orbit (LEO). After final system checks, solar array deployment and adjustment, the EUS will perform the Trans-Lunar Injection maneuver to place Orion and the Lunar View module on the transit trajectory to lunar orbit. Orion and Lunar View will then separate from EUS, Orion will dock to Lunar View and the Exploration Upper Stage will be expended. After the transit to lunar orbit, Orion and Lunar View will perform a lunar outbound transit and an outbound powered lunar flyby, followed by the insertion maneuver into the lunar NRHO, which will be carried out by the Orion spacecraft. In the Lunar Gateway orbit, Orion will perform the rendezvous and docking maneuver. After the arrival at the Gateway, the Lunar View module will be activated by the crew. The crew will transfer to the Gateway and activate the Human Landing System (HLS) in the following. Half of the crew will transfer to the HLS, undock from the Gateway and descent to the lunar surface. The surface exploration crew will stay there for one week, carrying out activities like Extravehicular Activities, surface science activities and rover operation. Meanwhile, the other half of the crew will stay on board of the Lunar Gateway and carry out setup tasks and science activities. After their mission on the lunar surface, the surface exploration crew will ascent back to low lunar orbit on board of the HLS, where they will rendezvous and dock with the Lunar Gateway. After docking, the crew will spend additional 5 days in lunar orbit. At the end of the mission, the crew will transfer back to the Orion spacecraft and undock from the Lunar Gateway. The Orion spacecraft will carry out the lunar orbit departure maneuver. Following this, Orion will perform a return powered flyby around the Moon, in order to place itself on the final return trajectory [8] [36]. When approaching the Earth, the European Service Module (ESM) will be separated from the Crew Exploration Module (CEM), to expose the CEM's heat shield to the Earth's atmosphere. Orion will directly enter the Earth's atmosphere and transit through it, followed by a splashdown landing in the Pacific Ocean.

2.3 Science and Engineering Challenges during Artemis Missions

Science activities will focus on a wide range of different disciplines. These will include the lunar field geology and sample collection. On the Moon, regions on the edges of craters exist, that are permanently shadowed and therefore feature unique local conditions. Drilled cores will enable helio science to study the records of the ancient sun. From lunar craters, the impact history in the Earth-Moon system can be studied. Instruments for the environmental characterization of the lunar surface will be installed on the surface. In addition to this, the understanding of planetary processes and volatile cycles will also be included in the scientific objectives of lunar missions.

The environmental conditions on the lunar surface lead to a number of engineering challenges. The lunar dust is electrically charged and therefore has adhesive properties. Dust mitigation methods have to be applied as preventive measures for astronaut health and in order to prevent mechanical and optical degradation of technical equipment. Powering a lunar base during the local night will also be an engineering challenge because the night period can last nearly two weeks in most locations. The In-Situ Resource Utilization (ISRU) of regolith and material from craters will also be a focus of engineering. Water could be extracted from icy regolith that can be found in permanently shadowed regions of lunar craters. Oxygen could be extracted from lunar rocks and also other minerals like Pyroxene, Anorthite, Olivine or Ilmenite could be extracted from lunar materials. Robotic systems will allow the access to areas on and below the lunar surface, which are inaccessible for astronauts. In addition to this, autonomous construction and manufacturing processes utilizing local resources will be a focus of engineering during lunar missions.

During missions on board of the Lunar Gateway, zero gravity science, explorational and technological risks and their mitigation, as well as the simulation of crewed missions to Mars can be a focus of research. Most of the science operations on board of the Lunar Gateway will be teleoperated [16]. Possible objectives of scientific experiments could be zero gravity science, heliophysics, space radiation, or space weather.

2.4 Vehicles and Systems

A number of different vehicles and systems will be employed within the Artemis program. The Space Launch System (SLS) is a Launch Vehicle that was developed under the leadership of NASA for missions within the Artemis program. Similar designs were developed by NASA for the Constellation program, a predecessor of the Artemis program that focused on the exploration of Mars and the Moon. The SLS Launch Vehicle was developed in order to enable the delivery of a crew on board of a Crew Exploration Vehicle and additional cargo directly to the Moon. It consists out of a core stage, powered by 4 RS-25 engines, two solid rocket boosters, and different upper stage configurations. An image of the first Space Launch System before the beginning of the Artemis 1 mission is shown in Figure 2.14.

Figure 2.14 Space Launch System of the Artemis 1 mission (image: © NASA, Kim Shiflett [41])

The RS-25 engines of the core stage were initially developed for the Space Transportation System (STS) program as engines for the space shuttle and have been upgraded for utilization in the SLS Launch Vehicle. Currently, a set of 16 RS-25 engines is left from the STS program, enabling the manufacturing of up to 4 SLS Launch Vehicles. Apart from these, the core stage also has 2 solid rocket boosters, that are providing 75% of the vehicles' thrust in the first 2 minutes of flight. These boosters were developed from the boosters utilized within the STS program as boosters for the space shuttle external tank. Versions of the SLS launch vehicle differ in their upper stage configuration, and, following this, also the amount of mass and volume the LV can launch into space is increasing with increasing versions of the vehicle. The SLS Block 1

Launch Vehicle utilizes the Interim Cryogenic Propulsion Stage (ICPS) that has been developed from the Cryogenic Upper Stage (CUS) of the Delta IV Launch Vehicle from United Launch Alliance. This version of the LV has been utilized in the Artemis 1 mission. Instead of the ICPS, the Block 1-B version of the SLS Launch Vehicle will make use of the Exploration Upper Stage (EUS), which will be powered by 4 RL10C-3 engines. The EUS will be more powerful than the ICPS, and therefore, it will allow to extend the payload capacity of the LV. Each version of the Block 1, Block 1-B and the Block 2 configuration of the SLS LV will be available in 2 configurations. The cargo configuration will be utilized for uncrewed launches. The absence of a crew and a Crew Exploration Vehicle will allow to extend the fairing in cargo missions in order to extend the cargo volume and mass. For crewed missions, the crew configuration of the SLS Launch Vehicle will be utilized. In the crew configuration, the Orion spacecraft will be mounted on top of a smaller cargo fairing. All key parameters for the Space Launch System launch capacity have been summarized in Table 2.1 and a visualization of the different configurations is shown in Figure 2.15.

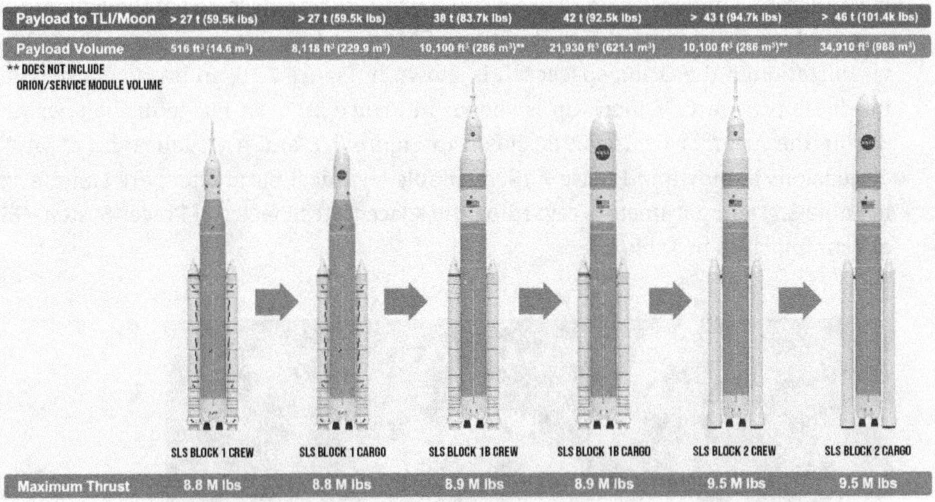

Figure 2.15 Different configurations and capacities of the Space Launch System (SLS) Launch Vehicle (LV) (image: © NASA [42])

Table 2.1 Cargo capacity of the Space Launch System (SLS) [42] [39] [40]

	Space Launch System (SLS)					
Configuration	**Crew**	**Cargo**	**Crew**	**Cargo**	**Crew**	**Cargo**
CEV included	yes	no	yes	no	yes	no
total payload to TLI	27 T	27 T	38 T	42 T	43 T	46 T
cargo volume	14.6 m^3	229.9 m^3	286 m^3	621.1 m^2	286 m^3	988 m^3

The SLS Block 1 Launch Vehicle has a cargo capacity of up to 38.1 T [39] [40] [42]. In the crew configuration, 14.6 m^3 are available for additional cargo, and in the cargo configuration, 229.9 m^3 are available [39] [40] [42]. The capacity of the SLS Block 1-B LV is greater than the capacity of the Block 1 LV [39] [40] [42]. In this version, the SLS LV can launch up to 38 T in the crew configuration, which allows the launch of the Orion spacecraft and additional 10 T of cargo to the lunar surface [39] [40] [42]. The available cargo volume of the SLS Block 1-B in crew configuration has a value of 286 m^3. In the SLS Block 1-B cargo configuration, the launch vehicle has a cargo volume of 621.1 m^2, whereby the cargo mass capacity has a value of 42 T [39] [40] [42]. Because of this, the SLS Block 1-B cargo launch vehicle will be capable of launching a complete habitat, additional equipment and a cargo lander to the lunar surface in a single launch. The SLS Block 2 has an even greater payload capacity. In the crew configuration, the SLS Block 2 LV is capable of launching of up to 43 T of cargo with a maximum volume of 286 m^3 [40] [42]. In the cargo configuration, the SLS Block 2 LV is capable to launch up to 46 T of cargo with a maximum volume of 988 m^3 [40] [42].

The Orion spacecraft was developed as Crew Exploration Vehicle (CEV) for the Artemis program. Similar designs were developed for the predecessor of the Artemis program, which focused on the exploration of Mars and the Moon in the early 2000s. A visualization of the Orion spacecraft is shown in Figure 2.16, an image of astronauts inside den Orion CEV mock-up is shown in Figure 2.17, an image of the spacecraft during the Artemis 1 mission is shown in Figure 3.9, and a visualization of Orion's dimensions is shown in Figure 2.18. Available technical parameters are summarized in Table 2.2, and parameters regarding the spacecraft's Electrical Power System (EPS) are summarized in Table 2.3.

Figure 2.16 Rendering of the Orion spacecraft in lunar orbit (image: © NASA [43])

Table 2.2 Technical parameters of the Orion Spacecraft [27] [28] [29] [36] [45] [46] [47]

Parameters	Dimensions
spacecraft diameter	5.2 m
span incl. Solar Array Wings (SAWs)	19.0 m
length of the Solar Array Wing	7375 m
width of the Solar Array Wing	2.0 m
height	5.2 m
mass	Crew Exploration Module: 10.39 T Service Module: 15.46 T total: 25.85 T
supplies	8.6 T propellant 240 kg portable water 30 kg nitrogen 90 kg oxygen
crew capacity	up to 4 astronauts
payload capacity	up to 0.57 m^3 up to 380 kg

Table 2.3 Electrical and power generation-related parameters of the Orion spacecraft [15] [46] [47]

Parameters	Dimensions
solar generator	4 Solar Array Wings in x-wing configuration, each consisting of 3 solar panels
mounting/tracking	2 axes of Solar Array Drive Mechanism (SADM) to swivel and rotate the Solar Array Wings inner axis: –35° to +25° outer axis: 0° to 360°
solar cell technology	3J-GaAs solar cells from Emcore
electrical power generation capacity	up to 11.2 kW
bus voltage	120 V, Direct Current (DC)
power transfer SAW to ESM	through 2-axis rotational joints
batteries	4 batteries

The Orion spacecraft will be utilized for the delivery of astronaut crews to lunar orbit and back and it will be launched on top of an SLS Launch Vehicle. It will be capable of housing a crew of up to 4 astronauts. Compared to the Crew Exploration Vehicles utilized for the Apollo program, the Orion CEV offers greater internal volume by one third, while also featuring a greater crew capacity by one third. During missions, Orion will be capable of housing the crew for up to 21 days [26].

Spacecraft generally consist of a payload and required service systems that are usually summed up under the term of the spacecraft bus. Typical functionalities of a spacecraft bus are the supply of the payload with electrical energy and other supplies, propulsion or communication. In the design of the Orion spacecraft, these two parts of the vehicle were integrated in different submodules. The Crew Exploration Module (CEM) acts as housing for the crew and it has additional systems, like an airlock for docking with the Gateway, crew interfaces, seats, or a waste removal system (Figure 2.17). Systems for the generation and the supply of electrical energy, propulsion, as well as the main air and water tanks are integrated into the European Service Module (ESM). The ESM was developed based on the Automated Transfer Vehicle (ATV) under the leadership of the European Space Agency and it is manufactured by more than 20 companies across Europe [45] [46] [51]. The Orion spacecraft has a diameter of 5.2 m and a height of 7.3 m [46]. The CEM accounts for 10,387 kg and the ESM accounts for 15,461 kg of the spacecraft mass [46] (Figure 2.18).

Figure 2.17 Astronauts inside the Orion Crew Exploration Module mock-up (image: © NASA [52])

The 4 Solar Array Wings (SAWs) of the ESM were integrated in an x-wing configuration. Each SAW has a length of 7.375 m and a width of 2 m (Figure 2.18), so that the total span of the solar array wings adds up to 19 m [46]. One SAW consists of 3 solar panels that have been manufactured utilizing triple-junction GaAs solar cells manufactured by

Emcore [46] [47]. In order to track the sun, the SAWs were mounted on Orion utilizing a Solar Array Drive Mechanism (SADM) that allows to swivel and rotate the Solar Array Wings in 2 axes. The inner axis allows a rotation between –35° and +25° and the outer axis allows a complete rotation in the magnitude of 360°. They are capable of generating up to 11.2 kW of electrical power [15] [46] [47]. The power is transferred from the SAWs to the ESM through the 2-axis rotational joints. Solar Array Drive Electronics (SADE) were implemented to control the solar arrays. Power regulation and the routing of electrical power to equipment was implemented utilizing a Power Conditioning and Distribution Unit (PCDU). Orion's power bus has a bus voltage of 120 V, utilizing Direct Current (DC) technology [15] [47]. A number of 4 batteries are used as energy storage for Orion's power bus [15]. The units of the ESM were connected to the CEV via an Onboard Data Network (ODN) that utilizes a 3-plane Time-triggered Gigabit Ethernet (TTGbE) network. In the CEV computers, the vehicle control and management software was implemented. Before reentering the Earth's atmosphere, the ESM has to be separated from the CEM in order to expose the heat shield to the Earth's atmosphere, which is integrated in between the CEM and the ESM.

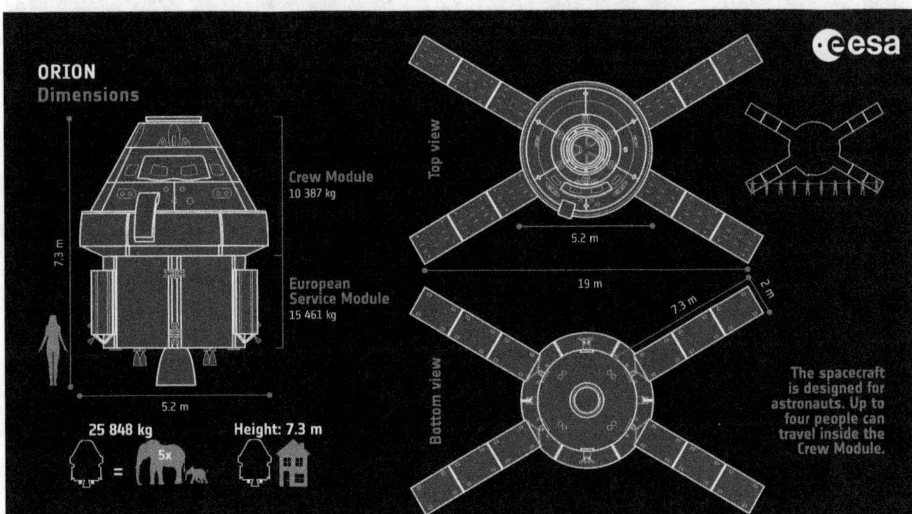

Figure 2.18 Dimensions of the Orion spacecraft (image: © ESA, K. Oldenburg [45])

During the Artemis 1 mission, a number of problems occurred [53]. These included structural degradation of the heat shield, melting and erosion of connection bolts between CEM and ESM, 24 events during which an opening of the latching current limiters of the Power Distribution Unit (PDU) were observed without a command, as well as damage at the launch systems after the Artemis 1 launch [53].

The Lunar Gateway will be a partially crewed space station in cislunar space. It is planned to utilize the Gateway as an intermediate step before a crewed landing on the lunar surface, whereby one half of an Artemis crew will decent to the lunar surface as a surface exploration crew, and the other half of the crew will stay on the Gateway and carry out logistical, operational or scientific activities. The Gateway will only be crewed for periods of 1 to 3 months. After the integration has been completed, the Gateway will be capable of housing up to 4 astronauts for a period of up to 90 days. The operation of the Gateway systems and science experiments will be remotely controlled during uncrewed periods. During the initial period, astronaut crews will stay on board of the Lunar Gateway for an estimated duration of around 40 days. In addition to its utilization as interim step for lunar surface missions, the Lunar Gateway will also be a zero gravity science outpost in deep space, outside the Earth's magnetic field. A visualization of the future Lunar Gateway is shown in Figure 2.19. Available technical parameters regarding the Gateway and its Electrical Power System (EPS) have been summarized in Table 2.4.

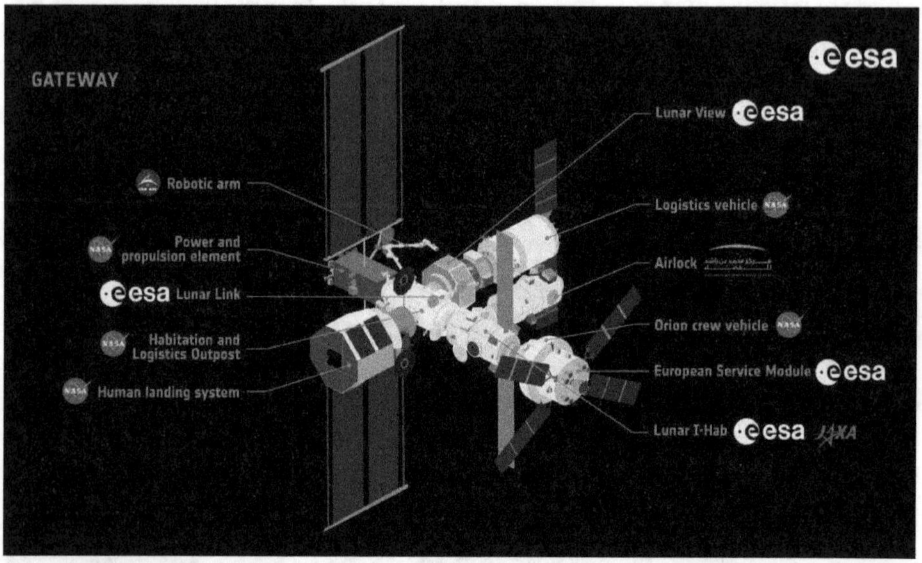

Figure 2.19 Rendering of the Lunar Gateway and possible modules (image: © ESA, K. Lochtenberg [54])

Table 2.4 Technical parameters of the Lunar Gateway [55]

Parameters	Measurements
total mass	63 T
dimensions	43 m tall, 19 m wide, 20 m long
electrical power generation capacity	60 kW from PPE module systems, generated utilizing photovoltaics

The Gateway will be contributed and operated by a conglomerate of different space agencies. Like the International Space Station (ISS), the Gateway will also need to be constructed sequentially in lunar orbit. After the construction phase, the Gateway will has a total mass of 63 T and an electrical power generation capability of 60 kW [55]. Gateway modules need to be delivered to lunar orbit utilizing uncrewed flights or utilizing Orion, by docking modules to Orion during transit from the Earth to lunar orbit. One of the first modules that will be launched to lunar orbit will be the Power and Propulsion Element (PPE). PPE was developed by Maxar Technologies and it will provide electrical propulsion utilizing solar power, thrust control, electrical power generation utilizing photovoltaics and optical communication based on lasers as a relay between the lunar surface and the Earth. The generative subsystems will be able to provide 60 kW of electrical power generation. The module Habitation and Logistics Outpost (HALO) was developed by Northrop Grumman and it will have crew cabins, an integrated basic Life Support System (LSS), a command, a control and data handling system, the energy storage and distribution system, the thermal control system, a communication system, as well as a tracking system. HALO will have 3 docking ports for visiting vehicles and the connection to other Gateway modules. It will have a length of 7 m and a diameter of 3 m [60]. The pressurized module of HALO will be provided by Thales Alenia Space.

The Lunar View module will be a module provided by the European Space Agency. It has been initially referred to as European System Providing Refueling Infrastructure and Telecommunication (ESPRIT). The Lunar View module will have a refueling submodule for the Gateway and docked Human Landing Systems, including an internal propellant storage facility for chemical and electrical propulsion systems featuring a capacity of up to 1.7 T [33]. Fuel transfer connections will be implemented at Lunar Views docking ports to transfer fuel to docked vehicles and NASA's HALO module. Lunar View will have an observation facility featuring 6 windows, enabling the astronauts to have a 360° view to the lunar surface and the space around the Gateway. In addition to this, it will also have cargo storage facilities in a size of 6.5 m^3 [33]. A docking port for visiting vehicles will be implemented on Lunar View. Attachment points for the CANADARM3 will be implemented on the outer structure of Lunar View. Lunar View is currently designed by Thales Alenia Space. In the final integration, it will have a mass of 8.5 T, a span of 4.6 m and a length of 6.4 m [34].

The module Lunar Link will be mounted on the HALO module. It will enable communication between the Gateway, systems on the lunar surface, as well as systems in lunar orbit. Lunar link will have 2 antenna dishes with a diameter of 125 cm, 2 low gain antennas, and computers [57]. It will provide communication speeds of up to 25 Mb/s while consuming up to 600 W [57].

The Gateway will have an additional habitat module referred to as Lunar International Habitat module (Lunar I-Hab), a logistics module, a science airlock for scientific experiments, an airlock for Extravehicular Activities and a gateway robotic system. Lunar I-Hab will have 2 docking ports for berthing vehicles, 1 docking port for visiting Crew Exploration Vehicles, 1 docking port connecting Lunar I-Hab to the Gateway, an experimental facility, 2 thermal control systems with 12 m long deployable radiators, an Environmental Control and Life Support System (ECLSS), batteries, exercise areas, medical areas, living and sleeping quarters for astronauts, as well as a dining table [58]. The internal habitable volume of Lunar I-Hab will have a value of around 10 m^3 and the total pressurized volume will have a value of around 36 m^3 [58]. Attachment points for the CANADARM3 were implemented on the outside of Lunar I-Hab. It was developed and is currently manufactured by Thales Alenia Space in Turin. In the final integration, Lunar I-Hab will have a mass of 10 T [58]. A photo of the Lunar I-Hab module mock-up at Thales Alenia Space is shown in Figure 2.20 and Figure 2.21 shows a photo of the rings of the Lunar I-Hab pressure vessel during manufacturing at Thales Alenia Space.

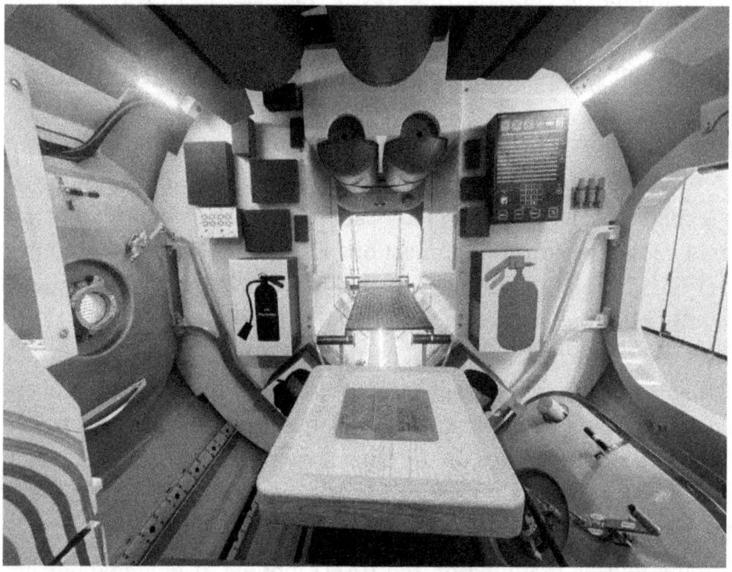

Figure 2.20 Image of the Lunar I-Hab mock-up (image: © Thales Alenia Space [16])

Figure 2.21
Rings of the pressure vessel of the Lunar I-Hab module during the manufacturing at Thales Alenia Space in Turin (image: © ESA, S. Corvaja [62])

The gateway robotic system will be the robotic arm CANADARM3, provided by the Canadian Space Agency (CSA). The Lunar I-Hab module, the Lunar View module, and the Lunar Link modules will be provided by the European Space Agency (ESA). The Japanese Space Agency (JAXA) will contribute components for the Lunar I-Hab module. An airlock module will be provided by the Mohammed Bin Rashid Space Centre (MBRSC) of United Arab Emirates (UAE) Space Agency.

HALO and PPE will be launched as first modules to the Lunar Gateway in a cargo mission. They will be launched together to lunar orbit in a single launch. Initially, the Space Launch System was scheduled to launch both, but the schedule has recently been changed towards launch services provided by a SpaceX Launch Vehicle. During the following Artemis 4 mission, Lunar I-Hab will be delivered to the Lunar Gateway, and Lunar View will be delivered to the Lunar Gateways with the Artemis 5 mission.

The Lunar Gateway will rely on periodic resupply logistics missions. Cargo spacecrafts will dock to the Lunar Gateway in advance of the arrival of a new crew. One logistics mission will be scheduled in advance of every crewed mission. Resupply missions need to include pressurized and unpressurized cargo. In the initial phase, cargo resupply missions will be carried out utilizing spacecraft provided by SpaceX. The Japanese Aerospace Exploration Agency (JAXA) is also discussing the development of an upgraded version of the HTV spacecraft to provide logistic resupply capabilities for Artemis missions.

The Gateway will be located in a lunar Near Rectilinear Halo Orbit (NRHO), which is a highly eccentric polar orbit with a perigee altitude of 3000 km and an apogee altitude of 70,000 km over the lunar surface. This orbit has a 7 day long period duration, so that a possible surface exploration crew can rendezvous and dock with the Gateway every 7 days, when the Gateway is located at the perigee location in its orbit [10]. A visualization of the gateway NRHO orbit is shown in Figure 2.22.

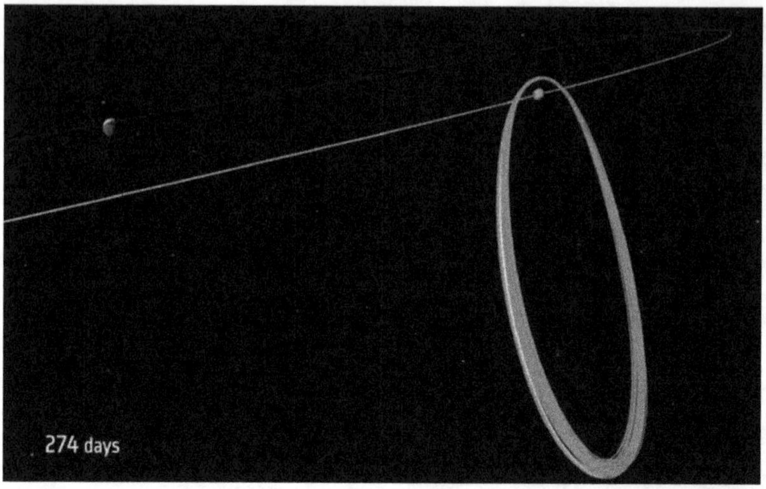

Figure 2.22 Near Rectilinear Halo Orbit (NRHO) of the Lunar Gateway (image: © ESA [10])

The Human Landing System (HLS) will be an incremental part of missions, where a landing on the lunar surface is included. It shall be docked to the Lunar Gateway in between lunar surface missions, it should be able to decent to the lunar surface and it should be able to ascent back from the lunar surface to the Gateway orbit. Docking shall be possible with the Lunar Gateway and with Orion. The HLS should be able to descent to the lunar surface with crew and additional cargo, like technology and science experiments, and it needs to be able to ascent with the crew and collected samples back to the Gateway orbit. Reusability is a key requirement for a lunar HLS and following this, an HLS should be capable of carrying out recurring missions to the lunar surface. Propellant shall be supplied to the HLS by the Lunar Gateway in between lunar surface missions. Currently, it is assumed that lunar surface crews will consist out of 2 astronauts, so that the minimum crew capacity requirement for the HLS is a number of 2. In advance of the first lunar surface mission within the Artemis program, the HLS with additional cargo for the lunar surface, including two Extravehicular Mobility Units (EMUs) and science equipment, needs to be delivered to lunar NRHO. Initial missions on the surface will have a duration of around 7 days, so that crew supplies for this duration need to be included in the HLS cargo. The destination of initial missions will be the lunar south pole [25]. During a 7 day long mission, between 3 and 5 extravehicular activities need to be enabled by the HLS [25]. It is expected, that it will be possible to extend the mission duration for missions to the lunar south pole to durations of up to 30 days. Missions with focus on non-polar locations featuring a duration of around 7 days shall also be enabled by a possible HLS. Within these missions, the HLS shall enable a minimum number of 5 EVAs and it should be able to sustain during the lunar night on the surface. For the development of the Human Landing System (HLS), SpaceX, Dynetics and a team lead by Blue Origin, also consisting of Lockheed Martin, Northrop Grumman, and Draper, have been selected [16]. Recently, the HLS concepts developed by SpaceX and Blue Origin have been selected by NASA [3].

The Starship spacecraft developed and manufactured by SpaceX is a fully reusable launch, landing and space transportation vehicle [16] [64]. It was designed for missions focusing on Earth orbit, Moon and Mars [64]. From the Earth, it is launched on top of a Falcon Super Heavy Launch Vehicle [16] [64]. Starship will transit to lunar orbit without a crew and rendezvous and dock with the Lunar Gateway or Orion in lunar NRHO autonomously [65]. The internal design of the Starship crew cabin will be adapted for a two-astronaut crew for missions within the Artemis program [65]. For missions outside of the Earth's orbit, refueling in Earth orbit will be required before departure and transit [16] [56]. A picture of Starship on the Launch Pad in Boca Chica is shown in Figure 2.23 and Figure 2.24 shows a rendering of Starship as HLS on the lunar surface.

Figure 2.23 SpaceX Starship on the Launch Pad in Boca Chica (image: © SpaceX [66])

Figure 2.24 Rendering of the SpaceX Starship as Human Landing System on the lunar surface (image: © SpaceX [65])

For these purposes, Starship will be available in different configurations [16] [56]. Different Starship configurations will vary in their payload bay design [16] [56]. The payload bay can be utilized as a room for cargo, for a crew cabin and for a combination of both and it can also be utilized for a tank [16] [56]. In the crew configuration, a crew compartment is integrated into Starship's payload bay [64]. The crew configuration will have spacious private crew cabins, solar storm protection, two airlocks, a gallery observation facility, as well as cargo and storage facilities [65] [64]. Theoretically, a single Starship can transfer up to 100 astronauts to Low Earth Orbit (LEO) or to the surface of the Moon or Mars [64]. A Starship in a tanker configuration will be utilized for in-orbit refueling of another Starship before its departure from Earth orbit [16] [56].

Starship's payload bay has a lower diameter of up to 8 m and a height of 17.24 m [64]. Between a height of 0 m and 8 m, the payload bay diameter is featuring a value of 8 m [64]. From there on to the top of the payload bay, diameter decreases to 3.6 m in the top part [64]. An extended version of Starship will be available, too, featuring a payload bay height of 22 m [64]. In the cargo configuration, Starship will be able to deliver more than 90.72 T to Low Earth Orbit (LEO) or to Geostationary Transfer Orbit (GTO) [64]. A fuel transfer in a LEO parking orbit from an additional Starship in tanker configuration enables to deliver up to 90.72 T to the lunar surface [64]. The Starship spacecraft will be able to land on the lunar surface [64]. It has been selected by NASA as a Human Landing System for the Artemis program [3] [65] [67]. Its utilization is scheduled for the Artemis 3 and Artemis 4 missions [3] [65] [67] [68]. Important technical details are summarized in Table 2.5.

Table 2.5 Technical parameters of the SpaceX Starship [64]

Parameters	Measurements		
available payload mass capacity	destination: LEO		more than 90.72 T
	destination: GTO		more than 90.72 T
	destination: Moon or Mars surface		more than 90.72 T (refueling in LEO required)
payload bay dimensions	diameter:	height 0.0 m – 8.0 m: 8.0 m height 8.0 m – 17.24 m: 8.0 m – 3.6 m	
	height:	17.24 m (extended version: 22.0 m)	
astronaut capacity	up to 100		
reusability	yes		
docking capabilities	yes, with Lunar Gateway and Starship		

The Integrated Landing System (ILS) has been developed as a Human Landing System (HLS) for lunar surface exploration missions by a team consisting of Lockheed Martin, Northrop Grumman, and Draper under the leadership of Blue Origin. The ILS was initially designed in a three-stage approach [16]. The descent stage was developed by Blue Origin, and the ascent stage, including the crew cabin, was developed by Northrop Grumman. The guidance, navigation, and control system, the avionics system, and additional software systems were developed by Draper. The ILS was designed in a way that launching it using a New Glen Launch Vehicle from Blue Origin or a Vulcan Heavy-Lift Launch Vehicle from United Launch Alliance (ULA) would be possible.

A later version of the ILS that is referred to as the Blue Moon Mark 2 spacecraft was selected as the Human Landing System in 2023 for the Artemis 5 mission. The Blue Moon Mark 2 spacecraft is a reusable spacecraft that allows to descent to the lunar surface and back to lunar NRHO with a crew of up to 4 astronauts. It will be 16 m tall and will have a dry mass of 14.52 T [68]. The spacecraft was designed for an unfueled launch on top of a New Glen Launch Vehicle. A space-tug vehicle will also be involved in missions utilizing the Blue Moon Mark 2 spacecraft, which will be loaded with fuel in Low Earth Orbit, travel to lunar orbit, and refuel the Blue Moon Mark 2 spacecraft there. The vehicle consists out of two propellant tanks mounted on top of each other in the top part and a Crew Exploration Module (CEM) implemented in the lower part. The CEM allows to house a crew of up to 4 astronauts for a duration of 30 days. A docking port was also integrated into the CEM, which enables docking with the Lunar Gateway.

In addition to this, the Blue Moon Mark 1 spacecraft was developed as an uncrewed single launch lunar Cargo Vehicle. It enables global access to the lunar surface, and has cargo capacity of 2.72 T and a height of 7 m [72]. The spacecraft was designed to be launched on top of a New Glen Launch Vehicle. A rendering of the Blue Moon Mark 1 spacecraft is shown in Figure 2.25 on the left side and a rendering of the Blue Moon Mark 2 spacecraft is shown in Figure 2.25 on the right side. The technical parameters of the two spacecraft are summarized in Table 2.6.

Figure 2.25 Rendering of the Blue Moon MK1 (left) and the Blue Moon MK2 spacecraft (right) (image: © Blue Origin [70] [71])

Table 2.6 Technical parameters of the Blue Moon Mark 1 and Mark 2 Spacecraft [68] [69]

Parameters	Blue Moon Mark 1	Blue Moon Mark 2
dry mass		14.52 T
available payload mass capacity	2.72 T	
astronaut capacity	-	up to 4
height	7 m	16 m
mission duration	stays on the lunar surface after the mission	up to 30 days on the lunar surface
reusability	no	yes
docking capabilities	no	yes, with Lunar Gateway and Blue Moon Space Tug

The Human Landing System (HLS) concept developed by Dynetics is a single stage Descent and Ascent Vehicle for the crew transfer between the lunar surface and the Lunar Gateway. In the HLS design, a crew habitat, as well as power, thermal control, and other subsystems have been included. It was designed to be launched on top of an SLS Block 1-B, a Vulcan Centaur or a Vulcan Super-Heavy Lift Launch Vehicle. The crew cabin of the Dynetics HLS has been designed for a two-astronaut crew, enabling a mission on the lunar surface that lasts up to 7 days. It also allows to extend the crew size to a number of 4 astronauts, but in this case the duration during which the HLS is in operation will need to be decreased.

A lunar base shall be utilized in sustaining lunar missions. This base shall be located close to the lunar south pole and it shall be designed to enable the permanent stay for a crew consisting of 4 astronauts for a mission duration of 1 to 2 months. Lunar base subsystems can be subdivided into static systems and mobile systems. Static systems can be defined as the non-moveable subsystems of a lunar base, such as the Electrical Power System (EPS), the lunar habitation module for the astronaut crew or logistic modules. Mobile lunar base systems are defined as those systems that can move on the lunar surface, such as the Unpressurized Rover Lunar Terrain Vehicle (LTV) or the Pressurized Rover Lunar Habitable Mobility Platform (HMP). In addition to these, autonomous systems for the offloading of vehicles or the construction of the lunar base are also conceivable. A lunar base will most likely rely on reoccurring cargo deliveries for lunar base systems and infrastructure, as well as on cargo and crew supplies delivery. The Electrical Power System (EPS) of a future lunar base will most likely employ solar generator systems based on photovoltaics and nuclear fission reactors. Therefore, the location of a lunar base will need to have relatively constant irradiance and temperature conditions. It is recommended to bury a future lunar base under a 1.5 m thick layer of the lunar surface material regolith in order to achieve sufficient shielding against space radiation and protection against meteoroid impacts. Solid structures can be manufactured from regolith with the employment of microwave-based processing technology. A visualization of a possible future lunar base is visualized in Figure 2.26. Important requirements for the design of a lunar base are summarized in Table 2.7.

A possible lunar base will most likely be constructed incrementally. A study from 2006 accounted a construction period between 15 and 20 years until the lunar base would be fully operational and autarkic without a need to rely on regular cargo supplies from the Earth. During Phase 0, autonomous robotic systems and a minimum amount of astronaut activities would be required for construction site preparation. First modules of the lunar base would be deployed in Phase 1. The operation would be carried out employing 3 to 4 astronauts who inhabit the base for periods of 4 to 6 months. Phase 2 has been defined as a 1-year-long built-up phase, during which the capabilities of the base would be extended to a capacity for housing up to 10 astronauts. Complete autarchy and a capacity for up to 100 astronauts would be reached during

Phase 3. Phase 4 has been defined by a lunar base state, where science and commercial activities are carried out. A Phase 4 lunar base would have the capacity to house more than 100 astronauts without a limitation for the duration of stay.

Figure 2.26 Rendering of a possible future moon village (image: © ESA, P. Carril [80])

Table 2.7 Mission specific requirements of a future lunar base [16] [25] [75]

Parameters	Measurements
crew size	4 astronauts
mission duration	1-2 months
location	lunar south pole

2.5 Research about Lunar Missions

Currently, a number of science projects is studying aspects of lunar bases. The think tank Spaceship EAC at the European Astronaut Centre (EAC) in Cologne conducts lunar mission-related research in the fields of energy production & storage, ISRU, additive manufacturing, virtual reality for training and exploration, radiation shielding, robotics, human factors, and exercising countermeasures. Exemplary outcomes are the lunar regolith simulant material EAC-1, the robotic system Marvin, the sensor system Sensor Confetti or the lunar energy systems simulation software LES-3.

The Moon and Mars Base Analog (MaMBA) is a project at the Center of Applied Space Technology and Microgravity (ZARM) of the University of Bremen. MaMBA is a lunar base concept consisting of 5 to 6 interconnected and independent cylindrical 2-story habitat modules. MaMBA focuses on the study of systems and human factors. A visualization of the MaMBA concept on the Moon is shown in Figure 2.27.

Figure 2.27 Moon and Mars Base Analog (MaMBA) (image: © Space is More [81])

The lunar analogue facility LUNA has recently started its operation at the European Astronaut Centre (EAC). This facility allows the simulation of missions on the lunar surface and can test new technologies for lunar exploration. An overview of the LUNA facility is shown in Figure 2.28 and an image of two astronauts in Extravehicular Mobility Units on the test bed of the facility is shown in Figure 2.29.

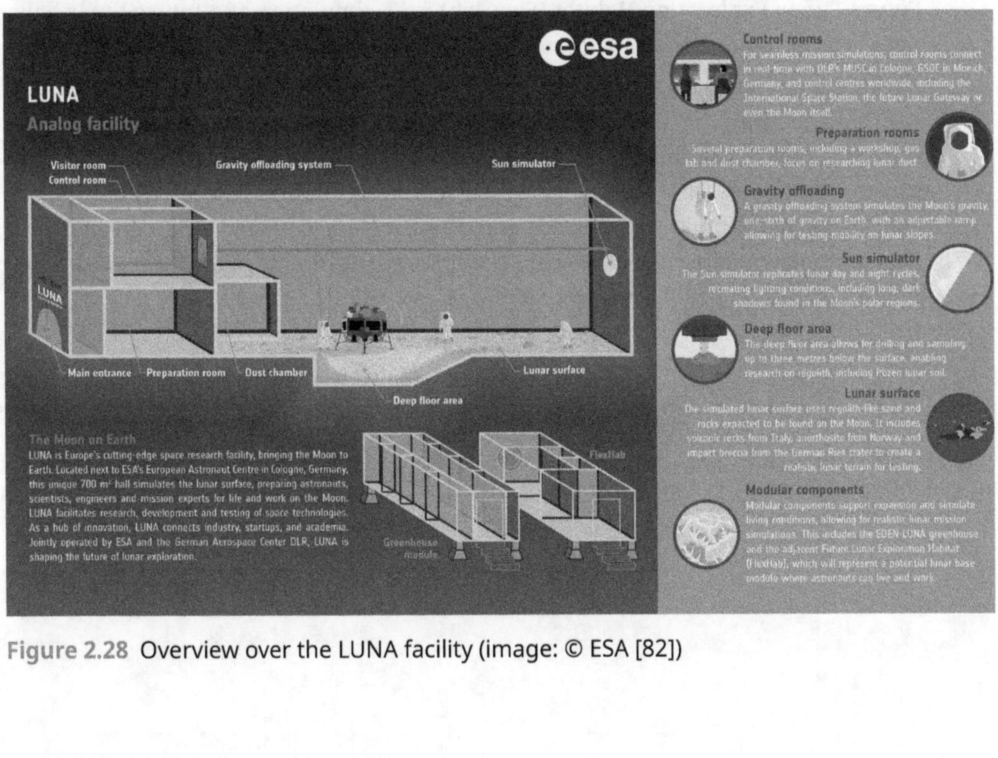

Figure 2.28 Overview over the LUNA facility (image: © ESA [82])

Figure 2.29 Astronauts simulating an Extravehicular Activity in the LUNA facility at the European Astronaut Centre (EAS) (image: © ESA/DLR [83])

The LUNA facility consists of a preparation room, a control room, a high dust facility, and a 700 m^2 hall filled with the regolith simulant material EAC-1. The hall includes a deep floor that allows to drill up to 3 m deep to collect samples. A sun-simulator has been implemented that allows to reproduce the lighting conditions at the lunar poles. In the future, the implementation of a gravity offloading system allowing the simulation of Extravehicular Activities on a lunar gravity level, the implementation of the Future Lunar Exploration Habitat (FlexHab) for a simulation of a future lunar habitat, as well the implementation of the greenhouse Eden-ISS are intended.

PANGEA is an analogue training campaign developed by the European Astronaut Centre (EAC) focusing on field geology in volcanic areas, for example on Lanzarote. During a PANGEA training, astronauts, operation engineers, and scientists work together to test new technologies and to train astronauts in field geology for the human exploration of the lunar surface and the surface of Mars. A photo of a simulated Extravehicular Activity during a PANGEA training campaign is shown in Figure 2.30.

Figure 2.30 ESA's PANGEA training campaign (image: © ESA, A. Romeo [84])

The Cooperative Adventure for Valuing and Exercising human behaviour and performance Skills (CAVES) is a training campaign developed at the European Astronaut Centre (EAC) [85]. CAVES is a 3-week-long training for astronauts in critical environments. During a CAVES training, astronauts have to conduct science and test new technologies, while exploring a cave for 6 days. They have to achieve their mission goal relying on ground control and their own skills. One of the focuses of the CAVES training is working in an intercultural team in critical environments. An image taken during a caves training is shown in Figure 2.31.

Figure 2.31 ESA's CAVES training campaign (image: © ESA, A. Romeo [85])

2.6 Landing Sites

Landing sites have to be evaluated based on scientific and operational objectives, like the access to permanently shadowed regions, the local geology, the launch date, the access to solar energy, the local temperature and temperature variations, the availability of a direct line of sight to the Earth and the surface configuration. In order to enable a save landing, the surface configuration needs to be as flat as possible without the presence of greater rocks.

The Schrödinger Basin within the Aitkin Basin at the lunar south pole has been classified as the location of the most scientific interest by the lunar and planetary institute at the Lyndon B. Johnson Space Center in an earlier study. For the upcoming landing on the lunar surface as part of the Artemis 3 mission, currently 13 regions within a circle of 6 degrees around the lunar south pole have been identified as candidates for possible landing sites. These landing sites have different geological properties of potential scientific interest, as well as the possibility of save landings enabled by relatively smooth surface properties and a continuous access to sunlight for a minimum period duration of 6.5 days. Possible regions are the Faustini Rim A, the Peak Near Shackleton, the Connecting Ridge and Connecting Ridge Extension, the de Gerlache Rim 1 & 2, the de Gerlache-Kocher Massif, Haworth, the Malapert Massif, the Leibnitz Beta Plateau, the Nobile Rim 1 & 2, and the Amundsen Rim [9]. These landing site candidates are visualized in Figure 2.32.

Figure 2.32 Visualization of 13 possible landing sites of Artemis 3 (image: © NASA [30])

3 Environmental Conditions

The lunar surface can be subdivided into two types of regions: The dark parts of the surface are referred to as lunar mare and the bright parts are referred to as lunar highlands. The lunar maria were originally formed from lava flows. As a result, these regions mainly consist of basalt. The lunar maria cover 17% of the lunar surface and the lunar highlands cover 83% [14]. Lunar maria can mainly be found on the lunar near side, where their share of the surface is 44% and where the lunar highlands share is 66% [14]. While the lunar maria have very smooth surface properties, the lunar highland terrain is very rough. This results from the age of the lunar highlands. These ancient parts of the surface have been exposed to meteoroid impacts for a long period of time. The different regions on the lunar surface are defined by different surface compositions. In the lunar mare, mostly basaltic volcanic rocks with significant contents of iron and titanium can be found. Pristine rocks with potassium, phosphorus, and pyroxene contents can mostly be found in the lunar highlands. Breccias resulting from meteoroid impacts can be found in lunar craters. In addition to this, formations of melted rocks can also be found in lunar craters that result from energy absorption during meteoroid impacts. The lunar surface is covered with the fine-grained material regolith. Regolith can also be defined as a highly abrasive material. It consists out of minerals, glass formed on the impacts of asteroids, rock fragments and volcanic ash. It also has a high iron content. The composition of the subsurface material and the fine-grained regolith on the lunar surface have similar material compositions, but differences can be observed between lunar highlands and lunar maria. The surface is generally very dry, with only a minority of locations where water resources can be assumed. Figure 3.1 shows images of the lunar near and far sides.

Figure 3.1 Photo of the lunar near side (left) and the lunar far side (right) (image: © NASA/Goddard Space Flight Center/Arizona State University [88] [89])

It takes the same time for the Moon to rotate one time around its rotational axis, as it takes for the Moon to complete one orbit around the Earth. As a result, only the lunar near side is constantly facing towards the Earth and the lunar far side is never facing the Earth. The areas on the lunar surface that are illuminated by direct solar irradiance change during the orbit of the Moon around the Earth. The change of the Moon's visibility from the Earth results from this condition, as only the illuminated parts of the lunar surface are visible from the Earth. The illumination conditions of the Moon during a lunar orbit around the Earth are visualized in Figure 3.2 for an observer who is located north of the rotational axes of the Earth and the Moon.

A lunar day/night cycle has a duration of 655.72 h [11] [20]. In most lunar locations, the lunar day and the lunar night have relatively equal durations. For locations between northern and southern latitudes below 70 degrees, around 14 days can each be accounted for the lunar day and the lunar night in a good approximation [93]. Illumination data from the lunar surface analyzed by Cook in 2000 suggests that close to the Shackleton crater nearly constant illumination conditions can be experienced. Resulting from its topology, the lunar night only has a duration of around 73 h in this location [93] [75]. The Peary crater has similar illumination conditions on the lunar north pole [87].

In addition to this, the Moon does not have a significant atmosphere, so the atmospheric condition on the lunar surface can, in a good approximation, be described as a vacuum. The total mass of the lunar atmosphere has a value of around $2.5 \cdot 10^3$ kg, which results in a surface pressure of $3 \cdot 10^{-15}$ bar [11].

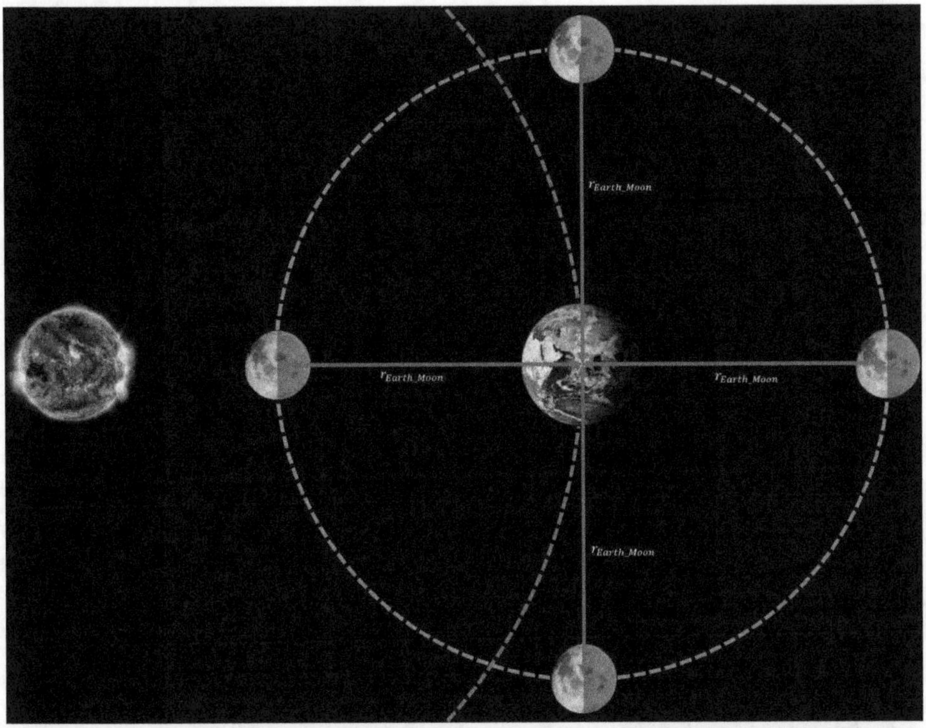

Figure 3.2 Illumination conditions on the lunar surface during a lunar orbit around Earth seen by an observer north of the lunar orbital plane: There is always only one lunar half-sphere illuminated by the Sun (used images: © NASA [90] [91] [92]).

On the lunar equator, temperature variations between minimum temperatures of –178.2 °C at the transition from night to sunrise and maximum temperatures between 113.9 °C and 123.9 °C can be observed [11] [134]. The temperature difference between shadowed and unshadowed areas can have a magnitude greater than 300 K [87]. As the distance between the Earth and the Sun varies during the Earth's orbit around the Sun, the distance between the Moon and the Sun also varies. Following this, the solar constant for the Moon has seasonal variations. Since the inclination of the Moon's rotational axis to the ecliptic only is around 1.5°, the seasonal variation in the progression of the Sun over the lunar surface and in the solar elevation angle only has very low values [87]. However, these temperature values have to be defined as the temperature of objects on the lunar surface. As the Moon does not have a significant atmosphere, the lunar temperature must not be mistaken for the ambient temperature that can be experienced on the Earth.

Due to its high iron content, regolith can be electrostatically charged. It can be detached from the lunar surface, for example from astronauts during Extravehicular Activities (EVAs), by the wheels of lunar rover vehicles, by impacts on the surface, by mining and In-Situ Resource Utilization (ISRU) activities or by the engine thrust during landing and launch maneuvers. In the following, regolith can form clouds of electrostatically charged dust. Resulting from this, electrostatically charged regolith can accumulate on surfaces, when it is detached from the ground. Accumulated regolith can alter the surface properties of systems like radiators, solar generators or optical systems. In addition to this, the direct exposure of astronauts to the lunar surface material poses significant health risks. To prohibit this, the application of dust mitigation methods, such as a grounding of surfaces, is highly recommended.

The high iron content of the lunar surface material can also result in conductive coupling between power lines that are implemented on the surface. The electrical conductivity of the lunar soil ranges from $1.0 \cdot 10^{-14} \frac{\Omega}{\mathrm{m}}$ up to $1.0 \cdot 10^{-9} \frac{\Omega}{\mathrm{m}}$ and its thermal conductivity ranges from $1.4 \cdot 10^{-4} \frac{\mathrm{W}}{\mathrm{cm \cdot K}}$ up to $3.0 \cdot 10^{-4} \frac{\mathrm{W}}{\mathrm{cm \cdot K}}$ [75].

The seismic activity of the Moon seems to be very low, since it does not have plate tectonics. Moonquakes have not been recorded by the Apollo surface experiments.

Since the Moon does only have a relatively small mass compared to the Earth, the gravitational force experienced on the lunar surface has a value in the magnitude of 1/6 of the gravitational force experienced on the Earth [14] [11]. In lunar orbit, only microgravity can be experienced.

The Moon does not have a significant magnetic field. Because of this, the lunar surface is also not shielded against space radiation. Ionizing radiation on the lunar surface mainly consists out of protons, electrons, and heavy ions. The biggest source of charged particles results from solar flares, which occur sporadically. Other sources of the ionizing radiation are Galactic Cosmic Rays (GCR), solar cosmic rays and the solar wind.

3.1 Natural Sources of Energy

Different sources of primary energy are conceivable for different space missions. If a source can be utilized by a space Electrical Power System (EPS) depends on its availability during the mission. Energy can be harvested from electromagnetic radiation emitted by a celestial body like the Sun. Chemical types of energy sources available on a celestial body could be mined and utilized in a process for the generation of electrical energy. It is also possible to harvest energy from the interplanetary magnetic field or the magnetic field of celestial bodies. Another possible approach would be the utilization of interplanetary radiation. When physical energy is available during a mission, for example in the form of wind or tides as mechanic sources of energy, or in the form of a geothermal heat as a thermal source of energy, these sources could also be utilized in order to generate electrical energy during a space mission. In addition to sources of energy that could be harvested from the environment of a space system, it is also possible to carry energy stored within an energy storage inside a space system and utilize it during the mission. For very short durations, primary batteries can be utilized, for intermediate durations, fuel cells can be utilized, and for the utilization during longer periods of time, nuclear sources of energy can be utilized. The types of energy sources available during a certain mission vary, since the availability of the different primary energy sources also varies.

For most missions, available sources of primary energy are limited to solar energy, nuclear energy and energy harvested from magnetic fields. When solar energy is available constantly or during reoccurring periods, this type of energy is usually chosen. The utilization of solar energy is suitable for missions roughly until the distance of Mars or the inner asteroid belt from the Sun. Since the solar intensity decreases with the square of the distance to a radiation source, from this distance to the Sun further outward, the required size of a solar generator would reach an order of magnitude that would be impossible to realize within a certain mission. Nuclear isotopes can be integrated into generative systems in order to enable a power generation independent from the external environmental conditions. Due to the long duration of the lunar night, nuclear energy systems have an advantage as a primary source of energy for the application in lunar missions. Energy systems utilizing the nuclear decay have a relatively steady output power, since the nuclear decay cannot be influenced, while systems utilizing nuclear reactions can increase or decrease their power output.

It is also often mentioned that Helium 3 could be mined on the lunar surface and utilized in a nuclear fusion process. However, since this technology is not yet ready developed for applications on the Earth, an application on a lunar base is highly unlikely. Spacecraft moving through space can also harvest energy from the intergalactic magnetic field. An approach has already been implemented on board of two American-Italian spacecraft, but there is no proof of concept yet because both missions failed.

3.2 Solar Energy in the Lunar Environment

There are different types of electromagnetic irradiance that can be received in space and on the lunar surface. Different sources can be differentiated by the spectrum of electromagnetic radiation they emit. The direction from which electromagnetic irradiance is received from a certain surface is depending on the geometrical configuration between the source of electromagnetic radiation and a receiving surface. The Sun is the main source of electromagnetic irradiance on the lunar surface. The lunar day/night cycle and its effect on the illumination conditions on the lunar surface causes the main difference in the availability of electromagnetic irradiance between the lunar surface and lunar orbit. During lunar day, the unaltered solar irradiance can be observed on the lunar surface, but during the lunar night, direct solar irradiance cannot be observed. A mean amount of direct solar irradiance in the magnitude of around 1360 $\frac{W}{m^2}$ is available from the Sun in perfect incidence conditions [11]. Seasonal variation caused by the Moon orbiting the Earth and by the Earth orbiting the

Sun can be observed in direct solar irradiance on the lunar surface. Around the 4th of January, direct solar irradiance in a magnitude of around 1410 $\frac{W}{m^2}$ can be experienced on the lunar surface, while around the 9th of July, solar irradiance in a magnitude of around 1310 $\frac{W}{m^2}$ can be experienced [87]. Starlight can be observed on the lunar surface, but its amount is neglectable compared to electromagnetic irradiance received from the Sun. Solar irradiance received by the Earth and partially reflected can be received on the lunar surface and in lunar orbit as the Earth's albedo irradiance in a mean magnitude of 0.07 $\frac{W}{m^2}$ in perfect incidence conditions, while the individual value varies between 0 $\frac{W}{m^2}$ and 0.145 $\frac{W}{m^2}$ in a sinusoidal progression during one lunar orbit [111]. The Earth is also emitting irradiance in the infrared spectrum, which can be received on the lunar near side and in lunar orbit in a mean magnitude of 0.07 $\frac{W}{m^2}$ in perfect incidence conditions, while the individual values varies in a sinusoidal movement around 0.07 $\frac{W}{m^2}$ during a lunar orbit [111]. The albedo irradiance of the Moon itself consists mostly of reflected direct solar irradiance and particles from cosmic radiation, such as neutrons, which result from the interaction of cosmic rays with lunar regolith. On the lunar surface, a maximum mean value of lunar albedo irradiance can be expected in a magnitude of 28.9 $\frac{W}{m^2}$ in perfect incidence conditions. The Moon is also emitting irradiance in the infrared spectrum. Close to the equator during noon, maximum values of lunar infrared irradiance can be received. During noon in perfect incidence conditions, a mean value of lunar infrared irradiance in a magnitude of 126 $\frac{W}{m^2}$ can be received [111]. For spacecraft calibration, lunar radiation has been well studied by Lane & Irvine (1973), followed up by the USGSs ROLO observation program. Figure 3.3 shows an overview of the different types of mean irradiance that can be received during a lunar orbit around the Earth on the lunar surface close to the equator by a photovoltaic generator in perfect incidence conditions for the respective type of irradiance. However, as different types of irradiance are usually received from different directions, being in perfect incidence conditions for one source of irradiance usually results in imperfect incidence conditions of the remaining types of irradiance.

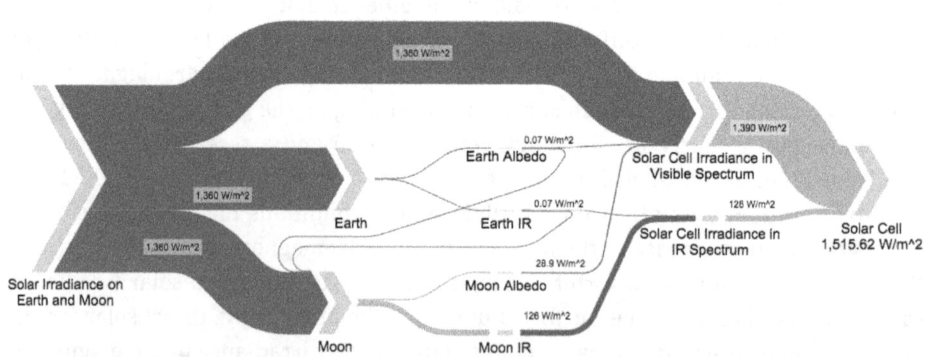

Figure 3.3 Energy transferred from the Sun and the Earth to a photovoltaic generator on the lunar surface

The Sun consists of 91% Hydrogen, 8% Helium and 1% other elements [97]. Inside the Sun, a constant nuclear fusion process is going on, during which Hydrogen reacts to Helium at a temperature of $15 \cdot 10^6$ K [97] [109]. The main reaction employs four hydrogen atoms that react to one helium atom. As side products, two positrons, two neutrinos, and a certain amount of energy result from the main reaction.

$$4\,{}^{1}_{1}p \longrightarrow {}^{4}_{2}\alpha + 2\,e^{+} + 2\,\nu_e + \Delta E_{Reaction}$$

The total mass of the reactant atoms is greater than the total mass of the reaction products by 0.0264 u [109]. This circumstance is usually referred to as mass defect. The mass–energy equivalence can be utilized in order to calculate the amount of energy that is released by the reaction.

$$\Delta E_{Reaction} = \Delta m_{reaction} \cdot c_0^2 = 24.687\ \mathrm{MeV}$$

The reaction and the resulting energy emission in the form of γ radiation is taking place in the center of the Sun. On the path from the center of the Sun to its surface, a constant absorption and reemission cycle is taking place, which results in the thermalization of the complete amount of γ radiation. The energy resulting from this process is transported to the surface of the Sun, where it is emitted into space as electromagnetic irradiance. The complete radiant power of electromagnetic radiation emitted by the Sun into space can be calculated utilizing the mass conversion rate $\frac{\Delta m}{dt}$ of the Sun and the mass–energy equivalence.

$$E_{Sun,total} = \Delta m_{total} \cdot c_0^2$$

$$\phi_{Sun,total} = \frac{E_{Sun,total}}{dt} = \frac{\Delta m_{total}}{dt} \cdot c_0^2 = 3.845 \cdot 10^{26}\ \mathrm{W}$$

The parameter radiant exitance is defined as the total amount of radiant flux, emitted through the surface of a source of irradiance, normalized to the surface area of the emitting source [109].

$$M_{Sun} = \frac{\phi_{Sun,total}}{A_{Sun,Surface}} = 63.3\ \frac{\mathrm{MW}}{\mathrm{m}^2}$$

Based on the Stefan-Boltzmann law, the surface temperature of a source of electromagnetic radiation can be determined utilizing the radiant exitance.

$$T_{Surface} = \sqrt[4]{\frac{M_{Sun}}{\sigma}}$$

The wavelength at which the spectrum of electromagnetic radiation reaches its maximum can be determined based on Wien's displacement law utilizing the surface temperature of the radiation source.

$$\lambda_{Emission,Max} = \frac{2897\ \mu\mathrm{m} \cdot \mathrm{K}}{T_{Surface}}$$

The spectrum of electromagnetic radiation emitted by the Sun can be described by the spectrum of a black body in a good approximation. A measure for the spectrum of radiation is the spectral radiance, which is defined as the radiant power emitted through the surface of a radiation source with reference to the wavelength of electromagnetic radiation. Figure 3.4 shows a visualization of the solar spectrum and the spectrum of a black body with reference to the radiation wavelength.

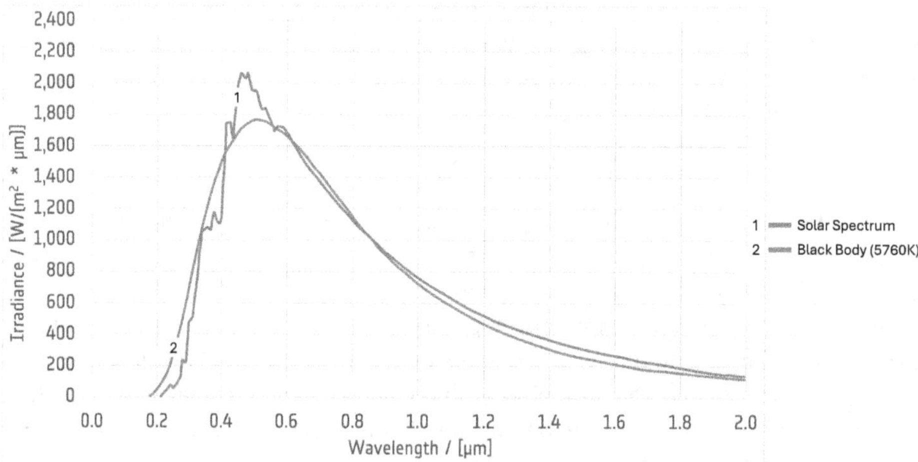

Figure 3.4 Solar spectrum and spectrum of a black body (image: data from NASA [110])

The spectral radiance of a black body can be calculated for a given temperature with reference to Planck's law. This temperature value can be determined employing Wien's displacement law or the Stefan-Boltzmann Law [97] [109].

$$L_{Planck}(\lambda, T_{Radiation}) = \frac{2 \cdot h \cdot c_0^2}{\lambda^5} \cdot \frac{1}{e^{\frac{h \cdot c_0}{k_B \cdot \lambda \cdot T_{Radiation}}} - 1} \cdot \frac{1}{\Omega_0}$$

The spectral radiant exitance is defined as the spectral radiance that is emitted through a certain element of the surface of a celestial body. It has to be determined, taking into account the radiation geometry of the radiation source. For sources radiating omnidirectional, it is sufficient to multiply the spectral radiance with π.

$$M(\lambda, T_{Radiation}) = L_{Planck}(\lambda, T_{Radiation}) \cdot \pi$$

The parameter spectral irradiance is defined as the amount of spectral radiant exitance that can be received by another surface in a certain distance to the surface of the source of radiation. It is determined by scaling the spectral radiant exitance by a factor relating the surface area of the radiation source to the surface area of a theoretic sphere. This theoretic sphere has a radius equal to the distance between the center of the radiation source and the surface receiving the radiation. Both the theoretic sphere and the source of radiation have to have coincident centers.

$$E_\lambda(\lambda, T_{Radiation}) \cdot A_{Sphere}(r_{Observer}) = M(\lambda, T_{Radiation}) \cdot A_{Sphere}(r_{Source})$$

$$E_\lambda(r_{Observer}, \lambda, T_{Radiation}) = M(\lambda, T_{Radiation}) \cdot \frac{A_{Sphere}(r_{Source})}{A_{Sphere}(r_{Observer})}$$

$$= M(\lambda, T_{Radiation}) \cdot \frac{4 \cdot \pi \cdot r^2_{Source}}{4 \cdot \pi \cdot r^2_{Observer}}$$

$$E_\lambda(r_{Observer}, \lambda, T_{Radiation}) = M(\lambda, T_{Radiation}) \cdot \frac{r^2_{Source}}{r^2_{Observer}}$$

The parameter irradiance is defined as the integral of the spectral irradiance over all wavelength.

$$E_0(r_{Observer}, T_{Radiation}) = \int_0^\infty E_\lambda(\lambda, T_{Radiation})\, d\lambda$$

$$E_0(r_{Observer}, T_{Radiation}) = \int_0^\infty M(\lambda, T_{Radiation}) \cdot \frac{r^2_{Source}}{r^2_{Observer}}\, d\lambda$$

$$E_0(r_{Observer}, T_{Radiation}) = \frac{r^2_{Source}}{r^2_{Observer}} \cdot \pi \cdot \int_0^\infty L_{Planck}(\lambda, T_{Radiation})\, d\lambda$$

During the orbital movement of the Earth and the Moon, the distance between the center of the Sun and a location on the surface of the Moon will be subject to variations. Following this, an observer on the surface of the Moon will also experience variations in the amount of irradiance it receives from the Sun. During one orbit of the Earth around the Sun, the value of direct solar irradiance on the lunar surface varies between around 1310 $\frac{W}{m^2}$ and around 1415 $\frac{W}{m^2}$ [111]. The influence of the Earth's orbital movement around the Sun on the amount of received direct solar irradiance on the lunar surface is much greater than the influence of the Moon's orbital movement around the Earth. The variation in direct solar irradiance received during a lunar day only has very small values. The same is true for variations caused by different lunar locations. Figure 3.5 shows the progression of direct solar irradiance received on the lunar surface in different latitudes during a lunar day/night cycle for both of the Earth's orbit extrema.

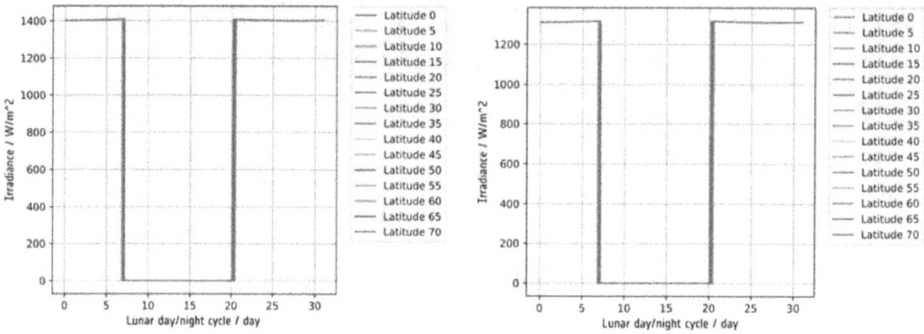

Figure 3.5 Direct solar irradiance received during a lunar day/night cycle on the lunar surface with reference to the latitude when the Earth is located at its perihelion (left) and when the Earth is located at its aphelion (right)

Ley et al. and Quaschning offer special equations to calculate the mean irradiance during one year on the Earth that can be received on the surface of the Earth or in Earth orbit [97] [109]. Both equations utilize the mean irradiance that can be received on the Earth at perigee together with a scaling factor dependent on the day in a year on the Earth. The equation featured in the work of Quaschning determines the irradiance received on the Earth for a day in an Earth year and the equation featured in the work of Ley et al. uses the number of days since the last perigee passage for the equation [97] [109]. In both equations, a day has to be defined by a value between 1 to 365, whereby 365 corresponds to the maximum number of days in a year.

$$E_{0,VQ}(d_{Year,Earth}) = E_{0,Perigee} \cdot (1 + 0.0334 \cdot \cos(0.9855° \cdot d_{Year,Earth} - 2.7198°))$$

$$E_{0,Ley}\left(d_{Year,Earth}\right) = E_{0,Perigee} \cdot \left(1 + 0.033 \cdot \cos\left(2\pi \cdot \frac{d_{Year,Earth}}{365}\right)\right)$$

These approaches deliver good approximations for the seasonal change in the mean irradiance observed on the Earths orbital track, but they assume that the observer is located in the center of the Earth. Contrary to this, the definition of the geometrical configuration of the solar system and the space systems allows to dynamically define the location of a solar generator on the surface of a celestial body or in orbit around it in order to determine the amount of received irradiance with reference to the distance between the solar generator and the irradiance source. A model for this purpose utilizing the concept of sequential reference frames can be found in Zünkler [111].

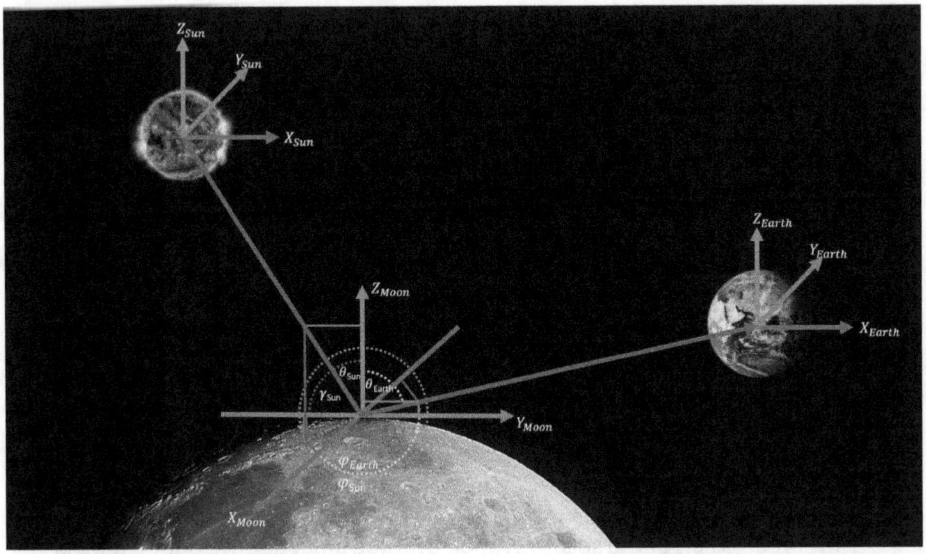

Figure 3.6 Elevation of the Sun and the Earth over a location on the lunar surface (used images: © NASA [90] [91] [92])

In the beginning of a lunar day, the Sun rises over the lunar horizon at a location on the lunar surface. From this point in time, it rises until noon and sinks from then on until the end of a lunar day, where the Sun sinks below the horizon of a lunar surface location. The solar elevation angle is a measure for the height of the Sun over the lunar horizon. It is defined as the angle between the lunar surface and the irradiance vector. Complementary to this, the surface incidence angle is defined as an angle between the irradiance vector and the surface normal.

During a lunar day, the solar elevation has maximum values during lunar noon. As the tilt of the lunar rotational axis only has a value around 1.5°, the progression of the solar elevation has similar progressions over the northern and southern lunar hemisphere. At the lunar equator, the Sun can rise up to a solar elevation of 90°. From the equator to the poles, the maximum solar elevation values decrease with the lunar latitude. The progression of the solar elevation during a lunar day/night cycle is visualized in Figure 3.7.

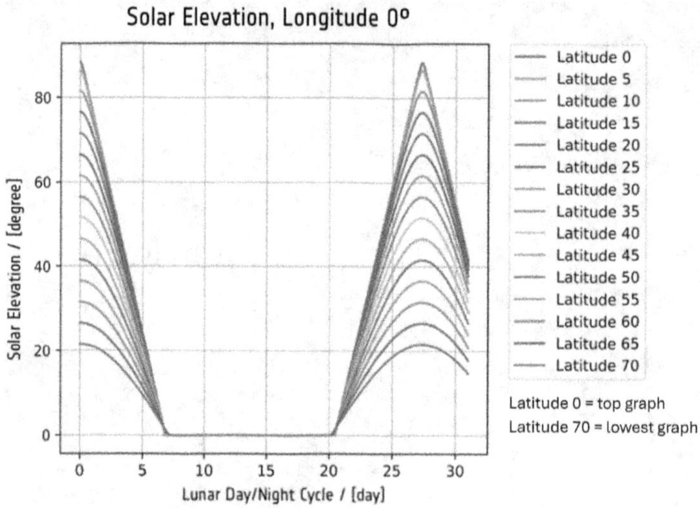

Figure 3.7 Progression of the solar elevation over the lunar surface during a lunar day/night cycle

Irradiance was defined normalized to the unit area. Following this, the total amount of radiant power received by a surface in space can be determined from the product of the irradiance value and the projected area of the receiving surface that is perpendicular to the irradiance vector. The size of this projected area can be determined by employing the cosines of the incidence angle as scaling factor for the total area of the receiving surface. The incidence angle is defined as the angle between the irradiance vector and the surface normal of the receiving surface.

$$A_{projected}\left(\theta_{incidence}\right) = A_{Observer} \cdot \cos(\theta_{incidence})$$

$$P\left(r_{Observer}, T_{Radiation}, \theta_{incidence}\right) = E_0\left(r_{Observer}, T_{Radiation}\right) \cdot A_{projected}\left(\theta_{incidence}\right)$$

The amount of albedo irradiance emitted by the Earth is highly dependent on the type of surface of the Earth underneath a spacecraft. Where glaciers and white clouds reflect a great share of the received irradiance into space, darker surfaces like forests, grasslands or oceans tend to absorb a significant share of the received irradiance. However, with an increasing altitude, the influence of the local variation of the albedo on the received Earth's albedo irradiance decreases. Especially for the distance of the Moon to the Earth, it is sufficient to calculate the amount of Earth's albedo irradiance as a mean value. Figure 3.8 shows a visualization of the progression of the Earth's albedo irradiance as received by a spacecraft in Earth orbit with reference to the spacecraft altitude.

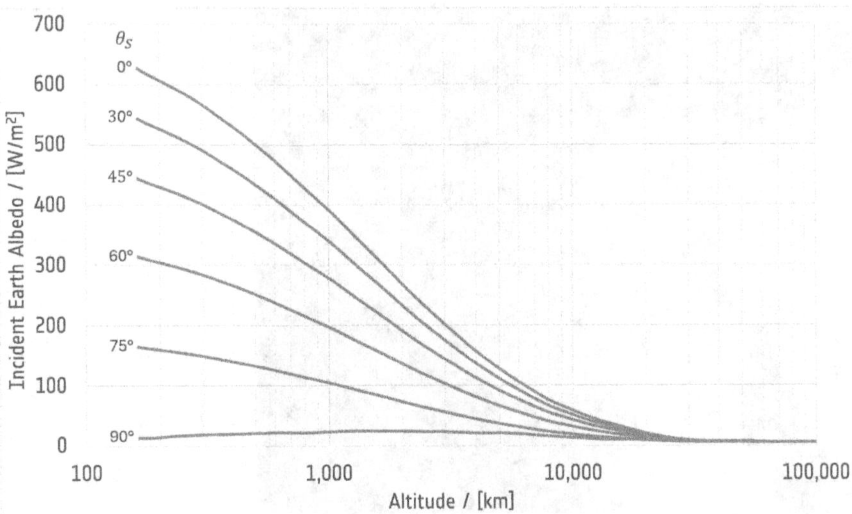

Figure 3.8 Earth's albedo irradiance for a satellite as a function of Earth orbit altitude (Data from NASA [110])

For the calculation of albedo irradiance, so-called albedo factors are employed that relate the total amount of received irradiance to the amount of reflected irradiance. The Earth's albedo factor varies between 0.05 and 0.6 [11] [97]. Although extensive models of the Earth's albedo exist, Ley et al. recommend to use a mean albedo factor of 0.3 and NASA indicates a mean albedo factor of 0.306 [11] [97]. The amount of albedo irradiance a satellite in orbit receives from the Earth is also dependent on the spacecraft altitude. Wertz and Larson offer an equation for the calculation of an altitude dependent scaling factor for the Earth's albedo factor [98] [112]. This equation depends on the angular radius of the Earth as experienced from the orbiting spacecraft, which is another measure for the spacecraft altitude.

$$\rho_{Earth} = \sin^{-1}\left(\frac{r_{Earth}}{r_{Earth} + H_{Sat}}\right)$$

$$K\left(\rho_{Earth}\right) = 0.664 + 0.521\,\rho_{Earth} - 0.203\,\rho_{Earth}^{2}$$

Only the day half-sphere of the Earth is receiving direct solar irradiance, and because of this, only the day half-sphere of the Earth is emitting the Earth's albedo irradiance. From space, the Earth also features visibility phases, similar to the Moon phases experienced from the Earth's surface. Figure 3.9 shows an image taken during the Artemis 1 mission from the Orion spacecraft, in which the visible half-spheres of the Earth and the Moon can be seen. These Earth phases are the reason for the variation of the Earth's albedo factor.

Figure 3.9 Orion spacecraft with the Earth and the Moon, with only a half-sphere of the Earth and of the Moon being illuminated (image: © NASA [113])

The phase parameter can be utilized as a measure for the "Earth phase." It features a value of 0 at "Full Earth," from there it increases over 0.25 at "Half Earth" to 0.5 at "New Earth." From "New Earth" onwards, the phase parameter increases over 0.75 at "Half Earth" to 1 at "Full Earth." The parameter can be calculated with reference to the true anomaly of a spacecraft or the Moon in an ecliptic orbit around the Earth. This equation assumes that the orbit perigee is located on the direct line between the centers of the Earth and the Sun, behind the Earth as seen from the Sun. For spacecraft in other orbits, an offset has to be implemented in the equation.

$$\xi_{Phase}(\nu) = \frac{1 + \frac{\nu_{Anomaly}}{180^\circ}}{2}$$

Based on the phase parameter, the length of the minor half axis of the visible Earth disc can be determined. For the calculation, it was assumed that the total area of the Earth disc features a value of π [114]. Following this, the length of the minor half-axis is equal to 1 at "Full Earth." During a lunar orbit, its value decreases over 0 at "Half Earth," where only half of the Earth disc is visible, to –1 at "New Earth."

$$s(\nu_{Anomaly}) = \cos(2\pi \cdot \xi_{Phase}(\nu_{Anomaly}))$$

Based on a geometrical representation, the area of the visible Earth disc can be calculated with reference to the following equation. The scaling factor for the share of the Earth disc that is visible in a certain position in Earth orbit or on the lunar surface can be calculated by normalizing the visible area to the maximum value of π for the visible area during "Full Earth."

$$A_{Earth}(v_{Anomaly}) = \frac{\pi}{2} + \frac{\pi}{2} \cdot s(v_{Anomaly})$$

$$\max(A_{Earth}) = \pi$$

$$\frac{A_{Earth}(v)}{\max(A_{Earth})} = \frac{\frac{\pi}{2} + \frac{s(v_{Anomaly}) \cdot \pi}{2}}{\pi} = \frac{1}{2} + \frac{s(v_{Anomaly})}{2} = \frac{1}{2} + \frac{\cos\left(\pi \cdot \left(1 + \frac{v_{Anomaly}}{180°}\right)\right)}{2}$$

In order to scale the Earth's albedo factor, this scaling factor has to be multiplied by the product of 2 and the mean value of the albedo factor. The resulting albedo factor varies between 0 and 0.612 around a mean value of 0.306. In order to include the altitude dependency, the altitude dependent scaling factor for the Earth's albedo also needs to be multiplied.

$$a_{Albedo,Earth}(v_{Anomaly}) = a_{Albedo,Earth,mean} \cdot 2 \cdot \frac{A_{Earth}(v_{Anomaly})}{\max(A_{Earth})} \cdot K(\rho_{Earth})$$

$$a_{Albedo,Earth}(v_{Anomaly}) = a_{Albedo,Earth,mean} \cdot \left[1 + \cos\left(\pi \cdot \left(1 + \frac{v_{Anomaly}}{180°}\right)\right)\right] \cdot K(\rho_{Earth})$$

Earth albedo radiation is defined as direct solar irradiance that was reflected by the Earth. Following this, the albedo radiant exitance emitted by the Earth's surface can be determined by calculating the direct solar irradiance received in the center of the Earth and multiplying the true anomaly and angular radius dependent albedo factor to it. In the next step, the amount of albedo irradiance can be calculated by scaling the albedo radiant exitance for the distance between the center of the Earth and an observer in lunar orbit or on the lunar surface in the same way as this has been done for solar irradiance.

$$E_{Earth,Sun}(r_{Sun,Earth}, T_{Sun}) = \frac{r^2_{Sun}}{r^2_{Sun,Earth}} \cdot \pi \cdot \int_0^\infty L_{Planck}(\lambda, T_{Sun})\, d\lambda$$

$$E_{Earth,Albedo}(r_{Earth,Observer}, T_{Sun}) = a_{Albedo,Earth}(v_{Anomaly}, \rho_{Earth}) \cdot \frac{r^2_{Earth}}{r^2_{Earth,Observer}} \cdot E_{Earth,Sun}(r_{Sun,Earth}, T_{Sun})$$

On the surface level of the Earth, it is emitting 237 $\frac{\text{W}}{\text{m}^2}$ of irradiance in the infrared spectrum [101]. The Earth's spectral infrared radiance can be determined in the same way as direct solar irradiance utilizing the value of the Earth's black body temperature. Based on this, the amount of infrared radiation reaching an observer on the lunar surface from the Earth can be determined employing the distance between the observer on the lunar surface or in lunar orbit and the center of the Earth. A visualization of the progression of received infrared irradiance in the Earth orbit can be found in Figure 3.10.

$$E_{Earth,IR}(r_{Earth,\ Observer}, T_{Earth}) = \frac{r^2_{Earth}}{r^2_{Earth,Observer}} \cdot \pi \cdot \int_0^\infty L_{Planck}(\lambda, T_{Earth})\, d\lambda$$

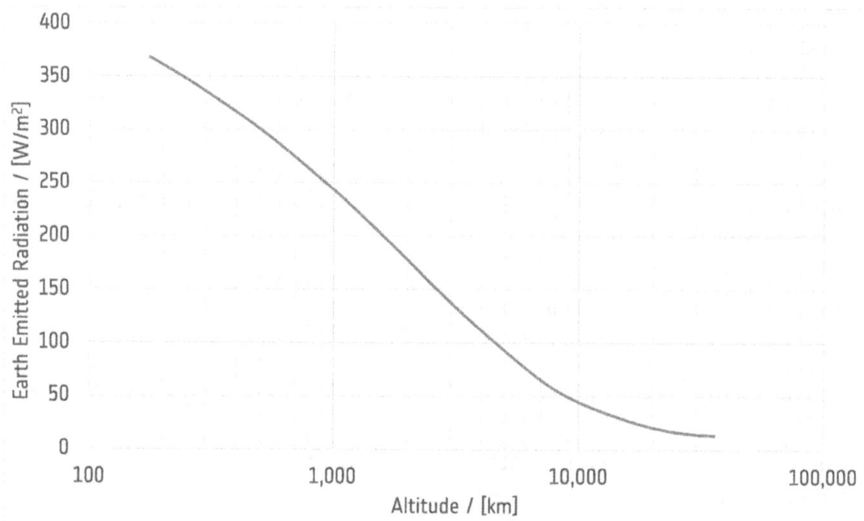

Figure 3.10 Earth's infrared irradiance as a function of the Earth orbit altitude (data from NASA [110])

The parameter horizontal irradiance is a measure for the amount of irradiance that a surface receives. In the case of perfect incidence conditions, a surface receives the total amount of available irradiance of a certain type, but in the case that the surface normal is perpendicular to the irradiance vector, the surface is not receiving irradiance of this type. The horizontal irradiance is defined as the product of irradiance and the cosines of the incidence angle. For the incidence on the lunar surface, the local surface incidence angle has to be employed and for the incidence on a photovoltaic generator, the incidence angle on the generator front side needs to be employed. Different values of the incidence angle have to be utilized for irradiance received from different sources like the Sun or the Earth.

$$E_{Hor.,Sun}(r_{Sun,Observer}, T_{Sun}, \theta_{incidence,Sun}) = \cos(\theta_{incidence,Sun}) \cdot E_{Sun}(r_{Sun,Observer}, T_{Sun})$$

$$E_{Hor.,Earth,Albedo}(r_{Earth,Observer}, T_{Sun}, \theta_{incidence,Earth}) = \cos(\theta_{incidence,Earth}) \cdot E_{Earth,Albedo}(r_{Earth,Observer}, T_{Sun})$$

$$E_{Hor.,Earth,IR}(r_{Earth,Observer}, T_{Earth}, \theta_{incidence,Earth}) = \cos(\theta_{incidence,Earth}) \cdot E_{Earth,IR}(r_{Earth,Observer}, T_{Earth})$$

For a surface that does not feature perfect incidence conditions for direct solar irradiance this circumstamce reduces the amount of available direct solar irradiance. For the lunar surface, or a solar array that would be lying directly on the lunar surface, perfect incidence conditions would only occur in equatorial locations during local noon. As an example, the progression of the horizontal direct solar irradiant for the lunar surface is visualized in Figure 3.11.

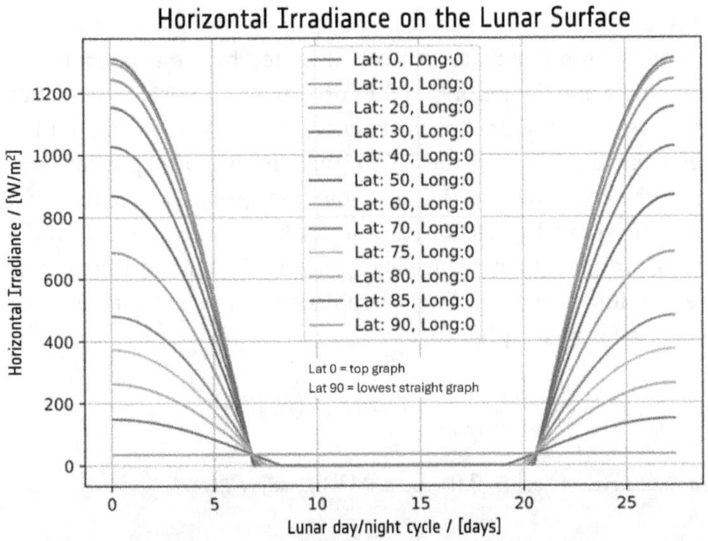

Figure 3.11 Progression of horizontal direct solar irradiance during a lunar day/night cycle for a location on the lunar longitude zero degree for different latitudes during an apogee passage of Earth

As a last step, the total radiant power a certain surface on the lunar surface or in lunar orbit receives has to be determined. For this operation, the projected area of the surface to the irradiance needs to be determined with reference to the individual incidence angles. Subsequently, the received radiant power can be determined by multiplying the projected area values with the different irradiance values.

$$A_{projected,Sun}(\theta_{incidence,Sun}) = A_{Surface} \cdot \cos(\theta_{incidence,Sun})$$

$$A_{projected,Earth}(\theta_{incidence,Earth}) = A_{Surface} \cdot \cos(\theta_{incidence,Earth})$$

$$P_{Earth,Sun}(r_{Earth,Observer}, T_{Sun}, \theta_{incidence,Sun}) = E_{Earth,Sun}(r_{Earth,Observer}, T_{Sun}) \cdot A_{projected,Sun}(\theta_{incidence,Sun})$$

$$P_{Earth,Albedo}(r_{Earth,Observer}, T_{Sun}, \theta_{incidence,Earth}) = E_{Earth,Albedo}(r_{Earth,Observer}, T_{Sun}) \cdot A_{projected,Earth}(\theta_{incidence,Earth})$$

$$P_{Earth,IR}(r_{Earth,Observer}, T_{Earth}, \theta_{incidence,Earth}) = E_{Earth,IR}(r_{Earth,Observer}, T_{Earth}) \cdot A_{projected,Earth}(\theta_{incidence,Earth})$$

The lunar albedo has also been studied extensively. It was determined that the lunar albedo varies highly with the surface composition of the Moon. The lunar surface properties and the lunar topography have an influence on the received lunar albedo irradiance. NASA indicates the mean lunar albedo with a value of 0.11 [11]. Lunar albedo irradiance can be calculated based on the direct solar irradiance and the Earth's albedo irradiance received on the lunar surface. The spectral horizontal irradiance received by a location on the lunar surface can be calculated by the product of the individual spectral irradiance and the cosines of the corresponding surface incidence angle. In order to calculate the total amount of irradiance a certain location on the lunar surface receives from the Sun and the Earth, the sum of both spectral horizontal irradiances has to be calculated. In the next step, the spectral irradiance of a certain lunar surface location can be calculated as a product of the resulting value and 0.5 and the albedo factor of the luar surface location. Based on the resulting spectral albedo irradiance, the lunar surface albedo irradiance can be calculated as an integral of this curve over all wavelengths.

$$E_{\lambda,Sun}(\lambda, r_{Sun,Observer}, T_{Sun}) = \frac{r_{Sun}^2}{r^2_{Sun,Observer}} \cdot \pi \cdot L_{Planck}(\lambda, T_{Sun})$$

$$E_{\lambda,Earth,Albedo}(\lambda, r_{Earth,Observer}, T_{Sun}) = a_{Albedo,Earth}(v_{Anomaly}, \rho_{Earth}) \cdot \frac{r_{Earth}^2}{r^2_{Earth,Observer}} \cdot E_{\lambda,Earth,Sun}(\lambda, r_{Sun,Earth}, T_{Sun})$$

$$E_{\lambda,Hor.,Sun}(\lambda, r_{Sun,Observer}, T_{Sun}, \theta_{incidence,Sun}) = E_{\lambda,Sun}(\lambda, r_{Sun,Observer}, T_{Sun}) \cdot \cos(\theta_{incidence,Sun})$$

$$E_{\lambda,Hor.,Earth,Albedo}(\lambda, r_{Earth,Observer}, T_{Sun}, \theta_{incidence,Earth}) = E_{\lambda,Earth,Albedo}(\lambda, r_{Earth,Observer}, T_{Sun}) \cdot \cos(\theta_{incidence,Earth})$$

$$E_{\lambda,Hor.,Surface}(\lambda, r_{Sun,Observer}, T_{Sun}) = E_{\lambda,Hor.,Sun}(\lambda, r_{Sun,Observer}, T_{Sun}, \theta_{incidence,Sun}) + E_{\lambda,Hor.,Earth,Albedo}(\lambda, r_{Earth,Observer}, T_{Sun}, \theta_{incidence,Earth})$$

$$E_{\lambda,Albedo,Surface}(\lambda, r_{Sun,Observer}, T_{Sun}) = \frac{1}{2} \cdot a_{Albedo,Moon} \cdot E_{\lambda,Hor.,Surface}(\lambda, r_{Sun,Observer}, T_{Sun})$$

$$E_{Albedo,Surface}(\lambda, r_{Sun,Observer}, T_{Sun}) = \int_0^\infty E_{\lambda,Albedo,Surface}(\lambda, r_{Sun,Observer}, T_{Sun})\, d\lambda$$

The horizontal lunar albedo irradiance that is received by a certain surface, such as the front side of a photovoltaic generator, can be calculated based on the angle between the surface normal of the lunar surface and the surface normal of the receiving surface. Employing this approach, a surface that is parallel to the lunar surface would not receive any lunar albedo irradiance, and a surface that is perfectly perpendicular to the lunar surface would receive the complete amount of available lunar albedo irradiance.

$$E_{Hor.Albedo,Surface}(\lambda, r_{Sun,Observer}, T_{Sun}, \theta_{Surface}) = E_{Albedo,Surface}(\lambda, r_{Sun,Observer}, T_{Sun}) \cdot (1 - \cos(\theta_{Surface}))$$

Infrared radiation is also emitted by the lunar surface, which has a mean black body temperature of 270.4 K [11]. The individual lunar surface temperature features a high variability that varies during lunar day/night cycle and with the lunar surface properties. It decreases from equatorial locations to polar locations. Following this, the amount of emitted infrared irradiance also varies with the lunar day/night cycle, surface properties, and lunar latitude. For the calculation of the lunar surface infrared irradiance, the temperature of the lunar surface in a certain location has to be determined. This can be done employing a thermal model based on the Stefan-Boltzmann law and the surface properties of the lunar surface material (see Section 3.4). Gaier et al. determined the absorptivity and emissivity properties for different lunar regolith simulants [115]. They indicated a mean absorptivity of 0.827 and a mean emissivity of 0.871 for the lunar mare. For the lunar highlands, a mean absorptivity of 0.502 and a mean emissivity of 0.895 was indicated by them.

$$T_{Surface} = f_{Thermal\ Model}(E_{Hor.,Sun}, E_{Hor.,Earth,Albedo}, E_{Hor.,Earth,IR}, \vartheta_{Sun}, \vartheta_{Earth}, \varepsilon_{LunarSurface}, \alpha_{LunarSurface})$$

Based on the resulting lunar surface temperature, the spectral lunar surface infrared irradiance can be calculated as black body irradiance in a similar way, as this has also been implemented for solar irradiance and the Earth's infrared irradiance.

$$E_{Moon,IR}(T_{Surface}) = \pi \cdot \int_0^\infty L_{Planck}(\lambda, T_{Surface})\, d\lambda$$

The amount of lunar infrared irradiance that is received by a certain surface, is referred to as horizontal lunar surface infrared irradiance. It can be calculated based on the angle between the surface normal of the receiving surface and the surface normal of the lunar surface, as this has also been done for lunar albedo irradiance.

$$E_{Hor.,Moon,IR}(T_{Surface}) = E_{Moon,IR}(T_{Surface}) \cdot (1 - \cos(\theta_{Surface}))$$

In addition to the radiation emitted by the Sun and the Earth, systems on the lunar surface also radiate infrared radiation with reference to their surface temperature and surface properties. The amount of infrared radiation radiated by a surface can be determined utilizing the Stefan-Boltzmann law. For the determination of the emitted radiant power, the emissivity parameter of the surface needs to be known.

$$P_{System,Emitted,IR}(A_{Surface}, \varepsilon_{IR}, T_{Surface}) = A_{Surface} \cdot \varepsilon_{IR} \cdot \sigma \cdot T^4_{Surface}$$

$$E_{System,Emitted,IR}(A_{Surface}, \varepsilon_{IR}, T_{Surface}) = \frac{P_{System,IR}(A_{Surface}, \varepsilon, T_{Surface})}{A_{Surface}} = \varepsilon_{IR} \cdot \sigma \cdot T^4_{Surface}$$

3.3 Radiation in the Earth-Moon System

The space radiation environment in the Earth-Moon system is characterized by high energy particle fluxes, high energetic electromagnetic radiation, and the microwave background radiation of space. The Sun is the main source for particle radiation in the Earth-Moon system and its sphere of influence reaches roughly until Jupiter's orbit. Electromagnetic radiation can be received on the lunar surface mainly from the Sun and for a minor extend from the Earth. To a great extent, particle radiation in space consists of electrons, protons and ions. The greatest share of particles is charged and only a small share of the particles is uncharged. The movement of charged particles in space is determined by magnetic field lines. A visualization of the different types of radiation in the solar system is shown in Figure 3.12.

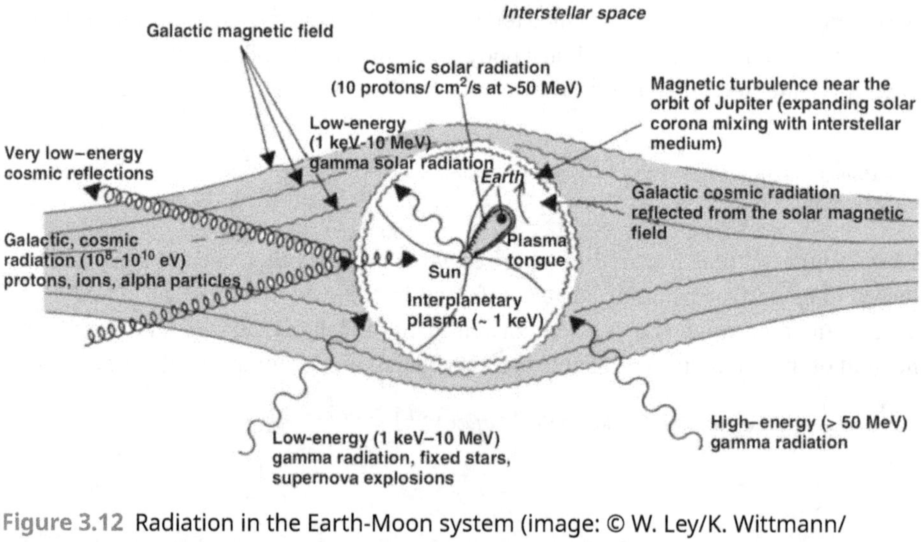

Figure 3.12 Radiation in the Earth-Moon system (image: © W. Ley/K. Wittmann/W. Hallmann [119])

Particles can be trapped inside the Earth's magnetic field, where particles can travel along the field lines between the magnetic north and south poles. Charged particle fluxes can be observed as trapped charged particle radiation in the Earth's magnetosphere, where the particle population reaches its maxima in the Van Allen radiation belts. The composition of the trapped radiation particles varies with the orbit altitude. In the radiation belts, protons with energies from 100 keV to more than 100 MeV as well as electrons with energies from 10 keV to more than 10 MeV can be observed [97] [98] [116]. The observed fluxes can be characterized as high and steady. The radiation belts do not reach the lunar orbit track and are therefore not of particular interest for studies about radiation in the lunar environment.

In the Earth-Moon system, charged particle fluxes constantly emitted by the Sun can be observed in the form of solar wind. With reference to the eleven-year-long solar cycle, the solar activity and the solar wind fluxes from the Sun vary over time. A measure for the solar activity is the total number of dark spots that can be observed on the Sun's surface. Sun spots are resulting from the Sun's magnetic activity. During a solar cycle, the maximum number of Sun spots occurs during solar maximum and the minimum number occurs during solar minimum. A visualization of the total number of Sun spots with reference to 5 solar cycles can be found in Figure 3.13 (top image).

Solar wind can be described as a plasma flux consisting of protons, electrons, and helium ions. It has a density of 6 cm^{-3}, a temperature range from 3500 K to 500,000 K, and a season-dependent velocity from 200 km/h to 2000 km/h [97] [101]. It travels from the Sun through the solar system and its trajectory is determined by the solar magnetosphere. Besides the regular solar flux, drastic peak fluxes can also be observed during Solar Particle Events (SPEs). These can basically happen any time, but the possibility for an SPE is higher during solar maximum. Solar Particle Events result from solar flares and coronal mass ejections. If the Earth-Moon system is close to the solar field line on which the solar flare is occurring on the Sun, SPEs can be observed in the Earth-Moon system. During an SPE, a huge number of charged particles featuring energies between 1 MeV and more than 1 GeV, as well as x-ray radiation, can be observed [97]. SPEs usually last from hours to days, but the greatest charges can be observed in their beginning. They have the potential to degrade solar arrays in the magnitude of a few percent. Figure 3.13 (bottom) shows a visualization of the proton flux for different particle energies during an SPE.

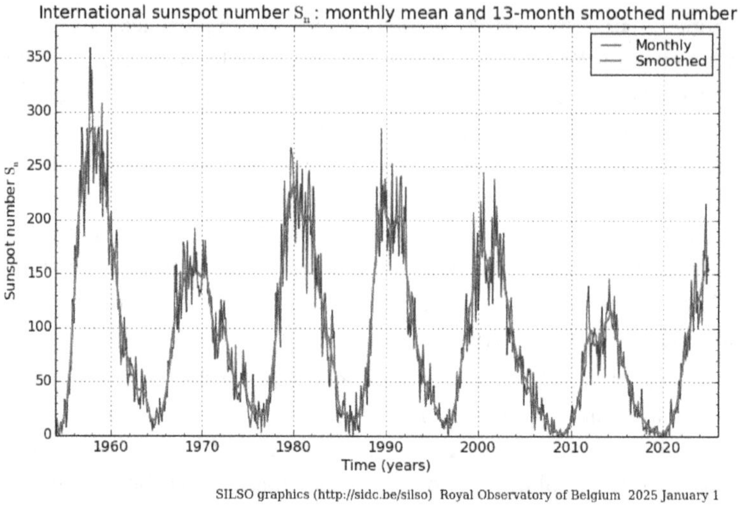

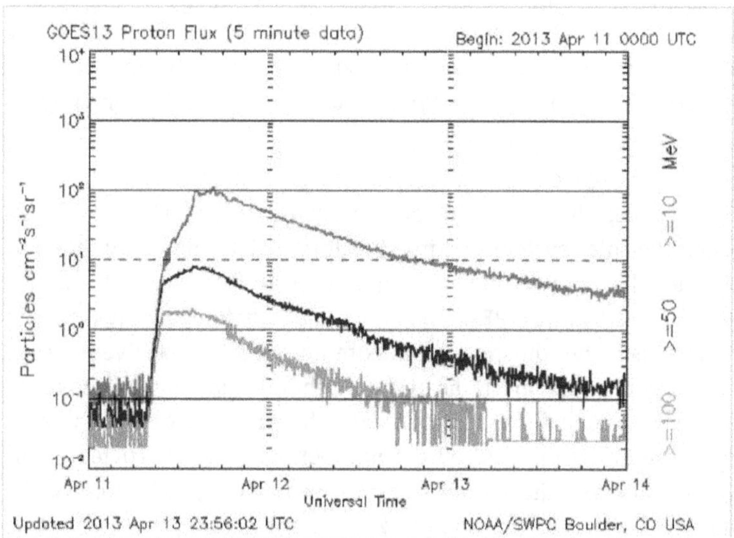

Figure 3.13 Sun spot number during a number of solar cycles (top), proton fluxes of a Solar Particle Event (SPE) on 13/04/2013 (bottom) (top image: © SIDC, Royal Observatory of Belgium; bottom image: © NOAA [116] [120])

Galactic Cosmic Rays (GCRs) have their origin outside the solar system and can be observed in the Earth-Moon system as omnidirectional space radiation. They have low fluxes, but individual particle energies can reach up to giga electron volts. The flux of GCRs varies with the solar cycle because the solar magnetic field modulates Galactic Cosmic Rays. GCRs mainly consist of hydrogen and helium ions.

The Earth's magnetosphere would have the shape of a perfect dipole in the absence of disturbances. When the solar wind is reaching the outer boundary of the Earth's magnetosphere, the so-called magnetopause, it slows down due to the impact into the Earth's magnetic field and a shock wave is traveling around the magnetosphere. Due to the impact, the solar wind is transferring a share of its kinetic energy into the Earth's magnetic field and, by this, deforms the shape of the field. Because of this, the side of the magnetic field facing the Sun is compressed and only reaches around 10 Earth radii into space. Its magnetic field on the side of the Earth facing deep space is reaching deep into space for more than 100 Earth radii. This end of the magnetic field is called the magnetotail. Due to the energy transfer from solar wind to the magnetic field, a share of this energy is stored inside the Earth's magnetic field. The magnetic field does not feature an infinite capacity, and because of this, the stored energy is dissipated in the form of magnetic storms from time to time, which can reach from the magnetotail until geosynchronous orbit.

Radiation in the lunar environment has been measured by a number of different experiments. On the lunar surface, the solar wind spectrometer that was part of the ALSEP (Apollo Lunar Surface Experiments Package), which was brought to the lunar surface with Apollo missions, measured electron fluxes featuring energies between 5 eV and 40 eV [121]. Local variations in the electron flux have been observed in connection with a weak global lunar magnetic field. More recently, the radiation during transit and in lunar orbit has been investigated by a number of experiments on the spacecraft Kaguya, Lunar Reconnaissance Orbiter, Chandrayaan-1 and Lunar Prospector. On board of Lunar Prospector, the electron reflectometer measured electrons utilizing 15 energy bins from 38 eV to 17 keV [117] [122] [125]. On the Lunar Reconnaissance Orbiter, the cosmic ray telescope for the effects of radiation measured the ionizing dose of Galactic Cosmic Rays and solar energetic protons featuring energies equal to or greater than 10 MeV [117] [122] [125]. On board of Chandrayaan-1, the RADOM-FM experiment measured particle fluxes of electrons with a minimum energy of 0.85 MeV and of protons with a minimum energy of 17.5 MeV [117] [123] [124]. Figure 3.14 shows the results from the measurements of the RADOM-FM experiment. During transit from the Earth to lunar orbit and in higher lunar altitudes, relatively constant fluxes were observed by the RADOM-FM experiment and a decrease in the flux rates was observed with decreasing lunar altitude. The weakening of the particle fluxes was related to a weak global lunar magnetic field. In addition to this, a particle upstream from the lunar surface was observed. The particle upstream was related to the impacting solar wind, the electrostatically charged surface and the formation of a weak plasma sheet around the Moon. Measurements indicate that fluxes caused by Galactic Cosmic Rays are relatively constant and that fluxes caused by solar particles are highly variable. Observed fluxes consisted of 2.0% electrons and positrons, 85.26% protons, as well as 11.76% Helium ions, and 0.98% consisted of higher order ions [127].

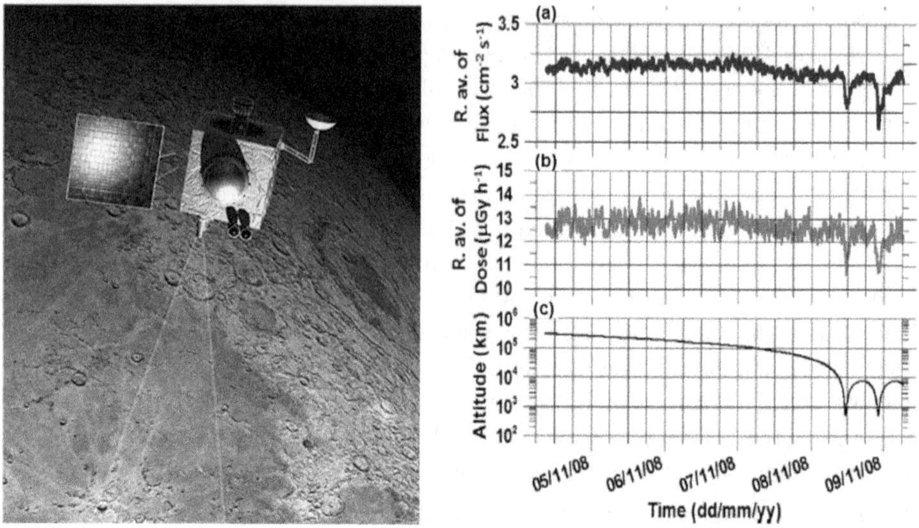

Figure 3.14 Visualization of Chandrayaan-1 spacecraft (left), lunar altitude, dose, and flux rates measured by RADOM-FM (right) (images: © D. Roam (spacecraft)/Dachev et al. (plots) [117] [124] [126])

Radiation can have problematic consequences for space systems in different ways. Low energy particles can accumulate on the surface of a spacecraft and high energy particles can penetrate spacecraft materials. These effects can cause system malfunctioning or damages, as well as health-related issues in crewed missions. Surfaces and materials can be altered, charges can be injected into semiconductors, and their internal structure can be altered by impacting charged particles.

Charged particle accumulation on surfaces can cause Electrostatic Discharge (ESD). Accumulated charges can build up over time and discharge in arches, once a critical charge is reached or a certain operational condition occurs. Arching can be external, for example from a solar array to a plasma flux, or internal, for example from the spacecraft or system structure to a subsystem. Electrostatic discharge can produce high currents and related spacecraft component damage. Figure 3.15 shows an example for ESD damage of solar cells on the right side.

Figure 3.15 SEB in a programmable logic (left) and ESD damage on a solar array (right) (image on the left: © NASA, K. A. LaBel; image on the right: © NASA [116] [128])

Charged particle fluxes impose hazardous risks to space missions as charged particles interact with matter. Absorbed radiation doses, bremsstrahlung, and ionization on spacecraft surfaces pose risks to spacecraft. Highly charged particles can penetrate space systems. During penetration, secondary particles can be formed inside the penetrated structures, leading to secondary radiation, whose effects can be even more severe than the effects of the original particles. The radiation effects can be differentiated between a constant radiation and sudden impacts of heavy charged particles. Constant radiation yields to an accumulation of radiation in space systems, which can be seen in the form of trapped charges. In electronics, trapped charges can yield to malfunctioning, such as leakage currents, threshold and time changes, as well as failures. Electronic parts can be damaged, when a charged particle is flying through an electronic part, by leaving an ionized trace in the part, or by the insertion of a charged particle into the electronic part. When a single particle is causing issues with electronics, this is often referred to as a Single Event Phenomenon (SEP). SEPs can be temporary, such as bit flips resulting from the deposited charge or as a latch-up, where a part stops working and draws excessive currents. These temporary effects can disappear over time or after a power cycle of the system. In addition to this, high charges can also cause total failures of systems by the deposition of high charges, which is often referred to as Single Event Burnout (SEB) or hard error. The effect of a SEB in a field programmable logic can be seen in Figure 3.15 on the left side. In addition to the effects radiation can have on space systems, it can also cause issues in the communication with spacecraft and it can be hazardous to an astronaut crew.

3.4 Temperature on the Lunar Surface

The temperature on the surface of the Moon has been studied for a long time. Even before the first lunar landings, J. C. Jaeger conducted studies about the surface temperature based on numerical calculations in the early 1950s. The first in-situ measurements were carried out during the Apollo missions. The solar cells employed during the Apollo 11 mission reached operating temperatures of 88 °C and the magnetometer employed in the Apollo 12 mission reached a maximum temperature of 78 °C [74]. Temperature measurements regarding the lunar surface were also conducted utilizing the Diviner experiment on board of the Lunar Reconnaissance Orbiter from 2009 onward. Regolith does not store a significant amount of thermal energy. Thermal anomalies can be explained by variations in the albedo and the surface composition, such as dense rocks or impact craters. Because the heat transfer from the lunar core is also very low, the surface temperature of the Moon is determined to a major extend by the balance of received and emitted electromagnetic radiation. Therefore, the temperature on the lunar surface is highly dependent on the available electromagnetic radiation during the lunar day/night cycle. Temperatures during the lunar night can be seen as comparatively constant in the domain from 117 K to 95 K [134]. Temperatures start to increase during the lunar dawn with an increase in the available irradiance. Surface temperatures at the end of lunar dawn feature values of around 133 K [134]. During the lunar day, the surface temperature can reach values of up to 397 K at the equator [134]. Surface temperatures at the end of lunar dusk feature values of around 163 K [134]. Higher standard deviations during the lunar dawn and the lunar dusk can be related to variations in the lunar albedo. The mean surface temperature during a lunar day/night cycle is visualized for different latitudes for lunar maria (left) and for lunar highlands (right) in Figure 3.16.

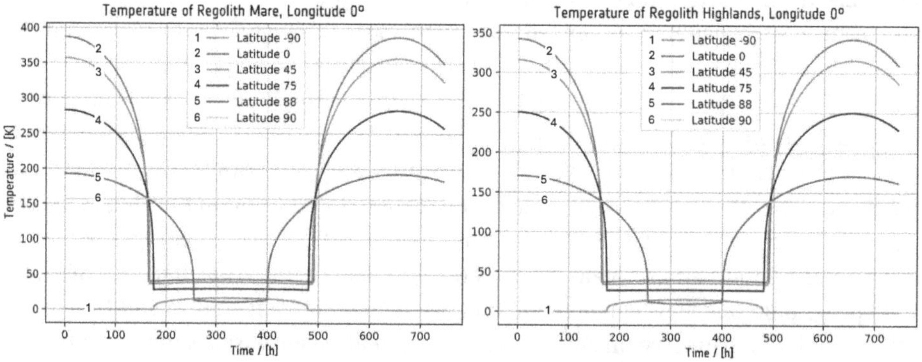

Figure 3.16 Simulated progression of the mean value of lunar surface temperature during a lunar day/night cycle for lunar maria and lunar highlands in an orbital condition where the Earth is located at its perihelion and the lunar rotational axis is pointing towards the Sun

The progression of the surface temperature during a lunar day is proportional to the progression of the solar incidence angle. The local incidence angle for solar irradiance determines the projected area of the lunar surface to direct solar irradiance. In the following, it also determines the amount of radiant power that a location on the lunar surface receives. A relation for the lunar surface temperature during local noon was developed by Williams et al. based on the analysis of remote sensing measurements [134]. Their equation allows to calculate the lunar surface noon temperature in a good approximation with reference to the latitude of a location, the solar constant, as well as the emissive and absorptive properties of the lunar surface utilizing the Stefan-Boltzmann law:

$$A_{Albedo,Moon}(\theta_{incidence}) = A_{Albedo,Moon,0} + a_{Albedo,Moon,0} \cdot \frac{\theta^3_{incidence}}{45°} + b_{Albedo,Moon,0} \cdot \frac{\theta^8_{incidence}}{90°}$$

$$T_{Surface,Moon}(\theta_{incidence}) = \sqrt[4]{E_0 \cdot (1 - A_{Albedo,Moon}(\theta_{incidence})) \cdot \frac{\cos(\theta_{incidence})}{\sigma \cdot \varepsilon_{Lunar\ Surface}}}$$

The temperature of the lunar surface during the lunar night has a different progression. At the end of the lunar day, the temperature drops to a very low value and from that point on, it decreases in a slow linear progression during the lunar night, after which temperatures rise again with the appearance of the Sun over the lunar horizon. Spacecraft are usually thermally designed, for example, in order to conduct thermal energy away from parts that would otherwise tend to overheat, such as power electronics, or in order to keep other parts, such as scientific instruments, within their working temperature range.

Thermal energy can be transmitted by convection, which is the thermal energy transfer mechanism employing gasses or liquids. Because of the absence of fluids and significant atmosphere in space and on the lunar surface, heat transfer by convection does only play a neglectable role in the heat transfer from the space environment to an object on the surface. Thermal energy can also be transported by conduction, which is a transfer mechanism through solids. It can be conducted inside a solid and between touching surfaces. The heat transfer results from the transfer of mechanical energy between atoms inside or on the surface of a solid. The thermal flux between two surfaces depends on different parameters, such as the roughness of the surfaces, the thickness of the gap between both surfaces, or the properties of a possible filling medium. The third mechanism of thermal energy transfer is thermal radiation. Surfaces featuring a certain temperature radiate and absorb energy in the form of electromagnetic radiation. The amount and the spectrum of radiated and absorbed electromagnetic radiation is depending on the surface properties and the surface temperature. Space systems usually radiate thermal energy in the infrared spectrum.

In the operation of space systems, internal power dissipation in subsystems usually leads to an increase in the system temperature, so that an amount of thermal energy usually has to be transported inside the space system and released constantly during operation. Simple mechanisms for the transfer of thermal energy inside the spacecraft are thermal conduction or thermal radiation. A widely applied but more complex method for the transport of thermal energy is provided by so-called heat pipes that employ a phase change from the liquid phase to the gaseous phase and back to transfer thermal energy from one point to another point. These systems feature a hot and a cold end. On the hot end, a liquid inside the heat pipe is evaporated, which will cool down the hot end to a certain extend. In the next step, this vapor will condensate on the cold end and thus increase the temperature of the cold end. In addition to this, the utilization of cooling loops employing a working medium is also a common method. The transport of thermal energy with cooling loops can be controlled by controlling the speed, at which a pump is pumping the working medium through the cooling loop. Thermal energy can be released from space systems utilizing technologies such as radiators on the outside structure of the spacecraft. Radiators are surfaces that have high radiating properties. They can also be combined with cooling loops, which enables them to control the amount of thermal energy that is radiated into space. Another method is the utilization of shutters that can be incrementally opened and closed to expose a hot surface inside the spacecraft to space.

The temperature of materials in space is subject to the Stefan-Boltzmann law. Every object featuring a certain temperature emits radiation with a certain radiant power, which can be determined by the objects surface area and its emissive surface property.

$$P_{Emission} = \sigma \cdot \varepsilon_{IR} \cdot A_{Surface} \cdot T_{Surface}^4$$

There are 3 mechanisms for the interaction between objects and electromagnetic radiation. The received electromagnetic radiation can be reflected at the surface of an object, absorbed by an object, and transmitted through it. For each mechanism, a specific factor can be defined in order to calculate the amount of received radiant power that is absorbed by, transmitted through, or reflected by it. The sum of these factors needs to be equal to one.

$$1 = \alpha_{Optic} + \rho_{Optic} + \tau_{Optic}$$

The absorptance defines the amount of received radiant power that is absorbed by an object, the reflectance defines the amount of radiant power that is reflected at the surface of the object, and the transmittance defines the amount of radiant power that is transmitted through the object. By multiplying the radiant power with the absorptance, reflectance, or transmittance, the amount of absorbed, reflected, and transmitted radiant power can be determined. The sum of these 3 types of radiant power has to be equal to the amount of received radiant power.

$$P_{Absorption} = \alpha_{Optic} \cdot P_{incoming}$$

$$P_{Reflexion} = \rho_{Optic} \cdot P_{incoming}$$

$$P_{Transmission} = \tau_{Optic} \cdot P_{incoming}$$

$$P_{incoming} = P_{Absorption} + P_{Reflexion} + P_{Transmission}$$

A surface emits and absorbs electromagnetic radiation at a certain wavelength in an equal magnitude. The property of a certain material to absorb and emit electromagnetic radiation is dependent on the radiation wavelength. Different from the 3 parameters, the spectral absorptance curve defines for each wavelength the magnitude in which a surface emits and absorbs electromagnetic radiation. Figure 3.17 shows the spectral absorptance of different materials with reference to the radiation wavelength.

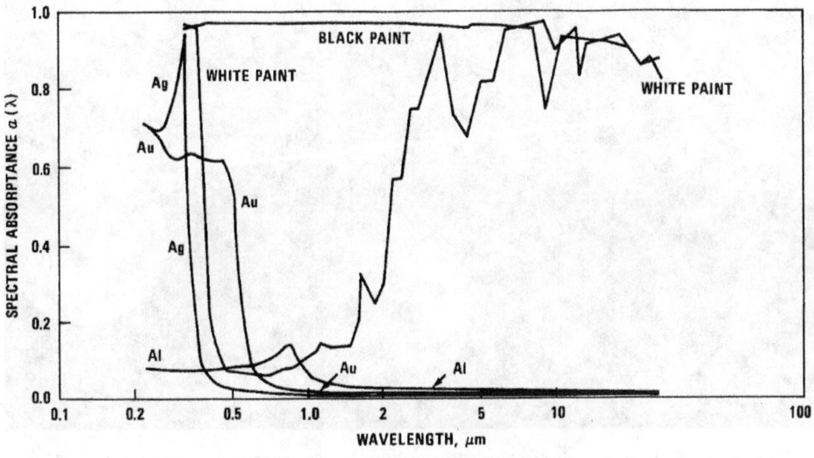

Figure 3.17 Spectral absorptance of typical surfaces (image: © NASA [110])

The spectral absorptance of black paint is relatively constant for different wavelengths and because of this, a black-painted surface emits and absorbs electromagnetic radiation in the visible and infrared spectrum. White paint has only small spectral absorptance values for visible light, but comparatively high values for shorter and longer wavelengths. In space, a surface painted in white will only absorb a little part of visible light, but emit electromagnetic radiation in the infrared spectrum. Because of this, white-painted surfaces are generally cooler than surfaces painted with a surface finish featuring a higher absorptance for electromagnetic radiation in the visible spectrum.

The spectral absorptance and emittance properties of surface materials can be employed in the thermal design of a space system. Subsystems can be cooled passively by the application of a surface finish like white paint, in order to increase the emissions in the infrared spectrum and in order to decrease the absorption in the visible spectrum. In the same way, subsystems can be passively heated by the utilization of a surface coating that absorbs a significant amount of the available electromagnetic radiation.

The equilibrium temperature of objects in space can be determined with reference to the balance of absorbed and emitted electromagnetic radiation employing the Stefan-Boltzmann law. In the Earth-Moon system, direct solar irradiance, the Earth's albedo and infrared irradiance, as well as lunar albedo and infrared irradiance are available. A visualization of the different types of electromagnetic irradiance in the Earth-Moon system are shown in Figure 3.18.

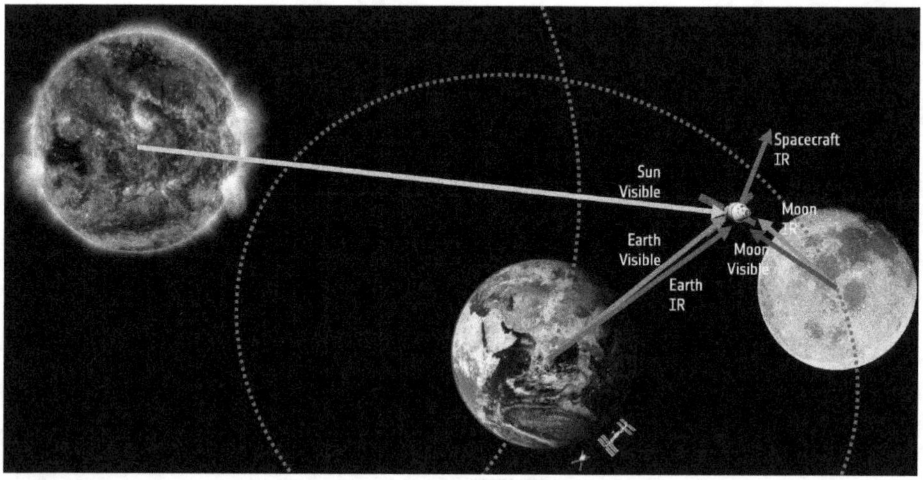

Figure 3.18 Sources of electromagnetic radiation in the Earth-Moon system (used images: © NASA [90] [91] [92] [141] [142])

For thermal models, the parameters absorptance and emittance are usually defined as static parameters instead of the utilization of the continuing spectral absorptance function. The mean value of the spectral absorptance function in the visible spectrum is usually referred to as solar absorptance and the mean value of this function in the infrared spectrum is usually referred to as infrared emittance. Since surfaces absorb and emit electromagnetic irradiance of a certain wavelength in an equal extend, the solar absorptance defines the mean amount of electromagnetic radiation a surface absorbs and emits in the visible spectrum, and the infrared emittance defines the mean amount of electromagnetic radiation a surface emits and absorbs in the infrared spectrum.

The radiant power a surface is absorbing can be defined for different sources and different spectra. It can be calculated as the product of the solar absorptance, the surface area and the horizontal irradiance for irradiance types in the visible spectrum. For irradiance types in the infrared spectrum, it can be calculated as the product of the infrared emittance, the surface area and the horizontal irradiance. The total amount of absorbed radiant power at a surface can be calculated as the sum over all available types of absorbed radiant power.

$$P_{Absorbed,Visible,i} = \alpha_{Vis} \cdot A_i \cdot E_{Hor.,Visible,i}$$

$$P_{Absorbed,IR,i} = \varepsilon_{IR} \cdot A_i \cdot E_{Hor.,IR,i}$$

$$P_{Absorbed} = \sum_{i=1}^{N} P_{Absorbed,IR,i} + \sum_{i=1}^{M} P_{Absorbed,Visible,i}$$

In the Earth-Moon system, electromagnetic irradiance in the visible spectrum emitted by the Sun, the Earth and the Moon is available. Infrared irradiance is available in the Earth-Moon system emitted by the Earth and the Moon.

The individual amount of radiant power absorbed by a surface of an object has to be calculated for each type of irradiance. The total amount of absorbed radiant power by an object can be calculated as the sum of the radiant power values absorbed at the different surfaces of an object in the following step.

$$P_{Abs.,Sun,i} = E_{Hor.,Sun} \cdot A_i \cdot \alpha_{Vis}$$

$$P_{Abs.,Earth,IR,i} = E_{Hor.,Earth,IR} \cdot A_i \cdot \varepsilon_{IR}$$

$$P_{Abs.,Earth,Albedo,i} = E_{Hor.,Earth,Albedo} \cdot A_i \cdot \alpha_{Vis}$$

$$P_{Abs.,Moon,IR,i} = E_{Hor.,Moon,IR} \cdot A_i \cdot \varepsilon_{IR}$$

$$P_{Abs.,Moon,Albedo,i} = E_{Hor.,Moon,Albedo} \cdot A_i \cdot \alpha_{Vis}$$

$$P_{Abs.,i} = P_{Abs.,Sun,i} + P_{Abs.,Earth,IR,i} + P_{Abs.,Earth,Albedo,i} + P_{Abs.,Moon,IR,i} + P_{Abs.,Moon,Albedo,i}$$

$$P_{Abs.} = \sum_{i=1}^{I} P_{Abs.,i}$$

The amount of radiant power a surface of an object emits in the infrared spectrum can also be defined utilizing the Stefan-Boltzmann law. Objects usually feature multiple surfaces. The total amount of emitted radiant power of an object can be calculated as the sum over the emitted radiant power of the individual surfaces of an object.

$$P_{Emission} = \sigma \cdot \varepsilon_{IR} \cdot A \cdot T^4 = \sum_{i=1}^{I} \sigma \cdot \varepsilon_{IR,i} \cdot A_i \cdot T_i^{\,4}$$

A share of the electromagnetic irradiance received by a surface will be absorbed by the surface and contribute to the temperature of the object. In turn, an increase in the temperature leads to an increase in the infrared emissions. Resulting from these two processes, an object pursues an equilibrium temperature defined by the radiation environment in space. In the equilibrium state, the amount of emitted and absorbed radiant power by an object have equal values. This relation can be utilized in order to determine the equilibrium temperature of a space system.

$$P_{Absorbed} = P_{Emission}$$

$$\sum_{i=1}^{N} P_{Absorbed,IR,i} + \sum_{i=1}^{M} P_{Absorbed,Visible,i} = \sigma \cdot T^4 \cdot \sum_{i=1}^{N} \varepsilon_{IR,i} \cdot A_i$$

$$T = \sqrt[4]{\frac{\sum_{n=1}^{N} \varepsilon_{IR,n} \cdot A_n \cdot E_{Hor.,IR,n} + \sum_{m=1}^{M} \alpha_{Vis.,m} \cdot A_m \cdot E_{Hor.,Visible,m}}{\sigma \cdot \sum_{i=1}^{I} \varepsilon_{IR,i} \cdot A_i}}$$

This relation allows to determine the equilibrium temperature of a passive object in space or on the lunar surface. A passive object could be something like a piece of rock or metal. Where a passive object would just heat up in space with reference to its electromagnetic irradiance environment, active objects have additional processes that influence the equilibrium temperature. These processes can be differentiated into processes converting a share of the received radiant power into a different type of power, so that this amount of converted radiant power does not contribute to the object temperature, and into processes supplying a certain amount of power to or drawing a certain amount of power from the object.

The process in solar cells converts a share of the received radiant power at the solar cell front site in the visible spectrum into electrical power. The converted share of the received radiant power does not contribute to the solar cell temperature. The magnitude of converted radiant power is dependent on the operational point at which a solar cell is operated. In open circuit conditions, no radiant power is converted into electrical power and drawn from the solar cell terminals. In Maximum Power Point (MPP) conditions, the maximum possible amount of radiant power is converted into electrical power and drawn from the solar cell terminals. The conversion of radiant power into electrical power results in a decrease in the solar cell temperature. The progressions of the solar cell temperature at the lunar equator for a solar cell in Maximum Power Point and open clamping operational state are visualized in Figure 3.19.

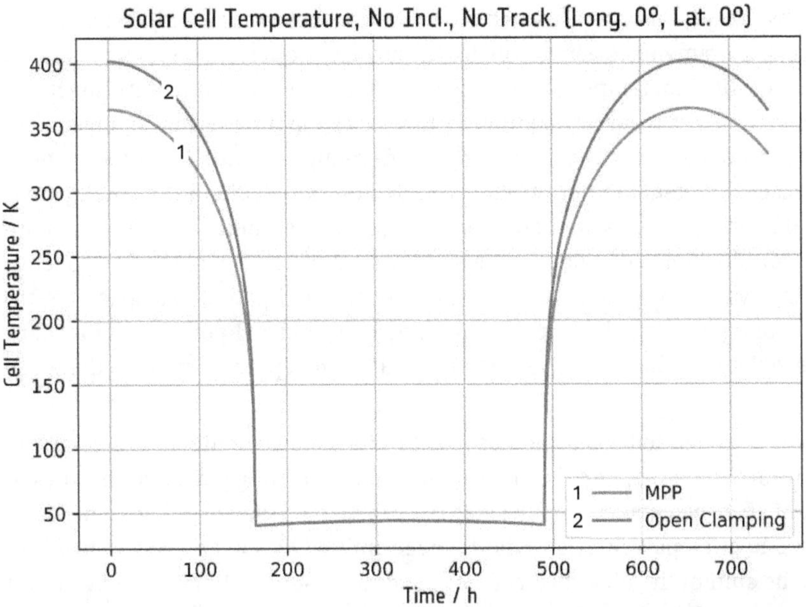

Figure 3.19 Solar cell temperature at the lunar equator during a lunar day/night cycle for the operation in Maximum Power Point (MPP) and open clamping conditions

The amount of radiant power converted into electrical power by a solar cell can be determined as the product of the solar cell efficiency and the radiant power of a certain type of irradiance in the visible spectrum absorbed at the solar cell front side. The utilization of the solar cell nameplate efficiency assumes the operation at the Maximum Power Point. The temperature determination in other operational states requires the dynamic determination of the solar cell efficiency. The electrical power converted from direct solar irradiance, the Earth's albedo irradiance, and the lunar albedo irradiance has to be taken into account for this analysis. The total amount of converted radiant power can be calculated as the sum of the individual converted power values.

$$P_{Conv.,Sun} = P_{Abs.,Sun} \cdot \eta(t) = E_{Hor.,Sun} \cdot A \cdot \alpha_{Vis} \cdot \eta(t)$$

$$P_{Conv.,Earth,Albedo} = P_{Abs.,Earth,Albedo} \cdot \eta(t) = E_{Hor.,Earth,Albedo} \cdot A \cdot \alpha_{Vis} \cdot \eta(t)$$

$$P_{Conv.,Lunar,Albedo} = P_{Abs.,Lunar,Albedo} \cdot \eta(t) = E_{Hor.,Lunar,Albedo} \cdot A \cdot \alpha_{Vis} \cdot \eta(t)$$

$$P_{Conv.} = \sum_{j=1}^{J} P_{Conv.,j} = P_{Conv.,Sun} + P_{Conv.,Earth,Albedo} + P_{Conv.,Lunar,Albedo}$$

Space systems usually feature active or passive measures to assure that the system temperature requirements are met. Due to the surface properties, a system is biased for a certain equilibrium temperature, which it will have in a certain electromagnetic radiation environment when no additional measures for the thermal management are integrated into the system. Systems can be differentiated into cold biased systems that feature a cold system temperature in equilibrium state without the employment of additional heating systems, and hot biased systems that feature a hot system temperature in equilibrium state without the employment of additional cooling systems. Active heating systems supply a certain amount of thermal power to a space system and active cooling systems draw a certain amount of thermal power from a space system. Internal power dissipation also results in a thermal power supply to the space system.

Electrical power generated from the absorbed radiant power reduces the complete amount of absorbed power, and the amount of supplied heating power contributes to the amount of absorbed power. Cooling systems can be modelled by utilizing negative values for the heating power. The radiant power emitted by a space system needs to be equal to the sum of the absorbed radiant power, the supplied thermal power and the negative value of the radiant power converted into electrical power. The equilibrium system temperature of active space systems can be determined in the same way as for passive objects in the next step.

$$P_{Absorbed} + P_{TM} - P_{Converted} = P_{Emission}$$

$$T = \sqrt[4]{\frac{\sum_{i=1}^{N} P_{Abs.,IR,i} + \sum_{i=1}^{M} P_{Abs.,Visible,i} + P_{TM} - \sum_{j=1}^{J} P_{Conv.,j}}{\sigma \cdot \sum_{i=1}^{I} \varepsilon_{IR,i} \cdot A_i}}$$

Especially in the case of a lunar base, there will be a great number of systems in the direct surrounding of other space systems on the lunar surface. These systems thermally interact with each other, and the interaction needs to be taken into account for accurate temperature determination. Such an analysis requires the simulation of the geometrical configuration of different space systems and the determination of the thermal interaction between different systems, which is a highly complex task.

Space systems feature special temperature requirements that can be defined by minimum and maximum temperature values. It can be differentiated between operational and survival temperature requirements. During the operation, a system has to be kept within its operational temperature domain in order to be functional. If a system is operated outside of its operational domain, this can result in the system losing its accuracy, in an increase in system noise, as well as in a temporarily or permanent malfunctioning or damage of the system. For optical systems, this requirement is usually very strict, for batteries, the operational temperature domain is broader than the requirement for optical systems, and for other systems the temperature requirement can be relatively broad. In addition to operational temperature requirements, there are also survivability temperature requirements for space systems. Survivability temperature domains are generally wider than operational temperature domains. Damage or malfunctioning does not need to be expected when a space system has temperatures outside the operational temperature domain but inside the survivable temperature domain, as long as it is out of operation when featuring such temperatures. Lower or hotter system temperatures not satisfying the survivability temperature domain can result in temporarily or permanent system malfunctioning and damage. Table 3.1 and Table 3.2 show the temperature domains for different space systems.

Table 3.1 Examples of typical thermal requirements for spacecraft components [112]

	Operational temperature domain	Survival temperature domain
battery	0 °C–15 °C	-10 °C–25 °C
solar arrays	-150 °C–110 °C	-200 °C–130 °C
power box baseplate	-10 °C–50 °C	-20 °C–60 °C
reaction wheel	-10 °C–40 °C	-20 °C–50 °C
gyro/IMU	0 °C– 40 °C	-10 °C–50 °C
star tracker	0 °C–30 °C	-10 °C–40 °C
hydrazine tank and lines	15 °C–40 °C	5 °C–50 °C
antenna gimbal	-40 °C–80 °C	-50 °C–90 °C
antenna	-100 °C–100 °C	-120 °C–120 °C

An example is the operation of batteries, which feature a relatively strict temperature domain in which the primary reaction is happening as expected. With decreasing temperatures, the battery reaction will slow down, so that the available capacity is decreasing with decreasing temperatures. There will also be a lower temperature limit from which the dielectricum will start to freeze and possibly damage the battery cell due to the deformation caused by the phase change. On the hot side, there will also be an upper limit until which the primary reaction happens safe and as expected. Above this limit, secondary reactions will increase, which could, for example, lead to the accumulation of gasses inside battery cells. Above a certain temperature, thermal run-away effects and battery cell explosions can be possible.

Table 3.2 Examples of typical thermal requirements for space certified batteries [97]

	NiCd and NiH_2	Li-Ion
operation	-10 °C–25 °C	-10 °C–35 °C
storage	-15 °C–35 °C	-20 °C–40 °C
charge optimum	-10 °C–10 °C	15 °C–25 °C
discharge optimum	10 °C–25 °C	0 °C–30 °C

Temperature can also have implications on the system and spacecraft structure. The dimensions of objects are usually defined with respect to a certain reference temperature, and below and above this temperature, deformations can occur. While small temperature variations are very likely to cause reversible deformations, greater variations can result in permanent deformations.

3.5 Lunar Orbits

Low lunar orbits can be defined as orbits featuring an altitude of up to 100 km [86]. These orbits are unstable and feature perturbations resulting from mass concentrations below the lunar surface, third-body effects and radiation pressure. Frozen low lunar orbits exist at inclinations of 27°, 50°, 76° and 86° that only have a minimum of perturbations over time [86] [144]. Higher lunar orbits can be defined by an orbit altitude of between 100 km and 225 km [86]. At orbit inclinations of 40°, 45° and 56°, frozen higher lunar orbits can be found [86]. While the utilization of orbit inclinations has advantages in the field of orbit stability, orbit inclinations can limit the access from a solar satellite to locations on the lunar surface. High lunar orbits are defined by an altitude of above 500 km and are subject to perturbations resulting from the non-spherical gravitational field in the Earth-Moon system [86]. Stationary orbits exist in Earth orbit and offer the advantage of a stable spacecraft position in the sky above the Earth. However, these do not exist in lunar orbit, as a spacecraft in the required altitude would be distracted by the Earth's gravitational force.

4 Technology and Design of Space Electrical Power Systems

Electrical Power Systems (EPSs) are incremental subsystems of all spacecraft and space missions. These subsystems provide power generation, power regulation, as well as energy storage functionalities, and supply electrical energy to subsystems and payloads during space missions. Typical spacecraft EPSs include generative systems, energy storages, power conversion and regulation systems, functional safety systems, a cable harness, and a Power Management And Distribution System (PMAD).

Generative subsystems of the EPS are utilized to convert the energy supplied by a primary energy source into electrical energy. Power regulators are employed in order to match the power generation to the power demand of loads and energy storages. The PMAD system provides functionalities like EPS control, power conditioning, power conversion, power transmission and distribution. Energy storages are integrated into the spacecraft EPS to supply the EPS loads during periods in which the employed primary energy source is not available, as well as during peak demand periods. All subsystems can be defined as EPS loads that are connected to the EPS and that are supplied by it with electrical energy but that are not part of the utilized EPS topology. In order to distribute energy within a spacecraft, usually one or multiple interconnected main buses are utilized. The Power Distribution Unit (PDU) is utilized in order to connect or disconnect different loads to the main bus.

Typical loads are payloads, communication systems, Environmental Control and Life Support Systems (ECLSSs), as well as Thermal Control Systems (TCS) or electrical propulsion systems. For a lunar base, also the supply of additional lunar base subsystems, science equipment, ISRU processes, transport facilities and transport systems need to be taken into account.

Spacecraft Electrical Power Systems must be able to handle the different loads that are connected to the EPS during all mission phases. Especially nominal and peak demand periods can be differentiated. An additional requirement is that the electrical input requirements of loads must be met. Spacecraft EPSs need to have a minimum of required maintenance. They need to be fault tolerant and stable in electrical error situations without interventions from operators.

Energy generation and storage systems can be differentiated according to the duration for which they can transfer a certain amount of power to the bus, according to the amount of power they can transfer to the bus, according to their power-to-mass and power-to-volume ratios, as well as according to the primary source of energy that is utilized by the generative subsystem. An overview visualization of different generative and energy storage systems is shown in Figure 4.1.

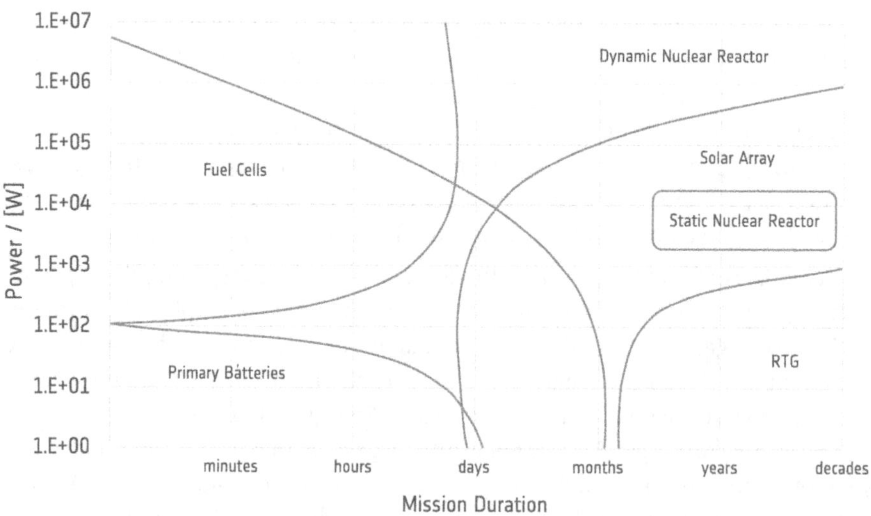

Figure 4.1 Comparison of different space power supply technologies regarding the power supply capabilities and the power supply duration (image: data from NASA, A. Baez [150])

Solar power generators are the quasi-standard generator type in space applications, that are currently utilized by 99% of all spacecraft [97] [98] [148]. Possible conversion processes for solar energy are photochemical, photoelectrochemical, photobiological, and photovoltaic processes. In addition to these, the radiation pressure of solar irradiance could also be utilized on the lunar surface. The conversion of solar irradiance to thermal energy as an interim step, together with the generation of electrical energy from the interim step of thermal energy, could also be a possible process. Related types of generators are usually referred to as solar thermal generators. The output power of photovoltaic generators is highly dependent on the environmental conditions in which a generator is operated. The electromagnetic irradiance a photovoltaic generator receives from natural sources of energy, the temperature of the generator, as well as the degradation influence the performance of photovoltaic generators. Photobiological and photochemical processes have not yet been demonstrated within space missions. Subsequently, space applications are not likely in the near future. When solar energy is utilized, tracking systems optimizing the generator orientation are beneficial subsystems because these allow to increase the energy yield by optimizing the incidence conditions.

Nuclear sources of energy utilize the nuclear decay or nuclear chain reactions in order to generate electrical energy. It is usually differentiated between Radioisotope Thermoelectric Generators (RTGs) and nuclear reactors. RTGs generate energy from the by-product thermal energy resulting from the nuclear decay. Nuclear reactors generate energy from the by-product thermal energy resulting from nuclear reactions. Static and dynamic reactors are available. Nuclear sources of energy can supply energy for long periods of time and are therefore utilized for long-term missions with limited access to solar energy.

Primary batteries are chemical energy storages. The internally employed chemical main reaction is not reversable and because of this, primary batteries cannot be recharged. Subsequently, primary batteries can only be employed for short durations. A possible application would be in Launch Vehicles. Secondary batteries are also chemical energy storages, but this type of batteries employs a reversable main reaction. Because of this, secondary batteries can be recharged. Nickel-hydrogen (NiH_2) batteries have been employed in space systems as a quasi-standard for many years. Recent developments employ lithium-ion batteries because these have higher energy densities.

Fuel cells require the supply of reactant chemicals that have to be stored in tanks in a space system. Because of this, the application of fuel cells is limited to a period of a certain number of days, until these chemicals have been consumed by a fuel cell system. Available technologies are PEM and alkaline fuel cells. Fuel cells can be utilized as a part of energy storages together with electrolyzers. This combination is usually referred to as a regenerative fuel cell. The electrolyzer is used in order to separate the product chemicals of the fuel cell again, so that these chemicals can be utilized in a closed cycle.

When high currents are demanded by bus loads, it can be beneficial to utilize additional supercapacitors for the supply of high currents and to utilize batteries for the supply of nominal currents. Supercapacitors have a self-discharge rate in the magnitude of $15 \frac{\%}{\text{day}}$ and because of this, this technology is not suitable as long-term storage and cannot replace other energy storage systems for longer periods of time [154].

For the exploration of the lunar surface, solar arrays and nuclear fission reactors are the most potential technologies as power generation systems. For most mission types, the utilization of solar energy offers advantages superior to other sources of energy. When nuclear generators are employed, the energy required during a mission needs to be integrated into these generators in the form of nuclear isotopes. Contrary to that, solar energy can be harvested with generators from the environment of a space system when it is available in a significant amount. Because of this, the employment of solar generators offers savings in the launch mass and also in the generator costs.

Electrical subsystems usually feature power dissipation, which contributes to the subsystem temperature. In this case, thermal control systems are usually required as support systems in most EPS designs in order to keep electrical components within their required temperature range.

Electrical Power System topologies can be differentiated by the stability of the bus voltage. Different topologies can have a relatively stable and controllable bus voltage, while other topologies can have a relatively great variation of the bus voltage. Following this, the bus voltage has different degrees of controllability and stability with different topologies. An EPS topology featuring a bus voltage that can be controlled to a certain steady voltage level is usually referred to as a regulated topology. Topologies in which the bus voltage is varying, for example, as a result of a directly connected battery-based energy storage, are usually referred to as unregulated topologies. Regulated topologies are utilized for space systems in the kilowatt EPS class. In addition to this, an EPS can also provide multiple voltage levels for different subsystems. EPSs for missions to the inner solar system need to be tolerant for the operation under high temperature conditions, due to the high solar intensity, and EPSs for missions to the outer solar system need to be tolerant to operate under low temperature conditions. Depending on the radiation environment, it is also possible, that radiation hardness can be a requirement of a spacecraft EPS.

The current state of the art for spacecraft EPSs are custom-designed EPSs for each space mission. These usually feature only limited expendability potential and automatization. For future space systems, the employed approach will most likely shift to a more modular approach, utilizing independent, smart and interconnected modules. Typical spacecraft EPSs were developed for supplying power in a magnitude between 10 W and 50 kW. An exception is the EPS of the International Space Station (ISS) that is capable of supplying 110 kW [97]. Typical power bus voltages lie in the domain between 20 V and 125 V [97] [112] [150]. An exception can be observed with the increasing popularity of electrical propulsion systems that require high voltage supply. A trend to higher voltages and power can also be observed so that it can be expected that future space EPSs have voltages greater than 300 V and power capacities greater than 100 kW [150]. Usually, Direct Current (DC) systems are employed for space EPSs. When Alternating Current (AC) supply is required by a load with a great power de-hand, it can also be expected that in the cases where the power dissipation with AC EPSs is lower than with a combination of a DC EPS and an additional conversion to AC in a load converter, AC EPSs will be utilized.

4.1 Photovoltaic Generators

Photovoltaic generators allow to convert solar irradiance into electrical energy utilizing the photovoltaic effect and have a high power to mass ratio. Around 99% of all spacecraft rely on photovoltaic generators, thus they are the most common type of generative systems utilized in space EPSs for missions with durations from a certain number of weeks to a certain number of years. Photovoltaic generators are commonly referred to as solar cells. An example for solar cells utilized to power the Orion Spacecraft are shown in Figure 4.2.

Figure 4.2 Solar Array Wings of the Orion spacecraft during the deployment test (image: © NASA [156])

The history of solar cells goes back to 1839, when the photo effect has been discovered by Henri Becquerel. Based on the photo effect, the theory of pn-junctions in semiconductors was developed by William Bradford Shockley in 1949. First solar cells were manufactured by the Bell laboratories for space exploration in 1954. Examples for the efficiencies achieved with different solar cells are shown in Figure 4.3.

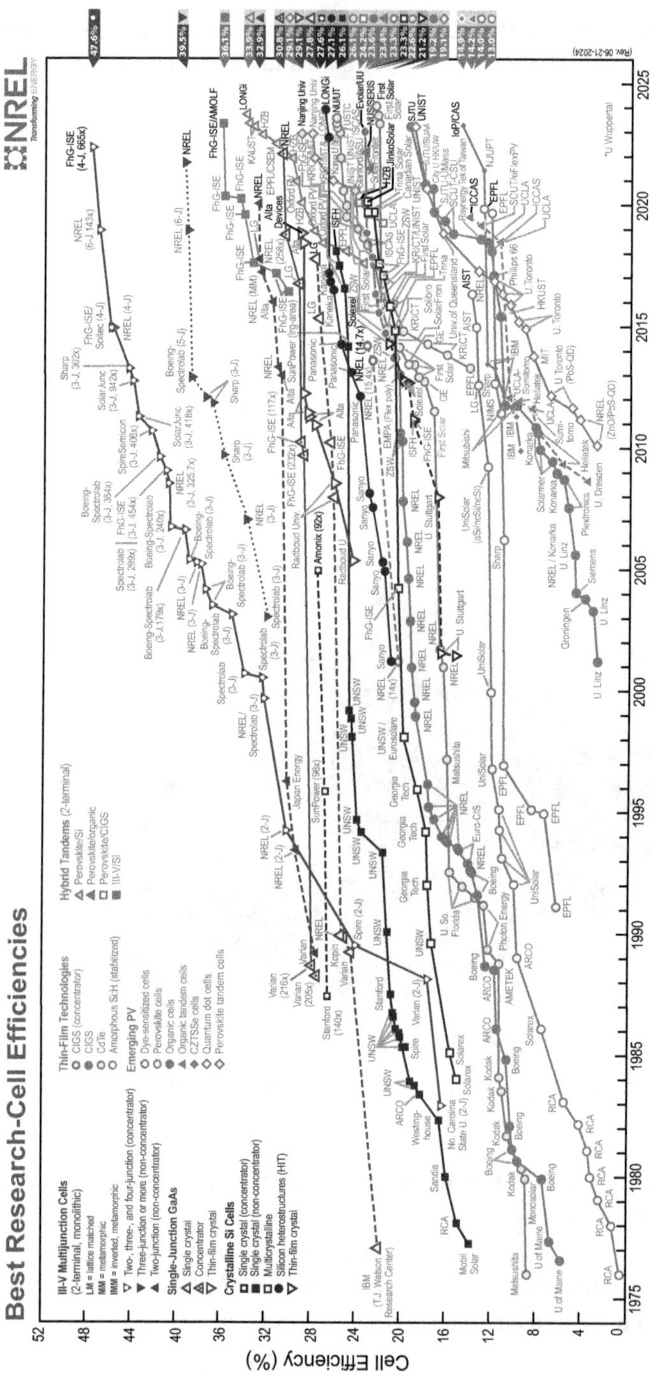

Figure 4.3 Achievements in the solar cell efficiency (image: © This plot is courtesy of the National Renewable Energy Laboratory, Golden, CO [158].)

Atoms consist of electrons, protons and neutrons. Neutrons are not electrically charged, protons are positively charged and electrons are negatively charged. Protons and neutrons can be found in the atomic nucleus and electrons are orbiting the nucleus. The atomic nucleus is positively charged [109] [151]. Since the number of negatively charged electrons in an atom is equal to the number of positively charged protons, atoms are usually neutrally charged. The atomic nucleus and the orbiting electrons attract each other by the Coulomb force.

$$F_{Coulomb} = \frac{1}{4 \cdot \pi \cdot \varepsilon_0} \cdot \frac{n_Z \cdot q_e^2}{r_{n_{Orbit}}^2}$$

Electrons orbiting the nucleus at a certain distance with a certain angular velocity will experience a centrifugal force resulting from their orbital movement.

$$F_{CF} = m_e \cdot r_{n_{Orbit}} \cdot \omega_{n_{Orbit}}^2 = \frac{m_e \cdot v_{n_{Orbit}}^2}{r_{n_{Orbit}}}$$

Both types of forces need to be equal in order to allow an orbital movement. Otherwise, electrons would either crash into the nucleus or depart from it.

$$F_{Coulomb} = F_{CF}$$

$$\frac{1}{4 \cdot \pi \cdot \varepsilon_0} \cdot \frac{n_Z \cdot q_e^2}{r_{n_{Orbit}}^2} = m_e \cdot r_{n_{Orbit}} \cdot \omega_{n_{Orbit}}^2$$

It is only possible for electrons to orbit atomic nuclei in a discrete number of orbits. Because of this, the orbital momentum of a certain electron orbit needs to be equal to an integer multiple of $\frac{h}{2 \cdot \pi}$.

$$L_e = m_e \cdot r_{n_{Orbit}}^2 \cdot \omega_{n_{Orbit}} = n_{Orbit} \cdot \frac{h}{2 \cdot \pi}$$

The equation of the force balance between Coulomb force and centrifugal force can be transformed together with the equation of the orbital momentum to the equation of the orbital radius. In a second step, the resulting orbital radius equation can be utilized together with the equation of the orbital momentum to derive the equation of the angular velocity of the orbiting electron.

$$r_{n_{Orbit}} = \frac{n_{Orbit}^2 \cdot h^2 \cdot \varepsilon_0}{\pi \cdot m_e \cdot n_Z \cdot q_e^2}$$

$$\omega_{n_{Orbit}} = \frac{\pi \cdot m_e \cdot n_Z^2 \cdot q_e^4}{2 \cdot n_{Orbit}^3 \cdot h^3 \cdot \varepsilon_0^2}$$

The possible electron orbits are referred to as shells or energy states. The different shells are indicated with different letters. The shell with the smallest radius is the k-shell and shells with greater radii are indicated employing the following letters in alphabetical order.

An electron in an orbit around an atomic nucleus needs to move with a certain velocity. Following this, the energy of an electron in a certain orbit can be calculated with reference to the electron mass and to its velocity. This energy value is also referred to as the energy state of the electron.

$$E_{n_{Orbit}} = \frac{1}{2} \cdot m_e \cdot v_{n_{Orbit}}^2$$

$$E_{n_{Orbit}} = \frac{1}{2} \cdot m_e \cdot r_{n_{Orbit}}^2 \cdot \omega_{n_{Orbit}}^2$$

$$E_{n_{Orbit}} = \frac{m_e \cdot n_Z^2 \cdot q_e^4}{8 \cdot n_{Orbit}^2 \cdot h^2 \cdot \varepsilon_0^2}$$

Electrons can be lifted from one shell to another shell. In order to lift an electron, the additional amount of energy required to reach the energy state of a higher shell needs to be supplied to the electron. This amount of energy can be supplied to it in the form of electromagnetic radiation at a certain frequency. It can be calculated employing the Planck-Einstein relation, as a product of the radiation frequency and the Planck constant. Electrons can also fall back from a higher shell to a lower shell [151]. In this case, the energy is emitted again.

$$\Delta E_{n_{Orbit}} = |E_{n_{Orbit},old} - E_{n_{Orbit},\mathrm{new}}| = h \cdot f$$

The atomic shells are inhabited by electrons from the lowest to the highest shell. The lowest shell is filled with electrons first, and from the lowest shell to the highest shell, shells are filled with electrons. The highest shell that can be filled with electrons, is determined by the total number of electrons orbiting the atom. The electrons in the highest shell that can be filled completely or partly with electrons, are referred to as valence electrons. These valence electrons are important for the formation of atomic bindings. Two valence electrons can form a covalent atomic bond. Covalent bonds are established between two neighboring atoms. In order to establish the covalent bond, one electron from each atom is utilized.

Of particular interest for the manufacturing of solar cells are materials featuring 4 covalent bonds because these materials form a regular crystalline lattice. Such materials can be manufactured from atoms featuring 4 valence electrons, by a combination of materials featuring 3 valence electrons and 5 valence electrons, or by a combination of materials featuring 2 valence electrons and 6 valence electrons. Semiconductors manufactured from atoms featuring 4 valence electrons are referred to as IV semiconductors, semiconductors manufactured from a combination of atoms featuring 3 valence electrons and atoms featuring 5 valence electrons are referred to as III-V semiconductors, and semiconductors manufactured from a combination of atoms featuring 2 valence electrons and atoms featuring 6 valence electrons are referred to as II-VI semiconductors. Exemplary materials for the manufacturing of IV semiconductors are silicon, germanium and tin. III-V semiconductors can be manufactured, for example, from a combination of gallium and arsenic (GaAs), from a combination of gallium and phosphor (GaP), from a combination of indium and antimony (InSb) or from a combination of indium and phosphor (InP). Examples of possible material combinations for the manufacturing of II-VI semiconductors are cadmium and telluride (CdTe), zinc and selenium (ZnSe), zinc and telluride (ZnTe), as well as mercury and selenium (HgSe).

After the formation of multiple atoms to a solid, the single shells or energy states are not clearly defined anymore. Furthermore, the energy states in the solid overlap to energy bands that result from side effects between the single orbiting electrons. The allowed energy states and bands of atoms and molecules are visualized in Figure 4.4.

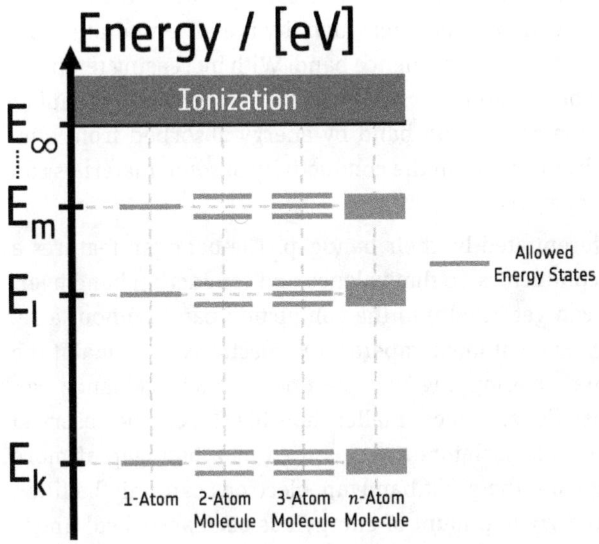

Figure 4.4
Allowed energy states for electrons orbiting an atomic nucleus

An energy band can be defined by a minimum and a maximum energy state. An electron in an energy band can have any value between the maximum and minimum value of the energy band. In a solid, the energy bands are filled from the lowest to the highest possible energy band. The highest possible energy band that can be filled with electrons, is referred to as the valence band. Electrons inhabiting the valence band are referred to as valence electrons. The next energy band featuring a higher minimum energy state as the maximum energy state of the valence band is referred to as the conduction band [109] [151]. In nominal conditions, the conduction band is not filled with electrons. However, it is possible for electrons to inhabit the conduction band.

Electrons can only have energy states between the minimum and the maximum state of an energy band. The maximum and minimum states of an energy band are also referred to as band edges. Different energy bands are separated by a bandgap that is sometimes also referred to as the forbidden zone. Electrons cannot inhabit the bandgap. The bandgap is located between the upper edge of the lower energy band and the lower edge of the upper energy band.

$$E_C = \min(E_{Conduction\ Band})$$

$$E_V = \max(E_{Valence\ Band})$$

$$\Delta E_{Gap} = |E_C - E_V|$$

In order to lift an electron from the valence band into the conduction band, it needs to absorb energy in the magnitude of the bandgap. Electrons in the conduction band can move freely within the solid. The conductivity of the material depends on the number of electrons that inhabit the conduction band. Their mobility is also temperature-dependent. At 0 Kelvin, all electrons inhabit the valence band. With increasing temperature, the temperature swinging of the atoms increases, so that an increasing number of electrons can be lifted into the conduction band by energy absorbed from temperature swinging. Subsequently, increases in the conductivity of some materials can be observed with increasing temperature.

Different materials can be differentiated by their bandgap. The bandgap features a value of zero electronvolts for conductors, so that valence and conduction band overlap. Because of this, electrons can get lifted into the conduction band without a requirement for absorbed energy. Even at low temperatures, electrons can inhabit the conduction band in conductors. Overlapping valence and conduction bands can mostly be found in metals. A specific resistance smaller than 10^{-5} Ωm can be observed in conductors [109]. Contrary to this, isolators are classified by a bandgap of more than 5 eV [109] [151]. Resulting from the great bandgap, electrons can only be lifted into the conduction band when a great amount of energy has been absorbed. Under normal conditions and even at high temperatures, only a small number of electrons or even no electrons at all inhabit the conduction band of isolators. Specific resistances greater than 10^{7} Ωm can be observed in isolator materials [109]. Materials that cannot either be classified as conductors or isolators are semiconductors. The bandgap of a semiconductor is defined by a value greater than the bandgap of conductors and smaller than the bandgap of isolators. At very low temperatures, semiconductors act in a similar way as isolators, and usually at temperatures of 200 °C and higher, semiconductors act similar as conductors [151]. In semiconductors, specific resistances between more than 10^{-5} Ωm and less than 10^{7} Ωm can be observed [109]. Figure 4.5 shows a visualization of the bandgap for the different types of material.

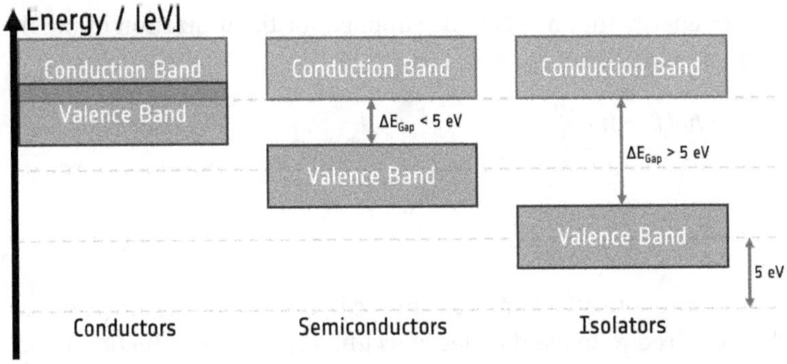

Figure 4.5 Valence band, conduction band and bandgap of different types of materials

A property that is utilized in photovoltaics, is that electrons in semiconductors can be lifted into the conduction band by the absorption of a relatively small amount of energy. The outer photo effect is defined as the detachment of an electron from an atom in response to the absorption of a photon. This effect is referred to as ionization.

The wavelength of the absorbed photon is related to its frequency by the speed of light. The amount of absorbed energy can be calculated employing the Planck-Einstein equation as the product of the radiation frequency and Planck's constant.

$$c_0 = \lambda \cdot f$$

$$\Delta E_n = h \cdot f = h \cdot \frac{c_0}{\lambda}$$

The required energy to detach an electron from an atom can be calculated employing a boundary analysis. This amount of energy is equal to the amount of energy that needs to be absorbed in order to lift the electron from the valence band to the infinite band. A photon featuring a certain wavelength can supply the required amount of energy.

$$\Delta E_n = \lim_{m_{Orbit} \to \infty} \left(E_{n_{Orbit}} - E_{m_{Orbit}} \right) \approx \frac{m_e \cdot n_Z^2 \cdot q_e^4}{8 \cdot n_{Orbit}^2 \cdot h^2 \cdot \varepsilon_0^2} - 0 = \frac{h \cdot c_0}{\lambda}$$

Resulting from the outer photo effect, the electrical resistance of a semiconductor material changes with the amount of absorbed electromagnetic radiation. This effect is utilized in photoresistors that change their resistance with reference to the amount of absorbed electromagnetic radiation. However, because relatively high energies are required for the utilization of the outer photo effect, this effect is not suitable for the utilization in generators.

The inner photo effect is defined as the lifting of an electron from the valence to the conduction band in response to the absorption of electromagnetic radiation (Figure 4.6). The amount of energy that needs to be supplied for the utilization of the inner photo effect is equal to the bandgap.

$$\Delta E_{Gap} = |E_C - E_V| = h \cdot f = h \cdot \frac{c_0}{\lambda}$$

$$\lambda = h \cdot \frac{c_0}{\Delta E_{Gap}}$$

Solar cells utilize the inner photo effect. In order to use it in a generator, the presence of an electric field is required to move the electrons lifted into the conduction band. The electrons moved in this way can be observed as electrical current at the solar cell terminals.

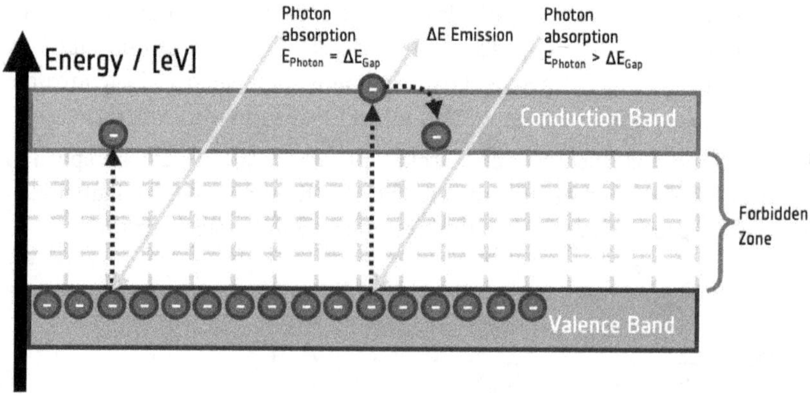

Figure 4.6 Visualization of the inner photo effect

In the dark state, with no presence of electromagnetic radiation, all electrons of a semiconductor crystal inhabit the valence band, and the conduction band is not inhabited by any electrons. By the absorption of electromagnetic irradiance, electrons can get lifted from the valence band to the conduction band. After an electron has been lifted into the conduction band, a hole remains in the valence band at the former location of the lifted electron. This process is referred to as electron-hole-pair generation. The process is visualized in Figure 4.6 and in Figure 4.7. The electrons that have been lifted into the conduction band can freely move there. Electrons can also fall back from the conduction band to the valence band. Both the generation of electron-hole-pairs and their recombination are constantly happening within semiconductors. Depending on the temperature, a mean number of electrons can be observed in the conduction band and a mean number of remaining holes can be observed in the valence band. The mean number of electrons in the conduction band and of holes in the valence band is referred to as the intrinsic conduction density.

$$n = N_C \cdot e^{-\frac{E_C - E_F}{k_B \cdot T}}$$

$$p = N_V \cdot e^{-\frac{E_F - E_V}{k_B \cdot T}}$$

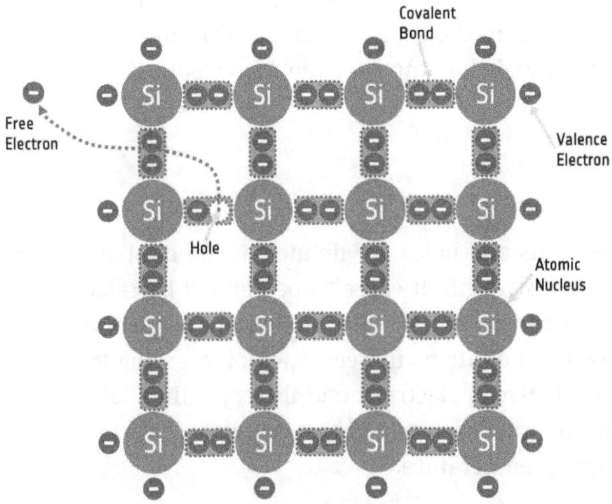

Figure 4.7
The crystalline structure of silicon and the photo effect

For every electron that has been lifted into the conduction band, a hole remains in the valence band. Because of this, the density of electrons in the conduction band in a semiconductor crystal is equal to the density of holes in the valence band.

$$n_{Conduction} = p_{Conduction}$$

Following this, the intrinsic charge carrier concentration can be calculated as the product of the intrinsic conduction densities of electrons and holes.

$$n_i^2 = n \cdot p = N_C \cdot e^{-\frac{E_C - E_F}{k_B \cdot T}} \cdot N_V \cdot e^{-\frac{E_F - E_V}{k_B \cdot T}} = N_C \cdot N_V \cdot e^{-\frac{E_C - E_F + E_F - E_V}{k_B \cdot T}}$$
$$= N_C \cdot N_V \cdot e^{-\frac{E_C - E_V}{k_B \cdot T}}$$

$$n_i^2 = N_C \cdot N_V \cdot e^{-\frac{\Delta E_G}{k_B \cdot T}} = n_{i0}^2 \cdot T^3 \cdot e^{-\frac{\Delta E_G}{k_B \cdot T}}$$

$$n_i = \sqrt{N_C \cdot N_V} \cdot e^{-\frac{\Delta E_G}{2 \cdot k_B \cdot T}}$$

Electrons and holes in a semiconductor move once an electrical voltage has been applied to the semiconductor terminals. The applied voltage induces an electrical field into the semiconductor that accelerates electrons in the direction of the positive electrical pole and holes in the direction of the negative one. The induced electrical field strength inside the semiconductor can be calculated with reference to the applied voltage at the semiconductor terminals and the length of the semiconductor crystal.

$$E_{El} = \frac{V}{l}$$

While electrons and holes move through the semiconductor, they collide with the atoms of the crystalline lattice. Resulting from these collisions, their velocity is limited. The mean velocity of electrons in the semiconductor is referred to as the drift velocity. The drift velocity can be calculated from the time that an electron needs to travel through the semiconductor crystal, and the length of the direct path through the semiconductor crystal.

$$v_{Drift}(T) = \frac{l_{Crystal}}{\Delta t}$$

The mobility parameters for electrons and holes are defined as the quotient of the drift velocity and the electrical field strength. It can also be defined by relating the mobility at a certain reference temperature to the mobility at another temperature value. Since the temperature swinging of atoms increases with increasing temperature, the probability for collisions between electrons and the crystalline lattice also increases. Because of this, both the drift velocity and the mobility of electrons and holes are decreasing with increasing temperature.

$$\mu_n(T) = \frac{v_{Drift}(T)}{E_{El}}$$

$$\mu_n(T) = \mu_{0n} \cdot \left(\frac{T}{T_0}\right)^{-\frac{3}{2}}$$

$$\mu_p(T) = \mu_{0p} \cdot \left(\frac{T}{T_0}\right)^{-\frac{3}{2}}$$

The total number of electrons that pass through a semiconductor, can be defined as the product of the electron density and the volume of the semiconductor lattice. The amount of charge that passes through a semiconductor during a certain time interval can be defined as the product of the charge of one electron and the total number of electrons that pass through it during the interval. In turn, the current through a semiconductor can be defined as the amount of charge that passes through it per unit time. This value is usually referred to as the field current.

$$N_E = n \cdot V_{Crystal} = n \cdot A_{Crystal} \cdot l_{Crystal}$$

$$Q_{El.,Semiconductor} = q_E \cdot N_E$$

$$I_F = \frac{Q_{El.,Semiconductor}}{\Delta t} = \frac{q_E \cdot N_E}{\Delta t} = \frac{q_E \cdot n \cdot A_{Crystal} \cdot l_{Crystal}}{\Delta t}$$
$$= q_E \cdot n \cdot A_{Crystal} \cdot v_{Drift}(T)$$

$$I_F = q_E \cdot n \cdot A_{Crystal} \cdot E_{El} \cdot \mu_n(T)$$

With reference to this, the current density in the semiconductor crystal can be defined as the division of the field current by the cross-sectional area of the semiconductor crystal.

$$j_F = \frac{I_F}{A_{Crystal}} = q_E \cdot n \cdot E_{El} \cdot \mu_n(T)$$

Holes can also move through semiconductor crystals. They move through it in the opposite direction of the movement of the electrons. However, the movement of the holes through the crystal is much slower than the movement of the electrons. Based on the mobility of the electrons and holes, as well as the density of both, the electrical conductivity and resistivity of the semiconductor can be defined. With decreasing temperature, a decrease in the conductivity and an increase in the resistivity can be observed.

$$\kappa = \frac{1}{\rho} = q_e \cdot (n \cdot \mu_n + p \cdot \mu_p) = q_e \cdot n_i \cdot (\mu_n + \mu_p)$$

The doping of a semiconductor is defined as the process of integrating atoms from another group of the periodic table into the semiconductor lattice. A semiconductor consisting of atoms from the fourth group of the periodic table can be doped employing atoms from the second, third, fourth and fifth group of the periodic table. The combination of semiconductors doped with atoms from the second group of the periodic table with semiconductors doped with atoms from the fifth group of the periodic table, as well as the combination of semiconductors doped with atoms from the third group of the periodic table with semiconductors doped with atoms from the fourth group of the periodic table, offers additional advantages that will be explained in the following sections. The crystalline structure of doped semiconductors is visualized in Figure 4.8.

In a regular semiconductor lattice, an integrated atom would form 4 covalent bonds with its neighboring atoms, utilizing 4 valence electrons. Since atoms from the second group of the periodic table only have 2 valence electrons, it can only form 2 covalent bonds with its neighboring atoms, and since atoms from the third group only have 3 valence electrons, it can only form 3 covalent bonds. Following this, bonds between 2 of the 4 neighboring atoms and the newly integrated atom from the second group have to consist of 1 electron from the neighboring atoms and 1 hole from the newly integrated atom from the second group, and bonds between 1 of the 4 neighboring atoms and the newly integrated atom from the third group have to consist of 1 electron from the neighboring atoms and 1 hole from the newly integrated atom from the third group. Such holes can move inside the semiconductor, while the electrons cannot move in the semiconductor.

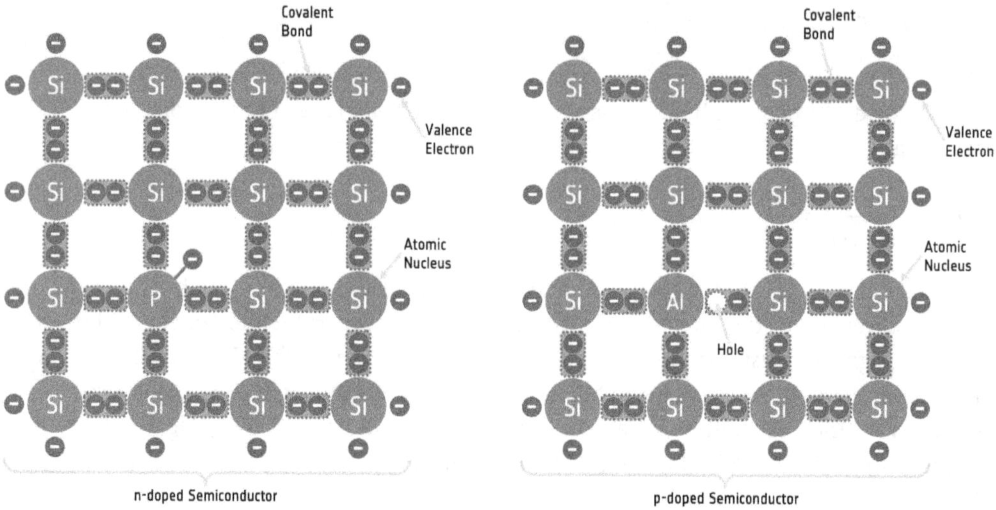

Figure 4.8 Crystalline structure of n-doped and p-doped semiconductors

A semiconductor that has been doped with atoms from the second or third group of the periodic table is referred to as a p-semiconductor. The process of doping a semiconductor with atoms from the second or third group of the periodic table is referred to as p-doping and the atoms utilized for the doping are referred to as the acceptor atoms or acceptors. In p-semiconductors, the mobile holes will recombine with static electrons, and following this, the number of holes is greater than the number of electrons. Therefore, holes are also referred to as majority charge carriers and electrons are referred to as minority charge carriers in p-semiconductors. The density of free holes in the p-semiconductor can be calculated with reference to the following equation.

$$p = \sqrt{\frac{n_A \cdot N_V}{2}} \cdot e^{-\frac{E_{Acc}}{2 \cdot k_B \cdot T}}$$

Because of the doping, a p-semiconductor has an additional energy state between valence and conduction band in the forbidden zone. This energy state allows holes to get lifted into the conduction band by the absorption of only a small amount of energy. As a result, hole conduction is possible in p-semiconductors.

A semiconductor that has been doped with atoms from the fifth or sixth group of the periodic table is referred to as n-semiconductor. The process of n-doping is defined as the integration of atoms from the fifth or sixth group of the periodic table into the semiconductor. Atoms from the fifth group of the periodic table have 5 valence electrons and atoms from the sixth group of the periodic table have 6 valence electrons. When these atoms are integrated into a crystalline lattice, consisting of atoms forming a crystalline lattice with 4 covalent bonds, 4 of the 5 valence electrons of the atoms from the fifth group, or 4 of the 6 valence electrons of the atoms from the sixth group will be utilized to form 4 covalent bonds with neighboring atoms. One electron will be left after the integration that is not utilized for forming a covalent bond, when atoms from the fifth group of the periodic table have been utilized for the doping. When atoms from the sixth group of the periodic table have been utilized for the doping, 2 electrons will be left after the integration that are not utilized for forming a covalent bond. These electrons are mobile and can move in the n-semiconductor, while holes in the n-semiconductor cannot move. The doping atoms integrated into the semiconductor lattice in this way are referred to as donators. The free electrons recombine with static holes, so that the number of electrons is greater than the number of holes in an n-semiconductor. Following this, electrons are referred to as majority charge carriers and holes are referred to as minority charge carriers in n-semiconductors. After the doping, an additional energy state between valence and conduction band has been formed that allows electrons to get lifted into the conduction band by the absorption of only a small amount of energy. The density of electrons in the n-semiconductor can be calculated with reference to the following formula.

$$n = \sqrt{\frac{n_D \cdot N_C}{2}} \cdot e^{-\frac{E_{Don}}{2 \cdot k_B \cdot T}}$$

Diffusion currents occur when the charge carrier concentration is greater in one part of a semiconductor than in another. In this case, a diffusion current from the zone of high charge carrier concentration to the zone of low concentration can be observed, until the charge carrier concentration has been equalized.

Interesting applications result from the combination of p- and n-semiconductors on top of each other. The interface between p-semiconductor and n-semiconductor is in this case referred to as the pn-junction. At the interface, diffusion currents of free holes from the p-semiconductor into the n-semiconductor, and of free electrons from the n-semiconductor into the p-semiconductor can be observed. The free holes of the p-semiconductor recombine with the electrons of the n-semiconductor in the n-semiconductor and the free electrons of the n-semiconductor recombine with the holes of the p-semiconductor in the p-semiconductor. The recombination starts at the interface between n-semiconductor and p-semiconductor and expands into both directions with time. After the recombination process of one electron and one hole, the atoms in the p-semiconductor have one hole less than in the equally charged state and atoms in the n-semiconductor have one electron less than in the equally charged state. As a result, atoms in the zone close to the pn-junction in the p-semiconductor are negatively charged and atoms in the zone close to the pn-junction in the n-semiconductor are positively charged. The free charge carriers recombine close to the interface between p-semiconductor and n-semiconductor. Subsequently, no free charge carriers can be observed in this region, which is referred to as the depletion zone [151]. This process is visualized in Figure 4.9.

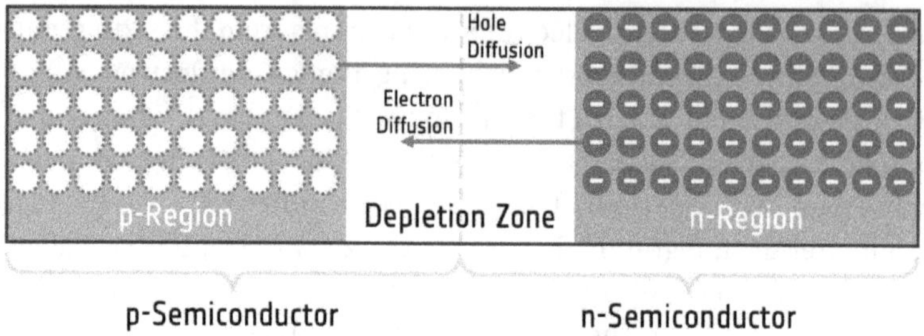

Figure 4.9 Charge carrier diffusion and depletion zone

The depletion zone is electrically charged. Between the positively charged zone of the n-semiconductor and the negatively charged zone of the p-semiconductor, an electrical field from n-semiconductor to p-semiconductor can be observed. This electrical field acts as a counterforce to the free charge carriers, forcing them away from the pn-junction. With increasing expansion of the depletion zone into the p-semiconductor and the n-semiconductor, the field strength of the electrical field increases. At a certain point, an equilibrium is reached between diffusion and field currents that equalize each other. Following this, the depletion zone will only expand until this equilibrium point. The electrical field between the n-semiconductor and the p-semiconductor also leads to a diffusion voltage. In the equilibrium state, the diffusion voltage can be calculated with reference to the following formula.

$$V_{Diffusion} = \frac{k_B \cdot T}{q_e} \cdot \ln\left(\frac{n_A \cdot n_D}{n_i^2}\right)$$

The total depth of the depletion zone can be calculated as the sum of the depth of the depletion zone in the n-semiconductor and the p-semiconductor.

$$d_{depletion} = d_{depletion,n} + d_{depletion,p} = \sqrt{\frac{2 \cdot \varepsilon_0 \cdot \varepsilon_r \cdot V_{Diffusion}}{q_e} \cdot \frac{n_A + n_D}{n_A \cdot n_D}}$$

The product of the individual depth of the depletion zone in the n-semiconductor and the density of the donator atoms in the n-semiconductor needs to be equal to the product of the individual depth of the depletion zone in the p-semiconductor and the density of the acceptor atoms in the p-semiconductor.

$$d_{depletion,n} \cdot n_D = d_{depletion,p} \cdot n_A$$

Electron-hole pairs can be generated from electromagnetic radiation absorbed in the semiconductor. After the generation, the electrical field of the pn-junction acts as a force to the electron-hole pair, forcing the electron in the direction of the n-semiconductor and the hole in the direction of the p-semiconductor. Following this, the electron-hole pair is separated, so that both cannot recombine. Electrons will accumulate in the n-semiconductor outside the depletion zone and holes will accumulate in the p-semiconductor outside the depletion zone. When an electrical load is connected to such a semiconductor, these electrons can move through the connected circuit and the load from the n-semiconductor to the p-semiconductor. At the solar cell terminals, these electrons can be observed as electrical currents. This process is visualized in Figure 4.10.

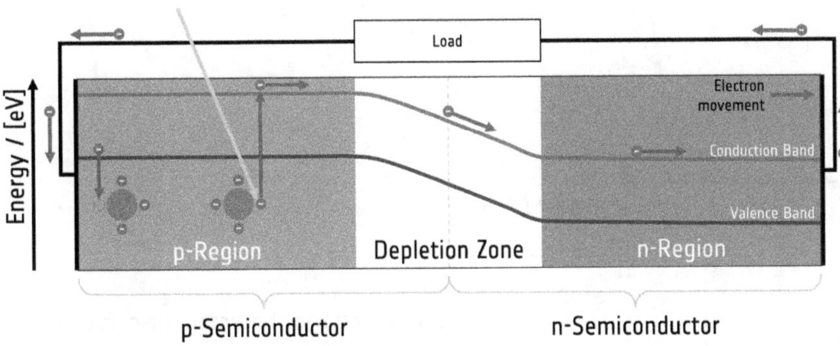

Figure 4.10 Working principle of a solar cell

Solar cells usually consist of a highly doped n-semiconductor as the top layer and a weakly doped p-semiconductor as the bottom layer. In case that a voltage is applied to the semiconductor in blocking direction, electrons and holes will be forced further away from the depletion zone, which leads to an expansion to the depletion zone. Following this, only very small currents can flow in the blocking state through the semiconductor. Contrary to this, in case that voltages in forward direction are applied to semiconductors, this voltage will force electrons and holes in the direction of the depletion zone around the pn-junction. In this case, the expansion of the depletion zone is decreased. It is also possible to equalize the depletion zone completely in this way, which enables higher current flows.

Different semiconductor materials can be differentiated into direct and indirect semiconductors. Direct semiconductors can directly absorb photons and generate electron-hole pairs in response to the absorption. Indirect semiconductors additionally require the presence of a phonon in order to absorb the photon and generate an electron-hole pair. The efficiencies of the direct semiconductor type of solar cells are higher than those of the indirect semiconductor type of solar cells. Silicon is defined as an indirect semiconductor material and gallium arsenide is defined as a direct semiconductor material.

In the dark state of a solar cell, no electron-hole pairs are generated and therefore the electrical behavior of a solar cell is similar to the electrical behavior of a diode. The current-voltage characteristic curve of a diode can be determined utilizing the Shockley equation in a good approximation. An example for the current-voltage characteristic curve of a solar cell in dark and in illuminated state is shown in Figure 4.11.

$$I_{Diode}\left(V_{Diode}\right) = I_0 \cdot \left(e^{\frac{V_{Diode}}{n_{Diode} \cdot V_{Temp}}} - 1\right)$$

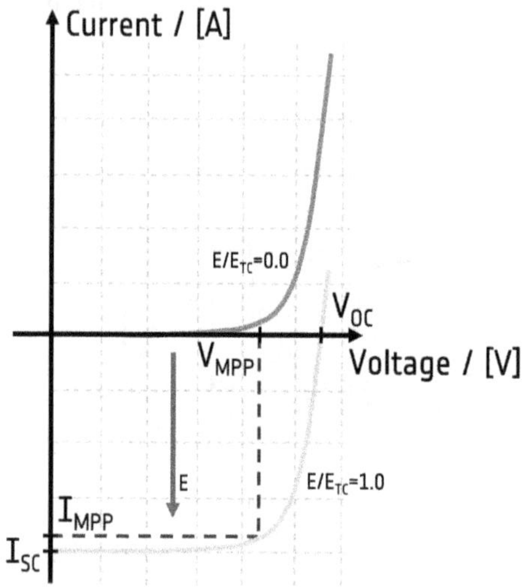

Figure 4.11
Current-voltage characteristic curve of a solar cell (consumer reference system) in different irradiance conditions

The saturation current is resulting from the movement of charge carriers inside the semiconductor lattice. It can be observed in blocking direction, when a diode is operated in the blocking state outside of the breakthrough domain. In solar cells, it can be observed in dark state. The value of diode temperature voltage can be determined with reference to the following equation.

$$V_{Temp} = \frac{k_B \cdot T}{q_e}$$

The generation of electron-hole pairs begins as soon as the solar cells absorb electromagnetic radiation featuring a wavelength that the solar cells are sensitive to. The generation of electron-hole pairs results in a photocurrent that increases with the amount of absorbed electromagnetic radiation.

$$I_{Ph}(E_0) = c_{Ph} \cdot E_0$$

In the illuminated case, the current-voltage characteristic curve can be determined by scaling the characteristic curve of the diode on the current axis. The characteristic curve of a diode is usually expressed in a consumer reference system (Figure 4.11), while the characteristic curve of a solar cell is usually expressed in generator reference system (Figure 4.12). Because of this, the current-voltage characteristic curve of the solar cell in a generator reference system can be determined from the characteristic in a consumer reference system by an inversion of the current-axis. The current-voltage characteristic curve of a solar cell is scaled on the current-axis in positive direction with an increasing amount of received irradiance. It is visualized for different irradiance conditions in Figure 4.12.

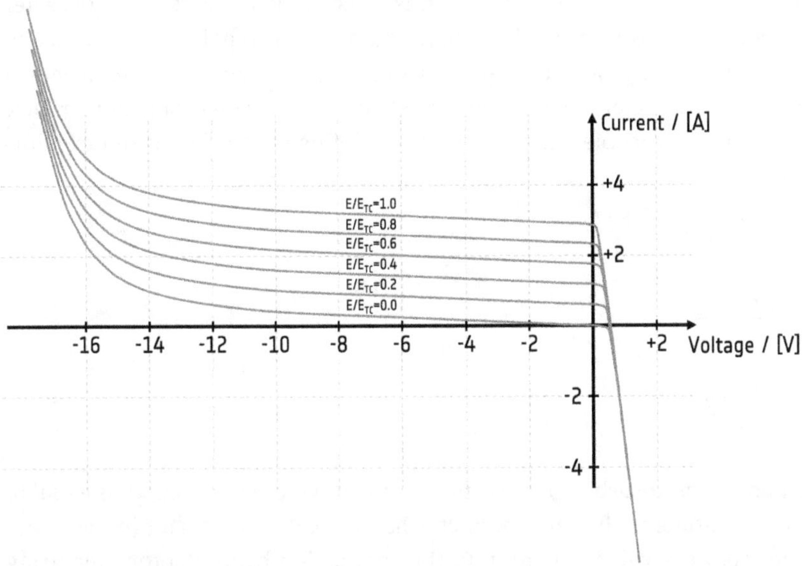

Figure 4.12 Current-voltage characteristic curve of a solar cell in different irradiance conditions in the generator reference system

The characteristic curve can be subdivided into different quadrants. Often, only the positive voltage and positive current quadrant, is visualized. Solar cells are usually operated in this quadrant because the output power features positive values. The solar cell power reaches its maximum value in the Maximum Power Point. From small cell voltages until the Maximum Power Point voltage, small and almost linear decreases in the solar cell current from the short circuit current to the Maximum Power Point current and an almost linear increase in the solar cell power can be observed. The solar cell current and power decreases to zero between cell voltages from the Maximum Power Point voltage to the open circuit voltage. For cell voltages featuring values greater than the open circuit voltage, the solar cell current and power turns negative. In the negative voltage domain, solar cells have positive output currents. At small and intermediate negative voltages, solar cells have relatively constant positive currents that increase with decreasing voltages. From a certain point on, the solar cell current rapidly increases. Even small increases in the cell current above the short circuit current can result in highly negative cell voltages. This characteristic is of particular importance in the case of solar array shadowing. With the exception of the positive voltage and positive current quadrant, where the solar cell output power has positive values, it has negative values.

In order to convert a photon into electrical energy, the photon energy needs to be equal to or greater than the bandgap of the solar cell. Because the photon energy decreases with increasing wavelength, a certain maximum wavelength can be defined, until which the electron-hole pair generation is possible. Photons featuring greater wavelengths cannot be utilized for the generation of electron-hole pairs because in this case, the photon energy would be smaller than the bandgap of the semiconductor. The bandgap is characteristic for a certain semiconductor material, and because of this, a maximum convertible wavelength can be defined for different semiconductor materials.

$$E_{Photon}(\lambda) = h \cdot f = h \cdot \frac{c_0}{\lambda}$$

$$E_G = h \cdot \frac{c_0}{\lambda_{Solar,max}}$$

$$\lambda_{Solar,\mathrm{max}} = \frac{h \cdot c_0}{E_G}$$

In case that a photon is absorbed in the semiconductor, whose wavelength is equal to energy in the magnitude of the semiconductor bandgap, this will result in the lifting of an electron from the valence band into the conduction band. Photons featuring energies greater than the bandgap will also lead to the lifting of an electron from the valence band to the conduction band. However, in this case, the higher amount of absorbed energy will bring the electron into an energy state higher than the lower edge of the conduction band. In the following, the electron will fall back to the lower edge of the conduction band by releasing the surplus of energy. The emitted energy will be absorbed by the crystalline lattice and contribute to the lattice oscillations. This process is referred to as thermalization. A visualization of this effect is shown in Figure 4.6.

Irradiance received by a solar cell is reflected at the solar cell front surface, absorbed inside the solar cell and transmitted through the solar cell (Figure 4.13). In order to account for these effects, different factors need to be defined. The factor absorptance accounts for the absorption of irradiance inside the solar cell, the factor transmittance accounts for the amount of irradiance that is transmitted through a solar cell and released at the solar cell backside, and the factor reflectance accounts for the amount of reflected irradiance at the solar cell front surface. These factors can be defined as a function with reference to the irradiance wavelength or as a mean factor that can be utilized for all wavelengths. The sum of these factors needs to be equal to one.

$$1 = \alpha_{Optic} + \tau_{Optic} + \rho_{Optic} = \alpha_{Optic}(\lambda) + \tau_{Optic}(\lambda) + \rho_{Optic}(\lambda)$$

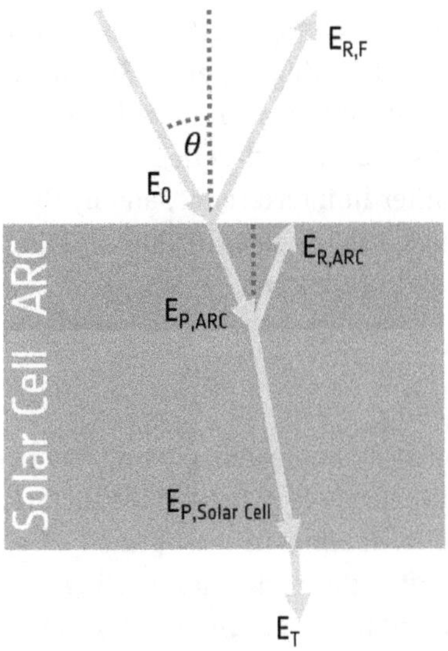

Figure 4.13
Reflection, absorption and transmission effects at a solar cell

Reflections at a surface are resulting from the difference in the speed of light between the environment of the surface and the surface material itself. A measure for this is the refractive index of a medium, which relates the lightspeed in the medium to the lightspeed in vacuum.

$$n_{Refractive,i} = \frac{c_0}{c_i}$$

The factor reflectance is a parameter accounting for reflections on a surface. Usually, the reflectance varies with the incidence angle of electromagnetic irradiance on the surface. It can be determined from the proportion of electromagnetic irradiance that has been reflected in the total amount of electromagnetic irradiance received. The reflectance factor for perpendicular incidence can be derived from the refractive indices of the medium surrounding the surface and of the surface material itself.

$$\rho_{Optic}(\theta_{incidence}) = \frac{E_{SolarCell,R}}{E_0}$$

$$\rho_{Optic}(\theta_{incidence} = 0°) = \left(\frac{n_{Refractive,space} - n_{Refractive,surface}}{n_{Refractive,space} + n_{Refractive,surface}}\right)^2$$

The utilization of anti-reflex coatings offers advantages as a possible countermeasure for reflections at a surface. Because of this, anti-reflex coatings can also increase the efficiency of a solar cell. Examples for possible coating materials are silicon dioxide SiO_2 or silicon nitride Si_3N_4. Anti-reflex coatings are implemented on top of a surface, in order to reduce its reflective properties. Received electromagnetic irradiance will first be received by the surface of the coating, then penetrate into the coating material, and after that it will be received by the coated surface itself. At both surfaces, reflections occur. The primary properties of an anti-reflex coating are the thickness of the coating layer and the refractive index of the selected coating material. The thickness has to be designed in a way that between the reflected irradiance at both surfaces, a phase shift in the magnitude of 180° occurs. As a result, both reflected types of electromagnetic irradiance will interact with each other in interference, and by this, dissolve each other. In order to achieve this, the thickness of the anti-reflex coating needs to be manufactured with reference to the following equation.

$$d_{ARC} = \frac{\lambda}{4 \cdot n_{Refractive,ARC}} \cdot (2 \cdot m_{Refractive} + 1)$$

The reflectance factor resulting from the application of an anti-reflex coating can be calculated for perpendicular incidence employing the refractive indices of the surface material, the surrounding medium and the anti-reflex coating. In the optimal case, the reflectance factor should have a value of zero after the anti-reflex coating has been applied. The required refractive index of a coating material can be determined utilizing the formula of the reflectance factor for perpendicular incidence, where a reflectance of zero should be achieved. For a certain surface, the employment of an anti-reflex coating featuring a refractive index in the magnitude of the square root of the product of the refractive indices of the surface material and the surrounding medium offers the advantage of a minimum of reflected electromagnetic irradiance in the case of perpendicular incidence conditions.

$$\rho_{Optic,ARC}(\theta_{incidence} = 0°) = \left(\frac{n^2_{Refractive,ARC} - n_{Refractive,space} \cdot n_{Refractive,surface}}{n^2_{Refractive,ARC} + n_{Refractive,space} \cdot n_{Refractive,surface}} \right)^2 = 0$$

$$n_{Refractive,ARC} = \sqrt{n_{Refractive,space} \cdot n_{Refractive,surface}}$$

The amount of irradiance that is reflected at a surface can be calculated as the product of the reflectance factor and the irradiance at the surface. The sum of the irradiance reflected at the surface and penetrating the surface has to be equal to the total amount of irradiance received by the surface. The remaining share of the received irradiance that is not reflected at the surface penetrates into the solar cell.

$$E_{SolarCell,R} = E_0 \cdot \rho_{Optic}$$

$$E_0 = E_{SolarCell,P} + E_{SolarCell,R}$$

$$E_{SolarCel,P} = E_0 - E_{SolarCell,R} = E_0 - E_0 \cdot \rho_{Optic} = E_0 \cdot (1 - \rho_{Optic})$$

In the case that electromagnetic irradiance is penetrating a solar cell, the greatest share of absorptions is happening in the beginning of the semiconductor material. The amount of electromagnetic irradiance penetrating deeper into the solar cell decreases with increasing penetration depth. The amount of electromagnetic irradiance penetrating a certain depth of a solar cell can be determined utilizing the absorptance factor.

$$E_{SolarCell}(d_{Penetration}) = E_{SolarCel,P} \cdot e^{-\alpha_{Optic,d} \cdot d_{Penetration}}$$

Electromagnetic irradiance is also transmitted through solar cells. The amount of the received electromagnetic irradiance that is transmitted through the cell can be determined from the amount of irradiance penetrating the solar cell at a depth equal to the total depth of the solar cell.

$$E_{SolarCell,T} = E_{SolarCel,P} \cdot e^{-\alpha_{Optic,d} \cdot d_{SolarCell}}$$

$$E_{SolarCell,T} = E_0 \cdot \left(1 - \rho_{Optic}\right) \cdot e^{-\alpha_{Optic,d} \cdot d_{SolarCell}}$$

The remaining share of the electromagnetic irradiance that penetrates the solar cell, but that is not transmitted through it, is absorbed inside the solar cell. Following this, the amount of absorbed electromagnetic irradiance can be calculated as the difference between penetrating and transmitted irradiance.

$$E_{SolarCell,A} = E_{SolarCell,P} - E_{SolarCell,T}$$

$$E_{SolarCell,A} = E_0 \cdot \left(1 - \rho_{Optic}\right) \cdot \left(1 - e^{-\alpha_{Optic,d} \cdot d_{Cell}}\right)$$

A solar cell needs to be designed thick enough, so that only a minimum amount of irradiance is transmitted through it. The different magnitudes of the absorptance factor in different materials lead to interesting implications for the manufacturing of solar cells. This factor has a value of 7.2 $\frac{1}{mm}$ in silicon and of 630.0 $\frac{1}{mm}$ in gallium arsenide [109]. Because of this, solar cells made from gallium arsenide can be manufactured thinner by magnitudes than silicon solar cells.

Inside a solar cell, processes occur limiting the solar cell efficiency. These processes lead to a recombination of generated electron-hole pairs. Parameters accounting for the magnitude in which recombination occurs inside a solar cell, are the charge carrier life duration and the diffusion length. The charge carrier life duration defines the mean duration a charge carrier exists until it recombines inside the solar cell [151]. The diffusion length is defined as the mean distance that a charge carrier travels through a solar cell until it recombines. A parameter connecting the charge carrier life duration and the diffusion length is the diffusion constant that can be defined for electrons and holes.

$$D_n = \frac{L_n^2}{\tau_n}$$

$$D_p = \frac{L_p^2}{\tau_p}$$

The electrons generated from photons penetrating a solar cell deeply have to travel a relatively long way until these reach the pn-junction. Following this, the chances are high that these electrons recombine inside the solar cell before reaching the pn-junction. In order to prevent these electrons from recombination, it is recommended to include countermeasures against their recombination in the cell design.

The total number of photons of a certain wavelength received by a solar cell can be calculated employing the time integral of the division of the received radiant power of a certain wavelength by the energy of a single photon of the same wavelength. The total number of photos received by a solar cell can be calculated by the integral of the number of photons received in a certain wavelength over all wavelengths.

$$N_{Photons}(\lambda) = \int \frac{A \cdot E_{\lambda}(\lambda)}{E_{Photon}(\lambda)} \, dt = \frac{\lambda}{h \cdot c_0} \cdot A \cdot E_{\lambda}(\lambda) \cdot \Delta t$$

$$N_{Photons} = \int_{0}^{\infty} N_{Photons}(\lambda) \, d\lambda = \frac{A \cdot \Delta t}{h \cdot c_0} \cdot \int_{0}^{\infty} \lambda \cdot E(\lambda) \, d\lambda$$

The efficiency of a solar cell can be defined as its ability to convert the received irradiance into electrical energy. Quantum efficiencies are defined with reference to the photon wavelength. The external quantum efficiency is defined as the ability of a solar cell to generate an electron-hole pair from a received photon. It is defined as the share of the received photons that can be utilized by the solar cell for the electron-hole pair generation. It can be calculated as the division of the total number of electron-hole pairs that have been generated by the total number of photons received in a certain wavelength.

$$\eta_{External}(\lambda) = \frac{N_{EHPairs}(\lambda)}{N_{Photons}(\lambda)}$$

The internal quantum efficiency also takes into account losses resulting from reflections of photons on the solar cell surface, and transmissions of photons through the solar cell. Because of this, the internal quantum efficiency is a measure for the ability of the solar cell lattice itself to generate electron-hole pairs from incident photons. Utilizing the absorptance, transmittance and reflectance parameters that are defined with reference to the photon wavelength, the equation of the internal quantum efficiency can be derived from the equation of the external quantum efficiency.

$$\eta_{Internal}(\lambda) = \frac{\eta_{External}(\lambda)}{\alpha_{Optic}(\lambda)} = \frac{\eta_{External}(\lambda)}{1 - \rho_{Optic}(\lambda) - \tau_{Optic}(\lambda)}$$

An example for the quantum efficiencies of the different subcells of a six-junction solar cell based on gallium arsenide technology are shown in Figure 4.14.

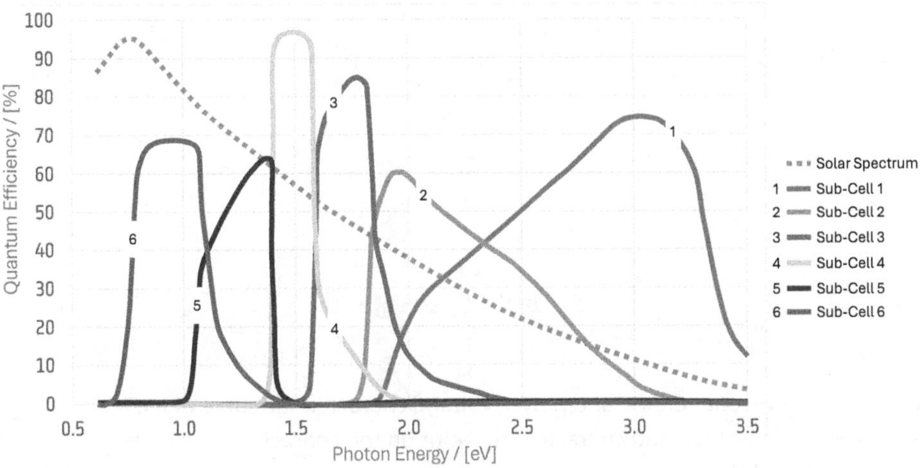

Figure 4.14 Quantum efficiencies of a six-junction gallium arsenide solar cell and solar spectrum (image: Data from R. R. King et al. [166])

The parameter spectral sensitivity is a measure for the ability of a solar cell to convert the energy of a photon featuring a certain wavelength into electrical energy. Following this, the spectral efficiency can be defined as a measure for the photocurrent that occurs in a solar cell in response to the absorbed spectral irradiance.

$$S(\lambda) = \frac{q_e}{h \cdot c_0} \cdot \lambda \cdot \eta_{External}(\lambda) = \frac{I_{Ph}(\lambda)}{A \cdot E_\lambda(\lambda)}$$

The photocurrent of a solar cell in certain irradiation conditions can be defined employing the spectral sensitivity. Usually, solar cells are not irradiated by monochromatic radiation. Because of this, the photocurrent needs to be defined for the total spectrum of the absorbed photons as the integral of the photocurrent for monochromatic radiation over all wavelengths to which a solar cell is sensitive.

$$I_{Ph}(\lambda) = S(\lambda) \cdot A \cdot E_\lambda(\lambda)$$

$$I_{Ph} = \int_0^{\lambda_{Max}} I_{Ph}(\lambda)\, d\lambda = \int_0^{\lambda_{Solar,max}} S(\lambda) \cdot A \cdot E_\lambda(\lambda)\, d\lambda$$

The electrical behavior of solar cells can be defined as voltage-dependent current sources. An example for a current-voltage and power-voltage characteristic curve of a six-junction gallium arsenide solar cell with indicated characteristic points are shown in Figure 4.15.

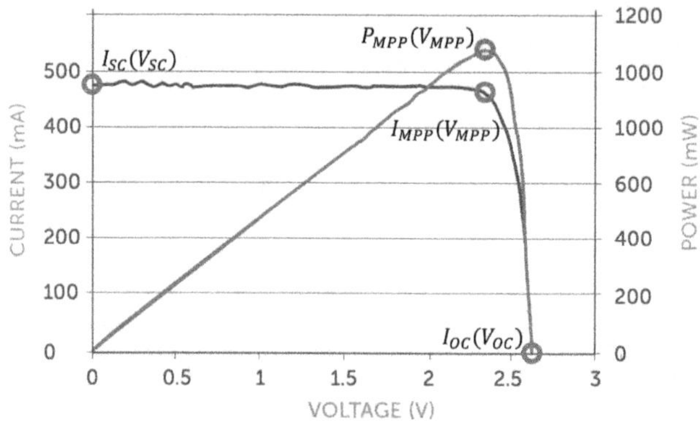

Figure 4.15 Current-voltage and power-voltage characteristic curve of a CTJ30 solar cell made by CESI and characteristic point on the characteristic curves (image: © CESI [167])

The current-voltage characteristic curve has 3 characteristic points in the positive voltage and positive current quadrant. In the open circuit characteristic point the solar cell output current has a value of zero ampere, the solar cell power has a value of zero watts and the solar cell voltage has its maximum value in the positive voltage and positive current quadrant. In the short circuit characteristic point the solar cell output voltage has a value of zero volts, the solar cell power has a value of zero watts and the solar cell current has its maximum value in this quadrant. The solar cell output power has maximum values in the Maximum Power Point. The corresponding current is positive but slightly smaller than in the short circuit characteristic point, and the corresponding voltage is also positive but slightly smaller than in the open circuit characteristic point.

Short circuit point:

$$V_{SolarCell} = 0\,\text{V},\ I_{SolarCell}(0\,\text{V}) = I_{SC},\ P_{SolarCell}(0\,\text{V}) = 0\,\text{W}$$

Maximum Power Point:

$$V_{SolarCell} = V_{MPP},\ I_{SolarCell}(V_{MPP}) = I_{MPP},\ P_{SolarCell}(V_{MPP}) = P_{MPP}$$

Open circuit point:

$$V_{SolarCell} = V_{OC},\ I_{SolarCell}(V_{OC}) = 0\,\text{A},\ P_{SolarCell}(V_{OC}) = 0\,\text{W}$$

Nearly the complete photocurrent generated inside the solar cell can be measured at the solar cell terminals in short circuit conditions. Its value can be determined employing a photocurrent factor that allows to determine the photocurrent with reference to the amount of horizontal irradiance at the solar cell front site.

$$I_{Ph} = c_{Ph} \cdot E_{Hor.,0} \approx I_{SC}$$

The solar cell has an output current of 0 A in open circuit conditions, where a voltage in the magnitude of the open circuit voltage occurs at the solar cell terminals. Its value can be defined with reference to the solar cell current equation for the simplified single diode model in the open circuit characteristic point.

$$I_{SolarCell}(V_{OC}) = I_{Ph} - I_0 \cdot \left(e^{\frac{V_{OC}}{n_{Diode} \cdot V_{Temp}}} - 1 \right) = 0A$$

$$e^{\frac{V_{OC}}{n_{Diode} \cdot V_{Temp}}} = \frac{I_{Ph}}{I_0} + 1$$

$$V_{OC} = n_{Diode} \cdot V_{Temp} \cdot ln\left(\frac{I_{Ph}}{I_0} + 1\right) = n_{Diode} \cdot V_{Temp} \cdot ln\left(\frac{c_{Ph} \cdot E_{Hor.,0}}{I_0} + 1\right)$$

The location of a characteristic point is depending on the amount of received irradiance, the cell temperature and the cell degradation. Each characteristic point has to be defined by one voltage and one current value. The value of a voltage or current is defined by the combination of two parameters, one parameter for the parameter value under test conditions and another parameter for the temperature variation of this parameter. Manufacturers of space solar cells usually defined these parameters as a function of the equivalent fluence in order to define the degradation of their products in response to the space radiation environment.

Open circuit voltage:

$$V_{OC}^0(\phi_{Ref}), \quad \frac{dV_{OC}}{dT}(\phi_{Ref})$$

Short circuit current:

$$I_{SC}^0(\phi_{Ref}), \quad \frac{dI_{SC}}{dT}(\phi_{Ref})$$

Maximum Power Point voltage:

$$V_{MPP}^0(\phi_{Ref}), \quad \frac{dV_{MPP}}{dT}(\phi_{Ref})$$

Maximum Power Point current:

$$I_{MPP}^0(\phi_{Ref}), \quad \frac{dI_{MPP}}{dT}(\phi_{Ref})$$

The value of one of these voltages or currents can be calculated employing the corresponding two parameters. Different standards and units can be utilized for the definition of these parameters. The temperature coefficient is usually given as a constant normalized to unit temperature. Subsequently, the temperature coefficient needs to be multiplied by the temperature deviation between the cell temperature and the reference temperature during test conditions.

$$\Delta T = T - T_{TC}$$

Solar cell currents respond linearly and solar cell voltages respond logarithmically to variations in irradiance. Irradiance-dependent factors are employed in order to implement the irradiance dependency of solar cell voltages and currents into the determination equations for the characteristic voltages and currents. The response of the electrical solar cell characteristics to variations in the available irradiance is visualized in Figure 4.16.

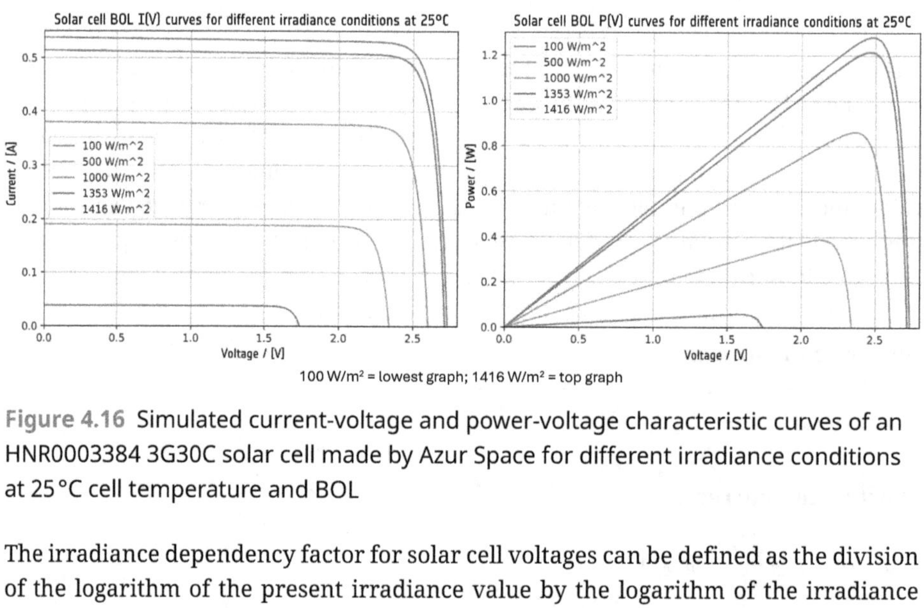

Figure 4.16 Simulated current-voltage and power-voltage characteristic curves of an HNR0003384 3G30C solar cell made by Azur Space for different irradiance conditions at 25 °C cell temperature and BOL

The irradiance dependency factor for solar cell voltages can be defined as the division of the logarithm of the present irradiance value by the logarithm of the irradiance value under test conditions. The irradiance dependency factor for solar cell currents can be defined as the division of the present irradiance value by the irradiance value under test conditions.

$$\alpha_{SolarCell,V}(E_0) = \frac{\ln(E_0)}{\ln(E_{0,TC})}$$

$$\alpha_{SolarCell,I}(E_0) = \frac{E_0}{E_{0,TC}}$$

A popular method for the determination of characteristic voltages and currents is to calculate them as the sum of the parameter value under test conditions and the product of the corresponding temperature coefficient and the temperature deviation from test conditions. The irradiance dependency can be implemented by multiplying the corresponding voltage or current irradiance factor. In this case, the parameter values under test conditions need to be defined employing the units volts and ampere. The corresponding temperature coefficients need to be defined in the units volts per kelvin or ampere per kelvin.

$$V_{OC}(E_0, T, \phi_{Ref}) = \alpha_{SolarCell,V}(E_0) \cdot \left(V_{OC}^0(\phi_{Ref}) + \frac{dV_{OC}}{dT}(\phi_{Ref}) \cdot \Delta T\right)$$

$$I_{SC}(E_0, T, \phi_{Ref}) = \alpha_{SolarCell,I}(E_0) \cdot \left(I_{SC}^0(\phi_{Ref}) + \frac{dI_{SC}}{dT}(\phi_{Ref}) \cdot \Delta T\right)$$

$$V_{MPP}(E_0, T, \phi_{Ref}) = \alpha_{SolarCell,V}(E_0) \cdot \left(V_{MPP}^0(\phi_{Ref}) + \frac{dV_{MPP}}{dT}(\phi_{Ref}) \cdot \Delta T\right)$$

$$I_{MPP}(E_0, T, \phi_{Ref}) = \alpha_{SolarCell,I}(E_0) \cdot \left(I_{MPP}^0(\phi_{Ref}) + \frac{dI_{MPP}}{dT}(\phi_{Ref}) \cdot \Delta T\right)$$

It is also a common method to define the temperature coefficients as factors featuring a value from zero to one in the unit one per kelvin. When temperature coefficients have been defined in this way, a factor needs to be calculated as the sum of one and the product of the temperature coefficient and the temperature deviation between cell temperature and reference temperature under test conditions. This factor needs to be multiplied with the product of the parameter value under test conditions and the corresponding irradiance factor in the next step. In the definition of temperature coefficients, the unit percent per kelvin is also commonly utilized. In this case, the temperature coefficient also needs to be deviated by 100%.

$$V_{OC}(E_0, T, \phi_{Ref}) = \alpha_{SolarCell,V}(E_0) \cdot V_{OC}^0(\phi_{Ref}) \cdot \left(1 + \frac{dV_{OC}}{dT}(\phi_{Ref}) \cdot \Delta T\right)$$

$$I_{SC}(E_0, T, \phi_{Ref}) = \alpha_{SolarCell,I}(E_0) \cdot I_{SC}^0(\phi_{Ref}) \cdot \left(1 + \frac{dI_{SC}}{dT}(\phi_{Ref}) \cdot \Delta T\right)$$

$$V_{MPP}(E_0, T, \phi_{Ref}) = \alpha_{SolarCell,V}(E_0) \cdot V_{MPP}^0(\phi_{Ref}) \cdot \left(1 + \frac{dV_{MPP}}{dT}(\phi_{Ref}) \cdot \Delta T\right)$$

$$I_{MPP}(E_0, T, \phi_{Ref}) = \alpha_{SolarCell,I}(E_0) \cdot I_{MPP}^0(\phi_{Ref}) \cdot \left(1 + \frac{dI_{MPP}}{dT}(\phi_{Ref}) \cdot \Delta T\right)$$

Characteristic voltages and currents can also be defined utilizing remaining factors. Remaining factors define the degradation of the parameters defining characteristic voltages and currents as functions with reference to the equivalent fluence. These factors can feature values between one and zero. In this case, the individual values of the parameter under test conditions and the corresponding temperature coefficient are defined as BOL values and not as a function with reference to the equivalent fluence. Remaining factors have to be multiplied with the corresponding parameter in order to determine a characteristic voltage and current value. In the next step, a factor has to be calculated as the sum of the product of a parameter under test conditions and the corresponding remaining factor, and as the sum of the product of the corresponding temperature coefficient, the remaining factor and the temperature deviation from test conditions. The value of a characteristic voltage or current has to be calculated as product of this value and the corresponding irradiance factor in the last step.

$$V_{OC}(E_0, T, \phi_{Ref}) = \alpha_{SolarCell,V}(E_0) \cdot \left(V_{OC,0}^{0} \cdot R_{V_{OC}}(\phi_{Ref}) + \frac{dV_{OC}}{dT} \cdot R_{\frac{dV_{OC}}{dT}}(\phi_{Ref}) \cdot \Delta T \right)$$

$$I_{SC}(E_0, T, \phi_{Ref}) = \alpha_{SolarCell,I}(E_0) \cdot \left(I_{SC,0}^{0} \cdot R_{I_{SC}}(\phi_{Ref}) + \frac{dI_{SC}}{dT} \cdot R_{\frac{dI_{SC}}{dT}}(\phi_{Ref}) \cdot \Delta T \right)$$

$$V_{MPP}(E_0, T, \phi_{Ref}) = \alpha_{SolarCell,V}(E_0) \cdot \left(V_{MPP,0}^{0} \cdot R_{V_{MPP}}(\phi_{Ref}) + \frac{dV_{MPP}}{dT} \cdot R_{\frac{dV_{MPP}}{dT}}(\phi_{Ref}) \cdot \Delta T \right)$$

$$I_{MPP}(E_0, T, \phi_{Ref}) = \alpha_{SolarCell,I}(E_0) \cdot \left(I_{MPP,0}^{0} \cdot R_{I_{MPP}}(\phi_{Ref}) + \frac{dI_{MPP}}{dT} \cdot R_{\frac{dI_{MPP}}{dT}}(\phi_{Ref}) \cdot \Delta T \right)$$

The parameters of characteristic currents are sometimes also defined as area-densities. In this case, also the solar cell front area size needs to be multiplied in the determination of characteristic currents.

The solar cell output power can be defined as the product of the solar cell voltage and current. Subsequently, the Maximum Power Point power can be defined as product of the Maximum Power Point voltage and current.

$$P_{SolarCell}(E_0, T, \phi_{Ref}) = V_{SolarCell}(E_0, T, \phi_{Ref}) \cdot I_{SolarCell}(E_0, T, \phi_{Ref})$$

$$P_{MPP}(E_0, T, \phi_{Ref}) = V_{MPP}(E_0, T, \phi_{Ref}) \cdot I_{MPP}(E_0, T, \phi_{Ref})$$

The voltage in the Maximum Power Point usually has a smaller value than the open circuit voltage and the current in the Maximum Power Point usually has a smaller value that the short circuit current. In a perfect solar cell, the Maximum Power Point voltage and the open circuit voltage, as well as the Maximum Power Point current and the short circuit current, would have equal values. Following this, the power observed in the Maximum Power Point of real solar cells is usually smaller than the product of the open circuit voltage and the short circuit current.

$$P_{MPP} = V_{MPP} \cdot I_{MPP} < V_{OC} \cdot I_{SC}$$

A parameter relating between the theoretically maximum output power of a solar cell and the real output power in the Maximum Power Point is the fill factor. It can be calculated as the division of the real output power in the Maximum Power Point by theoretical Maximum Power Point power that can be calculated as the product of the open circuit voltage and the short circuit current. A perfect solar cell would feature a fill factor of 1. However, internal power dissipation occurs in a real solar cell, and because of this, the fill factor usually features values smaller than 1.

$$FF = \frac{V_{MPP} \cdot I_{MPP}}{V_{OC} \cdot I_{SC}}$$

The maximum solar cell efficiency can be calculated as the division of the Maximum Power Point power by the received radiant power.

$$P_{El} = P_{MPP} = V_{MPP} \cdot I_{MPP} = FF \cdot V_{OC} \cdot I_{SC}$$

$$P_{Radiant} = \int_{0}^{\lambda_{Solar,max}} A_{Cross} \cdot E(\lambda)\, d\lambda$$

$$\eta_{SolarCell,Theo} = \frac{P_{El}}{P_{Radiant}} = \frac{FF \cdot V_{OC} \cdot I_{SC}}{\int_{0}^{\lambda_{Solar,max}} A_{Cross} \cdot E(\lambda)\, d\lambda} = \frac{FF \cdot V_{OC} \cdot I_{SC}}{A_{Cross} \cdot E_0}$$

For the simulation of the solar cell behavior in electrical circuits, equivalent circuit models were developed. Equivalent-circuit models replicate the electrical behavior of solar cells employing circuits of basic electrical parts, such as diodes, resistors or current sources. A visualization of different solar cell equivalent circuit models featuring different levels of complexity are shown in Figure 4.17.

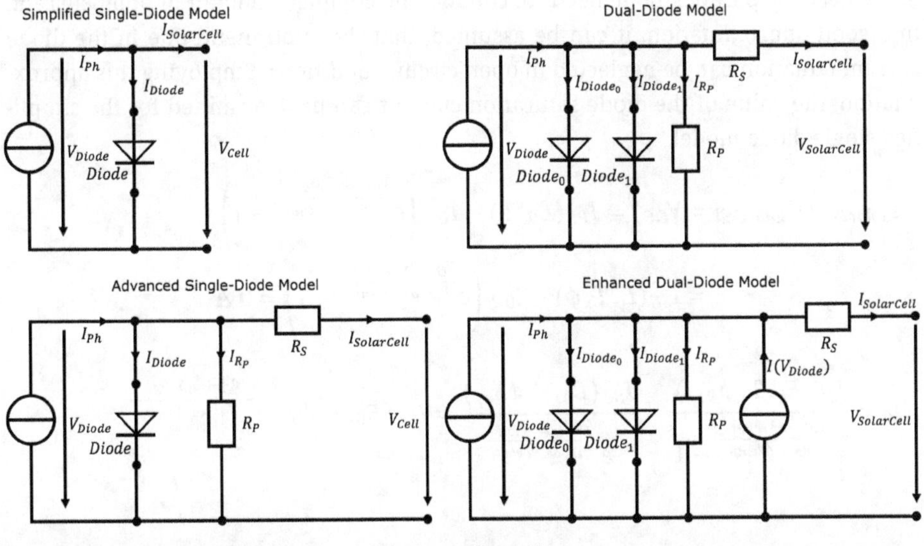

Figure 4.17 Equivalent circuit models for solar cells

The simplified single diode model is a parallel circuit consisting of an irradiance-dependent current source and a diode. It features a satisfying representation of the solar cell characteristic curve. The corresponding solar cell current equation can be determined employing Kirchhoff's law for the solar cell terminal current, the diode current and the photocurrent of the irradiance dependent current source.

$$I_{Cell}\left(V_{SolarCell}\right) = I_{Ph}(E_0) - I_{Diode}\left(V_{SolarCell}\right)$$

$$I_{SolarCell}\left(V_{SolarCell}\right) = c_{Ph} \cdot E_0 - I_0 \cdot \left(e^{\frac{V_{Solar\,Cell}}{n_{Diode} \cdot V_{Temp}}} - 1\right)$$

The voltage at the solar cell terminals has a value of zero volts in short circuit conditions, while the current at the solar cell terminals has a value in the magnitude of the short circuit current. In short circuit conditions, almost the complete amount of the available photocurrent can be drawn from the solar cell terminals. Subsequently, the current through the diode of the simplified single diode model can be neglected in a good approximation. Following this, the solar cell current equation of the simplified single diode model in the short circuit characteristic point can be employed for the definition of the photocurrent. When the simplified single diode model is employed, the photocurrent is equal to the short circuit current.

$$\begin{aligned} I_{SC}(E_0, T, \phi_{Ref}) &= I_{SolarCell}(V_{SolarCell} = 0\,V) = I_{Ph}(E_0, T) - I_0 \cdot \left(e^{\frac{0\,V}{n_{Diode} \cdot V_{Temp}}} - 1\right) \\ &= I_{Ph}(E_0, T) - 0 = I_{Ph}(E_0, T) \end{aligned}$$

In open circuit conditions, the current at the solar cell terminals has a value of zero ampere, while a voltage in the magnitude of the open circuit voltage occurs at the solar cell terminals. Employing the simplified single diode model, this results in a condition where the parallel diode needs to conduct the complete amount of photocurrent. In a good approximation, it can be assumed, that the subtrahend one in the diode current equation can be neglected in open circuit conditions. Employing this approximation, the value of the diode saturation current can be determined for the simplified single diode model.

$$\begin{aligned} I_{SolarCell}(V_{SolarCell} = V_{OC}) &= I_{Ph}(E_0, T) - I_0 \cdot \left(e^{\frac{V_{OC}(E_0, T, \phi_{Ref})}{n_{Diode} \cdot V_{Temp}}} - 1\right) \\ &= I_{SC}(E, T, \phi) - I_0 \cdot \left(e^{\frac{V_{OC}(E_0, T, \phi_{Ref})}{n_{Diode} \cdot V_{Temp}}} - 1\right) = 0\,A \end{aligned}$$

$$I_0 = \frac{I_{SC}(E_0, T, \phi_{Ref})}{e^{\frac{V_{OC}(E_0, T, \phi_{Ref})}{n_{Diode} \cdot V_{Temp}}} - 1} \approx \frac{I_{SC}(E_0, T, \phi_{Ref})}{e^{\frac{V_{OC}(E_0, T, \phi_{Ref})}{n_{Diode} \cdot V_{Temp}}}} = I_{SC}(E_0, T, \phi) \cdot e^{-\frac{V_{OC}(E_0, T, \phi_{Ref})}{n_{Diode} \cdot V_{Temp}}}$$

The advanced single diode model allows to simulate the current-voltage characteristic curve of a solar cell even more accurate than employing the simplified single diode model. This model consists of a parallel circuit consisting of an irradiance-dependent current source, a diode and a parallel resistor, as well as a series resistor. The corresponding solar cell current equation can be determined employing Kirchhoff's law and Ohm's law.

$$I_{SolarCell}(V_{SolarCell}) = I_{Ph}(E_0, T) - I_{Diode}(V_{Diode}, T) - I_{R_P}(V_{R_P})$$

$$I_{SolarCell}(V_{SolarCell}) = I_{Ph}(E_0, T) - I_0 \cdot \left(e^{\frac{V_{Diode}}{n_{Diode} \cdot V_{Temp}}} - 1\right) - \frac{V_{R_P}}{R_P}$$

$$I_{SolarCell}(V_{SolarCell}) = \frac{1}{1 + \frac{R_S}{R_P}} \cdot \left[I_{Ph}(E_0, T) - I_0 \cdot \left(e^{\frac{V_{SolarCell} + R_S \cdot I_{SolarCell}(V_{SolarCell})}{n_{Diode} \cdot \frac{k_B \cdot T}{q_e}}} - 1\right) - \frac{V_{SolarCell}}{R_P}\right]$$

The values of the parallel and the series resistance of the advanced single diode model can be determined with reference to the inclination of the current-voltage characteristic curve in the short circuit and the open circuit characteristic points. From the inclination of the characteristic curve in the open circuit characteristic point, the value of the series resistance can be determined, and from the inclination of the characteristic curve in the short circuit characteristic point, the value of the parallel resistance can be determined. Simplified equations have been developed by S. Vergura, that allow to define the series and parallel resistances of the advanced single diode model with reference to the characteristic points [168].

$$R_S = \frac{\frac{1}{4} \cdot [V_{OC}(E_0, T, \phi_{Ref}) - V_{MPP}(E_0, T, \phi_{Ref})]}{I_{MPP}(E_0, T, \phi_{Ref})}$$

$$R_p = \frac{V_{MPP}(E_0, T, \phi_{Ref})}{\frac{1}{2} \cdot [I_{SC}(E_0, T, \phi_{Ref}) - I_{MPP}(E_0, T, \phi_{Ref})]}$$

In order to utilize the advanced single diode model with datasheet parameters, a simplified exponential function can be utilized to define the diode current of the advanced single diode model in a good approximation. This exponential function has a value of one in open circuit conditions and a value of zero in short circuit conditions. Subsequently, the diode current is equal to the diode saturation current in open circuit conditions, and it is equal to zero in short circuit conditions.

$$I_{Diode} = I_0 \cdot e^{\gamma_{Diode} \cdot (V_{SolarCell} - V_{OC}(E, T, \phi_{Ref}))}$$

In short circuit conditions, the current through the parallel diode and the parallel resistor have minimum values and the majority of the photocurrent can be drawn from the solar cell terminals. Because of this, the equation for the photocurrent can be determined employing the short circuit characteristic point.

$$I_{SolarCell}(V_{SolarCell} = 0\,V) = \frac{1}{1 + \frac{R_S}{R_P}} \cdot \left(I_{Ph}(E_0, T) - I_0 \cdot e^{\gamma_{Diode} \cdot (0V - V_{OC}(E, T, \phi_{Ref}))} - \frac{0\,V}{R_P} \right)$$

$$= I_{SC}(E_0, T, \phi_{Ref})$$

$$I_{Ph}(E_0, T) - I_0 \cdot e^{-\gamma_{Diode} \cdot V_{OC}(E, T, \phi_{Ref})} = \left(1 + \frac{R_S}{R_P}\right) \cdot I_{SC}(E_0, T, \phi_{Ref})$$

$$I_{Ph}(E_0, T) = \left(1 + \frac{R_S}{R_P}\right) \cdot I_{SC}(E_0, T, \phi_{Ref}) + I_0 \cdot e^{-\gamma_{Diode} \cdot V_{OC}(E_0, T, \phi_{Ref})}$$

$$\approx \left(1 + \frac{R_S}{R_P}\right) \cdot I_{SC}(E_0, T, \phi_{Ref})$$

In open circuit conditions, the solar cell terminal voltage has its maximum value and the solar cell terminal current has a value of zero ampere. Following this, the voltage at the series resistor has a value of zero volts and the complete current is conducted by the parallel resistor and by the diode of the advanced single diode model. Subsequently, the diode saturation current can be defined employing the solar cell current equation of the advanced single diode model in the open circuit characteristic point.

$$I_{SolarCell}(V_{SolarCell} = V_{OC}(E_0, T, \phi_{Ref})) = \frac{1}{1 + \frac{R_S}{R_P}}$$

$$\cdot \left(I_{Ph}(E_0, T) - I_0 \cdot e^{\gamma_{Diode} \cdot (V_{OC}(E_0, T, \phi_{Ref}) - V_{OC}(E_0, T, \phi_{Ref}))} - \frac{V_{OC}(E_0, T, \phi_{Ref})}{R_P} \right) = 0\,A$$

$$\left(1 + \frac{R_S}{R_P}\right) \cdot I_{SC}(E_0, T, \phi_{Ref}) - I_0 - \frac{V_{OC}(E_0, T, \phi_{Ref})}{R_P} = 0\,A$$

$$I_0 = \left(1 + \frac{R_S}{R_P}\right) \cdot I_{SC}(E_0, T, \phi_{Ref}) - \frac{V_{OC}(E_0, T, \phi_{Ref})}{R_P}$$

In the last step of the model definition, the solar cell current equation in the Maximum Power Point can be utilized for the determination of the diode current exponential factor gamma.

$$I_{SolarCell}(V_{SolarCell} = V_{MPP}) = \frac{1}{1 + \frac{R_S}{R_P}}$$

$$\cdot \left(I_{Ph}(E_0, T) - I_0 \cdot e^{\gamma_{Diode} \cdot (V_{MPP}(E_0, T, \phi_{Ref}) - V_{OC}(E_0, T, \phi))} - \frac{V_{MPP}(E_0, T, \phi_{Ref})}{R_P} \right)$$

$$= I_{MPP}(E_0, T, \phi_{Ref})$$

$$I_0 \cdot e^{\gamma_{Diode} \cdot (V_{MPP}(E_0,T,\phi_{Ref}) - V_{OC}(E_0,T,\phi_{Ref}))} = I_{Ph}(E_0,T) - \frac{V_{MPP}(E_0,T,\phi_{Ref})}{R_P} - \left(1 + \frac{R_S}{R_P}\right) \cdot I_{MPP}(E_0,T,\phi_{Ref})$$

$$\gamma_{Diode} \cdot (V_{MPP}(E_0,T,\phi_{Ref}) - V_{OC}(E_0,T,\phi_{Ref})) = \ln\left(\frac{I_{Ph}(E_0,T) - \frac{V_{MPP}(E_0,T,\phi_{Ref})}{R_P} - \left(1 + \frac{R_S}{R_P}\right) \cdot I_{MPP}(E_0,T,\phi_{Ref})}{I_0}\right)$$

$$\gamma_{Diode} = \frac{1}{V_{MPP}(E_0,T,\phi_{Ref}) - V_{OC}(E_0,T,\phi_{Ref})} \cdot \ln\left(\frac{I_{Ph}(E_0,T) - \frac{V_{MPP}(E_0,T,\phi_{Ref})}{R_P} - \left(1 + \frac{R_S}{R_P}\right) \cdot I_{MPP}(E_0,T,\phi_{Ref})}{I_0}\right)$$

Even the simplified single diode model delivers satisfying results when it is accurately fitted. The advanced single diode model delivers slightly more accurate results. Subsequently, the utilization of a dual diode equivalent circuit model is only required in cases where highly accurate simulation results are necessary. However, the determination of the parameters for these models requires extensive laboratory measurements.

The losses occurring in solar cells can be subdivided into optical losses and electrical losses. Optical losses can be subdivided into losses resulting from reflections at the solar cell front side, shadowing resulting from front side contacts, and transmission losses. Electrical losses can be subdivided into ohmic losses, recombination losses, and losses resulting from cell mismatching. In addition to the loss mechanisms in the solar cell itself, the solar cell cover glass and possible darkening of the cover glass can also cause power losses. Reflections at the solar cell front side are resulting from the different refractive indices of the solar cell environment and the solar cell itself. These can be minimized by the utilization of anti-reflex coatings or by the texturization of the solar cell front side. A texturization does not decrease the amount of reflections itself, but due to the surface structure, the reflected share of the incident solar irradiance will encounter the solar cell material again and possibly penetrate the structured front side instead of being reflected again. The positive electrical contacts are usually directly printed on the solar cell front side utilizing screen printing processes. However, electrical contacts are not transparent and because of this, the share of the solar cell front side that is covered by electrical contracts cannot receive solar irradiance. Therefore, this share of the front side is shadowed by the contact material. In order to decrease this type of intrinsic shadowing, different approaches can be utilized. Usually, very thin transverse contacts with a thickness between 100 µm and

200 µm are implemented together with longitudinal bus bars that are utilized to collect the current from the transverse contacts [151]. The bus bars are usually only connected to the solar cell terminals on one side and not on the other. Following this, the current density in the bus bar will increase from the unconnected side to the connected side. This allows to design bus bars with tapered ends on the side not connected to the terminals. The front side contacts need to be designed to carry the solar cell currents. In order to enable the current conduction, the front side contacts require a certain diameter, but the geometric shape of the contacts is not an important parameter for the conduction of electrical currents. A possible approach to minimize the intrinsic shadowing is to manufacture high and narrow front side contacts. Front side contacts can also be buried into the solar cell frontside itself. In this case, narrow and high contact implementations are also advantageous. The penetration depth of long wavelength is relatively long and possibly longer than the thickness of the solar cell itself, so that a share of the incident solar irradiance can be transmitted through the solar cell. A countermeasure for these transmission losses is the utilization of back surface reflectors and a front surface texturization, so that long wavelength irradiance travels multiple times through the solar cell. Ohmic losses occur at the front side contacts, in the solar cell itself, and on the step from solar cell to front side contacts. Narrow and high contacts, as well as buried contacts, increase the diameter of the front-side conductors and, by this, minimize ohmic losses in the conductors. Solar cells are semiconductors and semiconductors are no good conductors at room temperature. Because of this, losses also occur in the solar cells themselves. An increase in the solar cell doping can decrease the internal losses in the solar cell. The potential step from solar cell front side to front-side contacts can act like a pn-junction. This effect can be reduced by implementing an extremely high doping around the front-side contacts. Recombinations at the solar cell surface and recombinations inside the solar cell can be differentiated. At the edges of a solar cell, the crystalline lattice is disrupted because the crystalline structure ends. Following this, edge atoms cannot form 4 covalent bonds with neighboring atoms. These can have open electron-pair bindings, which can cause recombinations. Recombinations can be prohibited by an oxide coating that is often referred to as passivation of the front surface. Anti-reflex coatings can also act as surface passivation. On the backside of a solar cell, a high p-doping can act as a back-surface field. This field applies a force to generated electrons, pushing them in the direction of the pn-junction. By this, a back-surface field is minimizing the recombination of electrons generated deep inside the solar cell. In addition to this, also impurities in the crystalline lattice impose additional possible energy states in the forbidden zone, which allow electrons to fall back from the conduction band to the valence band. Impurities in the crystalline lattice can result in recombination centers. These can result from disruptions in the crystalline lattice or from impurity atoms.

Solar cells made from mono-crystalline silicon are usually manufactured using the Czochralski or Floating Zone processes. They are also referred to as c-Si solar cells.

The p-doped wafer front side is textured and doped to form an n+ emitter. In the following, a surface passivation layer and an anti-reflex layer are coated on the solar cell front side. The front side contacts are usually manufactured by employing screen printing with aluminum paste. Mono-crystalline silicon cells can only generate electron-hole pairs from visible light. Common c-Si solar cells achieve efficiencies between 9% and 15% [97] [148] [155]. The theoretically maximum efficiency silicon solar cell has a value of 28.6% [151]. Space certified c-Si solar cells can be manufactured with a thickness of 0.18 mm [97]. In recent years, c-Si solar cells gained popularity again in space applications, due to a satisfying efficiency and relatively low costs. The efficiency of c-Si cells can even be increased by increasing the thermal radiative properties and by decreasing the internal parasitic resistance. After 15 years in geostationary orbit, a degradation in the magnitude from 17% efficiency at BOL to 12% efficiency at EOL can be expected for c-Si solar cells [97].

Hi-ETA solar cells are made from c-Si solar cells that have been textured with a pyramid-type structure on the front side, a back surface reflector, a front side passivation layer and an anti-reflex coating. Efficiencies of up to 17% have been achieved with this type of solar cells. An example for the internal structure of a Hi-ETA silicon solar cell is shown in Figure 4.18.

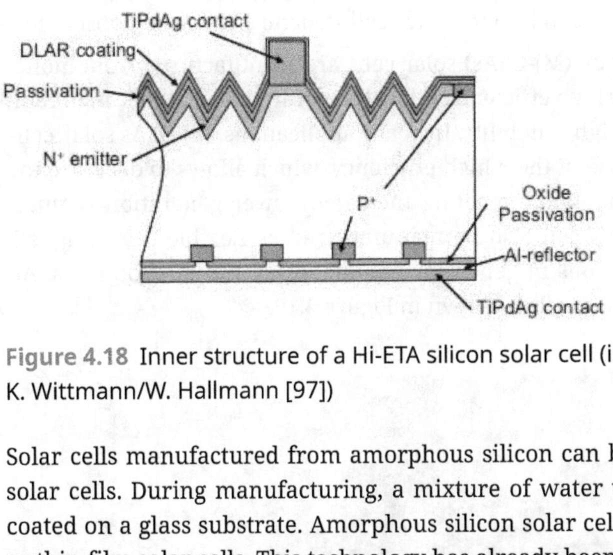

Figure 4.18 Inner structure of a Hi-ETA silicon solar cell (image: © W. Ley/K. Wittmann/W. Hallmann [97])

Solar cells manufactured from amorphous silicon can be defined as polycrystalline solar cells. During manufacturing, a mixture of water vapor and gaseous silicon is coated on a glass substrate. Amorphous silicon solar cells are also referred to as a-Si or thin-film solar cells. This technology has already been applied in space.

CIGS solar cells are manufactured on a glass substrate with molybdian coating as a backside contact layer. As cell materials, $CuInSe_2$, $CuInS_2$, $CuGaSe_2$ and $CuGaSe_2$ can be utilized [151]. A common cell material is $CuIn_xGa_{1-x}Se_2$ that is usually referred to as CIGS material. The bandgap can be designed by a variation in the indium content that is indicated with the "x." Bandgaps between 1.0 eV and 1.7 eV can be manufactured in this way [151]. On the Molybdian back contacts, a p-doped CIGS layer is first coated as an absorber, followed by an n-doped CIGS layer as emitter. The front contacts are manufactured by a tin oxide layer. Utilizing CIGS solar cells, efficiencies of up to 15% were achieved [151].

Heterojunction solar cells with intrinsic thin layer are manufactured from a combination of amorphous and crystalline silicon. This type of solar cells is usually referred to as HIT solar cell. As a substrate for HIT solar cells, n-doped crystalline silicon is utilized. On the front side of the substrate, a p-doped layer of amorphous silicon on top of an intrinsic layer of amorphous silicon is implemented, and on the backside an intrinsic amorphous silicon layer, followed by an n-doped layer made from amorphous silicon, is implemented. Efficiencies of up to 25.6% and cell voltages of 0.7 V were achieved with HIT solar cells [151].

Solar cells made from cadmium telluride (CdTe) are III-VI semiconductors and feature a bandgap of 1.45 V [151]. For the manufacturing, glass is usually employed as a substrate and it is usually coated with Indium Tin Oxide (ITO) as a front contact layer. On the ITO layer, an n-doped layer of CdTe is coated first, followed by a p-doped layer of CdTe and a back contact layer. CdTe solar cells can theoretically reach efficiencies of up to 29.7%, but currently the maximum efficiency has a value of 22.0% [151]. In some space applications, CdTe solar cells were utilized. In orbit, only efficiencies between 11% and 13% have been demonstrated [201]. Additional aging effects resulting from backside contact material diffusion into the solar cell structure have been observed.

Multi-junction gallium arsenide (MJ-GaAs) solar cells are manufactured from monocrystalline material and offer high efficiencies, low temperature variations, high radiation hardness, as well as a high reliability. In space applications, MJ-GaAs solar cells are commonly utilized because of their high efficiency, which allows to decrease the solar array mass and volume, while meeting the same power generation requirements. This type of solar cells can also be manufactured as flexible thin film cell, which enables further reductions in generator volume, mass and mission costs. An example for flexible space solar cells is shown in Figure 4.19.

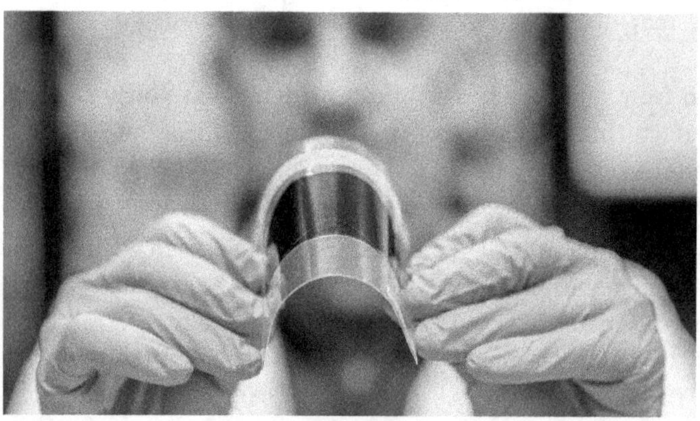

Figure 4.19 0.02 mm thick solar cell made by Azur Space (image: © ESA, SJM Photography [206])

Implementing additional Bragg reflectors, forces the irradiance on a path that lets it travel through the solar cell multiple times. Because of this, the utilization of this effect allows to reduce the thickness of MJ-GaAs solar cells even further. In addition to this, GaAs solar cells can operate at significantly higher temperatures than silicon-based solar cells. In MJ-GaAs solar cells, multiple individual solar cells are integrated into one physical solar cell on top of each other. An example for the internal structure of a triple-junction gallium arsenide solar cell is shown in Figure 4.20.

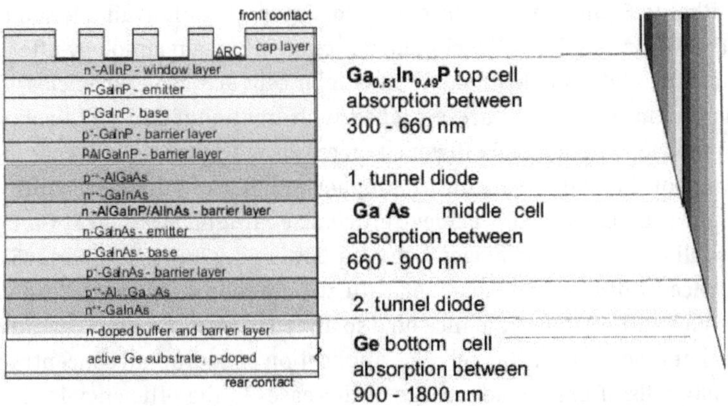

Figure 4.20 Inner structure of a triple junction gallium arsenide solar cell on germanium substrate with anti-reflex coating (image: © W. Ley/K. Wittmann/ W. Hallmann [97])

With every additional sub-cell, the total cell voltage increases, resulting from the internal series connection of the different sub-cells. Because of the series connection, the cell current is determined by the weakest subcell. Following this, decreases in the cell current can be observed with an increasing number of subcells. As the individual subcells of a multi-junction cell are usually connected in series, the parasitic capacitance of multi-junction solar cells is significantly increased compared to single-junction solar cells and the magnitude can even reach a factor of 5. During operation, the increased capacitance results in current spikes during regulator switching. An increased sensitivity to reverse voltage and currents can be observed in MJ-GaAs solar cells. Because of this, the utilization of bypass diodes is recommended for this type of solar cells. Bypass diodes can be integrated externally and they can also be integrated directly into the crystalline structure of a MJ-GaAs cell. Gallium arsenide solar cells have a bandgap of 1.42 eV, which is nearly perfect for the conversion of solar irradiance into electrical energy [151]. This allows a very good absorption of the different wavelengths of the solar spectrum. Different subcells have different quantum-sensitivities for different wavelengths. The top layer absorbs the irradiance with the highest photon energies and the lowest layer absorbs the lowest photon energies. It is possible to convert visible, ultraviolet and infrared irradiance into electrical energy utilizing MJ-GaAs solar cells. Because of this, multi-junction solar cells can achieve higher efficiencies in total. Triple-junction GaAs cells can theoretically achieve 33.5% efficiency, quadruple-junction cells can theoretically achieve 56%, and quintuple-junction cells can theoretically achieve 62% [202]. The maximum possible efficiency achievable with a gallium arsenide solar cell featuring an infinite number of junctions is 68.2% [202].

The majority of the available MJ-GaAs solar cells is currently employing 3 junctions, but quadruple GaAs solar cells are also available. In future missions, the application of MJ-GaAs cells with 5 to 6 junctions can be expected. Comparing all available solar cell technologies, the MJ-GaAs solar cell technology is currently the technology offering the highest efficiencies [48]. Single-junction GaAs solar cells achieve efficiencies of up to 19%, but maximum efficiencies are achieved with multi-junction GaAs solar cells [151] [155]. In theory, the solar cell efficiency increases with the total number of integrated junctions, but the manufacturing of a higher number of junctions is difficult. Available 3J-GaAs solar cells usually achieve efficiencies around 30% and 4J-GaAs to 6J-GaAs cells usually achieve efficiencies between 33% and 37% [97] [155] [166] [202] [207] [208]. Concentrator types of solar cells utilize parabolic mirrors or Fresnel lenses in order to concentrate solar irradiance on a solar cell to maximize the amount of solar irradiance that reaches the solar cell. The application of additional concentrators on MJ-GaAs solar cells offers the advantage of increases in the efficiency in the magnitude of up to 10% to efficiencies in the magnitude of 40% [205].

A high radiation tolerance can be achieved with MJ-GaAs solar cells. Experimental results show that MJ-GaAs solar cells experience less radiation degradation in orbit than other types of solar cells in many cases and are also not worse than the degradation that single-junction GaAs solar cells experience. After the exposure to radiation featuring a damage equivalent fluence in the magnitude of $1 \cdot 10^{15} \frac{\mathrm{MeV_e}}{\mathrm{cm^2}}$, a remaining efficiency in the magnitude of 83% to 87% of the initial efficiency has been observed in MJ-GaAs solar cells [210] [211]. TJ-GaAs solar cells initially featuring efficiencies of 30%, usually still achieve 26.5% of efficiency after 15 years of operation in geostationary orbit [97].

A possible solution in order to increase the efficiency and radiation hardness and in order to decrease the solar cell thickness is the utilization of light trapping structures. The light trapping structures allow to reduce the thickness of the middle cell of a triple-junction solar cell and, by this, enable reduction in the overall solar cell thickness. In addition to this, it is expected that the radiation-induced degradation is decreased utilizing this approach [213].

The manufacturing processes of MJ-GaAs solar cells is highly complex and the production is expensive. Because of this, the utilization of MJ-GaAs solar cells is currently limited to the applications in space. There are two different approaches for the manufacturing of multiple junctions in a single solar cell. The individual subcells can be grown on top of each other to manufacture monolithic solar cells and the individual cells can also be manufactured individually to manufacture mechanically-stacked solar cells. During the manufacturing of the monolithic solar cells, the upper subcells are grown on top of a base layer. In order to enable this, the solar cell material needs to be lattice-matched. The manufacturing of mechanically stacked solar cells offers several advantages, since the individual lattice constants do not need to be taken into account for the manufacturing of the individual subcells and since the electrical contacts of the individual subcells can also be brought out separately, so that the individual cells do not necessarily need to be connected in series. Following this, individual terminal contacts for the different sub-cells can be implemented as a countermeasure for current-mismatching losses. However, after the manufacturing of the individual cells, a significant share of the substrate needs to be removed in order to manufacture thin mechanically stacked solar cells.

During the manufacturing of the individual subcells, additional tunnel-junctions need to be implemented in between the individual pn-junctions. Lattice-matched solar cells can be manufactured with the epitaxial growth method and potentially offer high efficiencies. Inverted metamorphic (IMM) and upright metamorphic (UMM) solar cells are manufactured utilizing metamorphic growth methods. Metamorphic materials allow a greater freedom in the bandgap selection and wider bandgap top cell materials. In the following, this allows to achieve higher efficiencies with metamorphic materials. IMM solar cells are grown from the top subcell to the bottom subcell and UMM solar cells are grown from the bottom subcell to the top subcell. IMM solar cells allow further reductions in the cell thickness, since the germanium substrate can be removed after the growth process.

The front side contacts of a buried contact solar cell is manufactured by engraving structures in the front side material and filling them with contact material instead of printing the contacts on top of the cell front side. Usually, the engravings are additionally n+ doped in order to reduce the electrical resistance of the transition from the solar cell to the contacts. In addition to this, usually p+ doping at the backside is also implemented in order to manufacture a back surface field. The paths of light through the buried contact cell are enlarged by an additional texturization of both the front and the backside.

Integrated back contact IBC solar cells feature the positive and the negative electrical contacts on the backside. Subsequently, this type of solar cell does not feature intrinsic shadowing resulting from the implementation of front contacts. For IBC solar cells, front surface passivation and structuration are employed and the backside is manufactured passivated. In order to maximize the number of electrons reaching the solar cell backside without recombination, silicon substrate with a high degree of purity is required for the manufacturing. For the manufacturing of the backside contacts, the backside passivation layer is removed at the location of the electrical contacts. At the positive contacts, a local p+ doping is implemented and at the negative contacts, a local n+ doping is implemented.

Quantum dot solar cells are possible candidates for future space photovoltaic generators. This technology combines thin film technology with quantum dot absorption and photonic nanostructures. Quantum dot solar cells offer weight reductions by 90%, while achieving efficiencies greater than 30% [216].

For the exploration of the inner solar system, special so-called High Intensity High Temperature (HIHT) solar cells, and for the exploration of the outer solar system, special Low Intensity Low Temperature (LILT) solar cells are available. Both types of solar cells require a huge amount of testing. LILT solar cells use cells that have high internal parallel parasitic resistances. HIHT cells need to pass laboratory tests in order to ensure that those cells are tolerant to temperatures up to 230 °C, to extreme ultraviolet radiation and to high incidence angles [97].

For the application on board of spacecraft, photovoltaic generators need to meet certain requirements. For the application in Earth orbit, they need to be tolerant to thermal cycling in a temperature range between –150 °C and 120 °C during one sunlit/eclipse period. Examples for typical operating temperatures of gallium arsenide solar cells in space are shown in Table 4.1.

Table 4.1 Temperature of a gallium arsenide solar cell in space [74] [97]

	Average temperature of GaAs cells
High Intensity High Temperature (HIHT)	230 °C
Sun synchronous (LEO)	70 °C
Low Earth Orbit (LEO)	80 °C
Medium Earth Orbit (MEO)	65 °C
Geostationary Earth Orbit (GEO)	60 °C
Low Intensity Low Temperature (LILT)	-130 °C
Lunar surface	78 °C–88 °C

In near Sun environment GaAs solar cells can heat up to 230 °C, and in the outer solar system GaAs solar cells are usually operated at –130 °C [97]. In Low Earth orbit (LEO) cell operating temperatures between 70 °C and 100 °C can be observed during sunlit periods and during eclipse periods cell temperatures can drop to values of around –80 °C [97] [98]. In medium Earth orbit GaAs solar cells usually have an operating temperature of 65 °C, and in geosynchronous orbit GaAs solar cells usually have an operating temperature of 60 °C [97]. The employed solar cell technology should minimize the temperature-induced variations in the output parameters. The output parameters of gallium arsenide and indium phosphide solar cells vary less with temperature than those made out of c-Silicon. An example for the variation of the electrical solar cell characteristics with reference to temperature variations is shown in Figure 4.21.

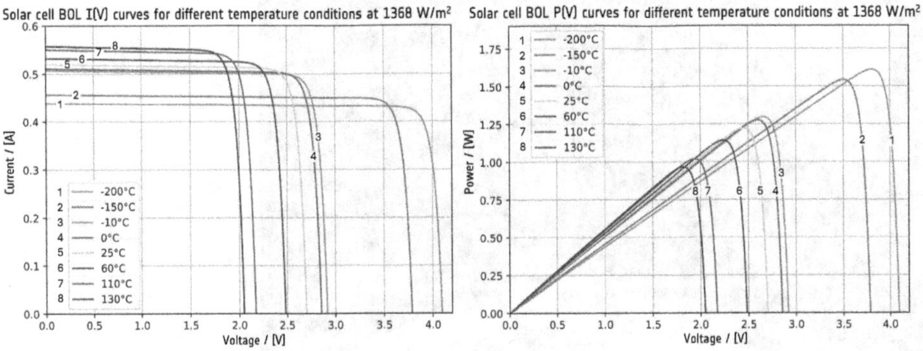

Figure 4.21 Simulated current-voltage and power-voltage characteristic curve of an HNR0003384 3G30C solar cell made by Azur Space for different temperature conditions at 1368 $\frac{W}{m^2}$ and BOL

Two different methods for the integration of solar cells can be differentiated. Solar cells can be body-mounted directly on the outer structure of a spacecraft or they can be mounted on external Solar Array Wings (SAW). Usually, spacecraft utilize either one of these approaches, but scientific missions sometimes also utilize both approaches together. Tracking systems are usually employed for SAWs featuring an output power of 2 kW and more [97] [98]. In order to optimize the solar incidence angle, tracking systems employ one or two motors to rotate the SAWs around one or two of their unit axes [48] [98]. Spacecraft with a generator power of 7 kW and more usually employ two Solar Array Wings (SAWs), each consisting of 4 to 6 solar arrays [97]. SAWs can be subdivided into rigid arrays and flexible arrays. Flexible solar arrays are manufactured by laminating solar cells into flexible foils. For example, a MJ-GaAs solar cell on a germanium waver can be manufactured with a thickness of 10 µm and laminated into 50 µm flexible sheets made from Kapton [166]. Other possible polymer materials are EVA, PVB, and TPU. Compared to rigid arrays, flexible arrays enable increases in the power to mass ratio. In the implementation on the lunar surface, fixed solar array inclinations and motorized solar array tracking can be utilized. A visualization of the optimization of the solar array incidence conditions on the lunar surface are shown in Figure 4.22.

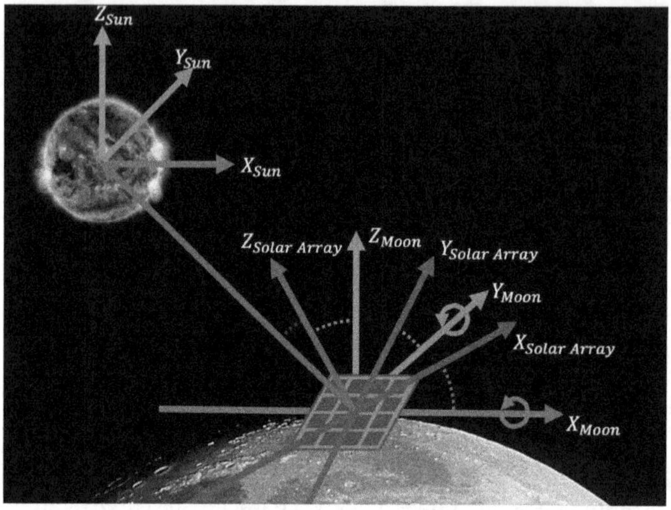

Figure 4.22 Improving the incidence conditions for direct solar irradiance with mounting and tracking methods

On the lunar surface, Solar arrays can be tilted towards the equator in order to improve the incidence conditions during noon. In this case, a tilt-angle in the magnitude of the latitude enables nearly perfect incidence conditions during noon. A combination of solar arrays tiled in western and eastern direction allows to optimize the solar array output power during the lunar day. A motorized tracking around the north-south axis allows to achieve nearly perfect incidence conditions in equatorial regions. Similar tracking results can be achieved in non-polar locations, when a tracking in the north-south axis is combined with a static inclination towards the equator. In polar locations, vertical solar arrays that track the Sun around their vertical axis achieve nearly perfect tracking results. Up to perfect tracking results can be achieved by a tracking around the 2 unit axes of the solar array plane. However, the achieved tracking results are only slightly better than with a combination of a static inclination around the east-west axis and a motorized tracking around the north-south axis. In Figure 4.23, the achieved horizontal direct solar irradiance by a solar array in a lunar polar location was visualized for the duration of a lunar day/night cycle during a perigee passage of the Earth for different ways to incline and track a solar array.

Single solar cells have relatively small voltages and currents. Following this, solar cells need to be integrated into solar arrays in order to increase voltage, current and power at the array terminals. Solar cells are connected in series to solar cell strings in order to increase the voltage, and multiple solar cell strings are connected in parallel in order to increase the current and power. Depending on the solar array regulator topology, a solar array needs to be designed in a way that the string voltage is lower, higher or equal to the bus voltage. The schematic of a solar array is visualized in Figure 4.24.

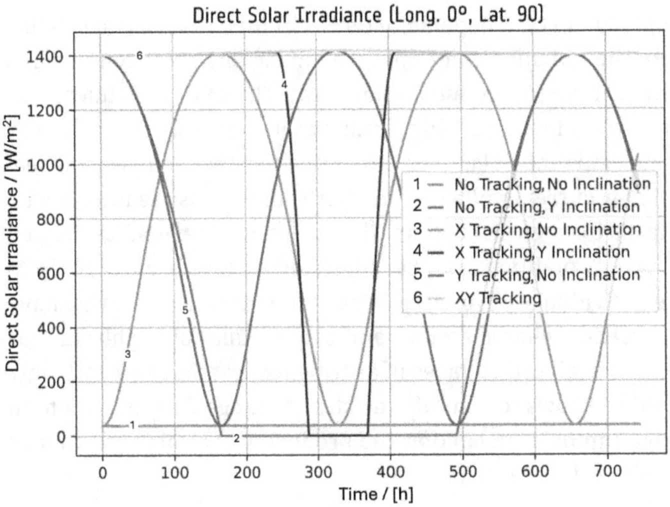

Figure 4.23 Received horizontal direct solar irradiance by a solar array in a lunar polar location during a lunar day/night cycle when the Earth is passing its perigee: The unit axes of the solar array plane are indicated with the letters X (north-south axis) and Y (east-west axis) (see Figure 4.22).

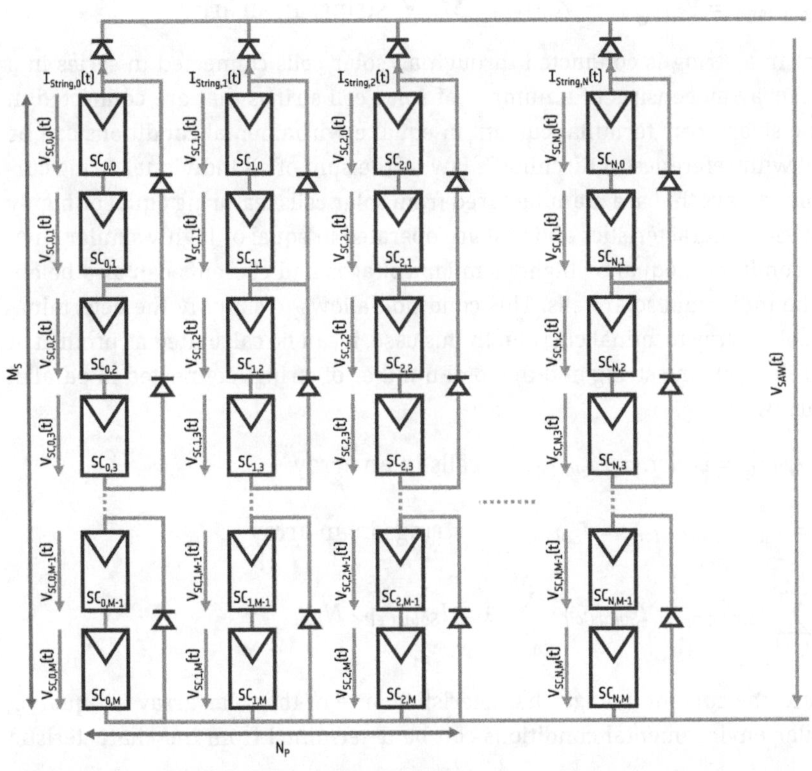

Figure 4.24 Schematic of a solar array with implemented string diodes and bypass diodes

The performance of a single solar cell is dependent on environmental parameters like irradiance, temperature or the radiation conditions. Following this, solar cells need to be characterized before their integration into a solar array. The solar cells integrated together in a solar array need to feature highly similar characteristics. All solar cells in a string conduct the same current. Solar cells integrated together into a string featuring equal or highly similar current-voltage characteristics will also feature highly similar or equal cell voltages for a given current. The terminal voltage of a solar cell string in equal environmental conditions can be calculated with reference to Kirchhoff's law as a sum of the individual cell voltages. When all solar cells in a string have equal or highly similar electrical characteristics and are in equal or highly similar environmental conditions, then all cell voltages in a string feature the same or highly similar values. This condition allows to simplify the string voltage determination. In this case, the string voltage can be calculated as the product of the solar cell voltage and the total number of solar cells in a string.

$$V_{String\ n_P,\ Cell\ m_S} = const. = V_{SolarCell} \ \forall \ \text{cells in an array}$$

$$V_{String\ n_P} = \sum_{m_S=1}^{M_S} V_{String\ n_P,\ Cell\ m_S} = \sum_{m_S=1}^{M_S} V_{SolarCell} = V_{SolarCell} \cdot \sum_{m_S=1}^{M_S} 1 = V_{SolarCell} \cdot M_S$$

$$V_{SAW} = V_{String\ 1} = V_{String\ n_P} = V_{SolarCell} \cdot M_S \ \forall \ \text{strings in an array}$$

The current in a string is conducted through all solar cells connected in series in a string. A solar array consists of a number of solar cell strings that are connected in parallel. The solar array terminal current in equal environmental conditions can be determined with reference to Kirchhoff's law as the sum of all individual string currents. In solar arrays that are manufactured from solar cells featuring equal or highly similar electrical characteristics and that are operated in equal or highly similar environmental conditions, equal or highly similar voltages and currents can also be observed at the individual solar cells. This condition allows to simplify the determination of the solar array terminal current. In this case, it can be calculated as product of the current through one string and the total number of strings connected in parallel in a solar array.

$$I_{String\ n_P,\ Cell\ m_S} = const. = I_{SolarCell} \ \forall \ \text{cells in an array}$$

$$I_{String\ n_P} = I_{String\ n_P,\ Cell\ m_S} = I_{SolarCell} \ \forall \ \text{strings in an array}$$

$$I_{SAW} = \sum_{n_P=1}^{N_P} I_{String\ n_P} = I_{SolarCell} \cdot \sum_{n_P=1}^{N_P} 1 = I_{SolarCell} \cdot N_P$$

Subsequently, the current-voltage characteristic curve of the solar array in equal or highly similar environmental conditions can be determined from the characteristic curve of a single cell by multiplying the cell current with the number of parallel strings and by multiplying the cell voltage with the number of solar cells in a string.

This approach cannot be applied in the case, that a part of the solar cells in a solar array is shadowed and, following this, does receive less irradiance than the rest of the cells. It is sufficient that one solar cell is shadowed in order to decrease the total output power of a solar array drastically. The output power of a solar array that has been shadowed by 2% can be reduced by 70% [109]. Less irradiated solar cells have lower values of the electrical output parameters than fully irradiated ones. For a given value of the string current, this can result in a condition where the string current has a smaller value than the short circuit current of the fully irradiated solar cells, while the string current has a greater value than the short circuit current of the shadowed solar cells. In this case, the fully irradiated solar cells in a string would have positive voltages, while the shadowed solar cells would feature negative voltages. Small increases in the string current can result in a drastic decrease of the voltage at shadowed solar cells because the current-voltage characteristic curve of a solar cell has a low inclination in the negative voltage and positive current quadrant. The string voltages in shadowed solar cell strings can be determined with reference to Kirchhoff's law as the sum of the individual solar cell voltages. In the shadowed case, string voltages can be calculated as the sum of the products of the number of solar cells receiving the same amount of irradiance and the cell voltages of solar cells in the corresponding irradiance condition.

$$V_{String\ n_{SolarArray}}(I_{String\ n_P}) = \sum_{m_S=1}^{M_S} V_{String\ n_P,\ Cell\ m_S}(I_{String\ n_P}, E_{String\ n_P,\ Cell\ m_S})$$

$$V_{String\ n}(I_{String\ n_P}, E_0) = M_{E_{0,0}} \cdot V_{SolarCell}(I_{String\ n_P}, E_{0,0}) + M_{E_{0,1}} \cdot V_{SolarCell}(I_{String\ n_P}, E_{0,1}) + \ldots$$

Solar cells only feature small open circuit voltages in the magnitude between less than one volt and a small number of volts in the positive voltage and positive current quadrant. In the negative voltage and positive current quadrant solar cells can feature highly negative voltages and act as consumers featuring negative power values. It is possible that the voltage occurring at one shadowed solar cell for a given value of the string current have a greater negative value than the sum of the positive voltages occurring at all of the remaining fully irradiated solar cells in a string. This is the reason, why a single shadowed solar cell can cause the breakdown of the complete string voltage. The string voltages occurring at the different solar cell strings in a solar array are also subject to Kirchhoff's law and have equal values. Subsequently, the breakdown of the string voltage in one string can result in a voltage breakdown in all parallel strings in a solar array. In addition to this, the negative power is dissipated by the solar cell and contributes to its temperature. Resulting from the high amount of dissipated power, a drastic increase in the solar cell temperature can be observed that can result in damages such as melting of the cover materials, melting of conduction materials, or melting of the solar cell itself. Because of this, solar arrays shall only be operated in the positive voltage and positive current quadrant on the characteristic curve and voltage breakdowns have to be prohibited. As a result, the most shadowed cell in a solar cell string limits the amount of string current.

Diodes can be integrated into solar arrays in order to limit the effect of shadowing and imbalances between the individual solar cells. It has to be differentiated between the bypass diodes and string diodes. Bypass diodes are implemented anti-parallel with a number of solar cells in a string in order to reduce the effects of shadowing. The voltage at a bypass diode can be calculated with reference to Kirchhoff's law as the sum of the individual cell voltages and the diode voltage.

$$V_{Diode,\ j_{Segment}} = -1 \cdot \sum_{k_{Cells/Diode}=1}^{K_{Cells/Diode}} V_{Segment\ j_{Segment},\ Cell\ k_{Cells/Diode}}\left(I_{Segment\ j_{Segment}}, E_{String\ n_P,\ Cell\ m_S}\right)$$

Solar cells feature positive voltages in nominal generative operating conditions, and because of this, the voltages at the anti-parallel bypass diodes have negative values in these conditions. Subsequently, bypass diodes are operated in the blocking domain when solar cells are operated in the nominal generative domain and only conduct a neglectable amount of current. In the case of shadowed solar arrays, this is different. The voltages occurring at a string segment connected anti-parallel to a bypass diode have maximum values in open circuit conditions and decrease with increasing string current. Once the string current has a value at which the voltage occurring at a string segment featuring a shadowed solar cell is equal to the negative diode threshold voltage, the diode will transition into a conductive state. Diodes typically feature threshold voltages of around 0.7 V [109] [151]. In order to minimize the effect of the diode voltage on the solar array terminal voltage, it is recommended to integrate bypass diodes featuring small voltages, such as germanium diodes or Schottky barrier diodes, into solar arrays. After the diode has transitioned into conduction state, the string current can be increased further. The voltage occurring at the segment anti-parallel to the conducting diode will remain at a small negative value instead of increasing further in an implementation without implemented bypass diodes. The effect of shadowing on the current-voltage characteristic curve is in this case reduced for a small number of shadowed solar cells and increases with the total number of solar cells and strings that are affected by the shadow conditions. A visualization on the resulting changes in the current-voltage characteristic curve of a solar array resulting from a shadowed solar cell are shown in Figure 4.25.

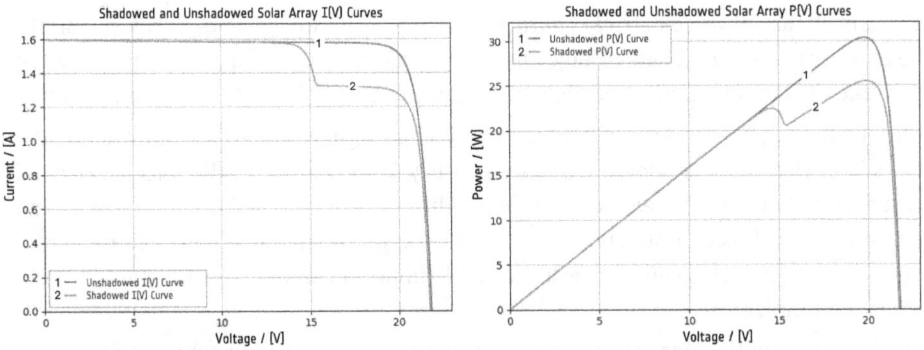

Figure 4.25 Simulated shadowed and unshadowed current-voltage and power-voltage characteristic curves

The power dissipated by the bypass diode can be calculated by the product of the diode current and the diode voltage. Dissipated power is converted into thermal energy and contributes to the diode temperature. After a bypass diode was turned into conductive state, the diode voltage has to turn to a negative value again in order to bring the diode back into blocking state. For this, the string current has to be reduced to a value small which is enough to allow the segment solar cells connected anti-parallel to the conducting bypass diode to have a positive voltage. Perfect protection against shadowing conditions can be achieved by implementing a single bypass diode anti-parallel to every single solar cell, but usually more than one solar cell is connected anti-parallel to a bypass diode [109]. Solar cells featuring integrated bypass diodes are also available. Another interesting approach are bypass diodes that are manufactured in a way that they fit in the cut corners of solar cells.

String diodes can be implemented in series with solar cell strings in order to protect them against problematic operational conditions like back-currents. Back-currents are experienced as negative currents by the integrated solar cells. Following this, solar cells are operated in the positive voltage and negative current quadrant, where solar cells act dissipative. String diodes also allow the voltage to be maintained at the remaining strings in case that a solar cell string is affected by an electrical fault condition. However, string diodes are operated in the forward domain, where positive voltages and currents occur. Subsequently, also the power value is positive and the string diodes are operated dissipative. In order to minimize the amount of power dissipated in string diodes, it is recommended to utilize string diodes featuring small voltages such as germanium diodes or Schottky barrier diodes. The utilization of string diodes also results in a decrease of the solar array terminal voltage, which is resulting from the diode voltage. Electrical fuses can also be utilized for the protection of solar cell strings instead of string diodes. Their utilization is advantageous because electrical fuses have a lower power dissipation than string diodes.

The simulation of the electrical characteristics of a shadowed solar array is complex because the individual cell voltages have to be simulated for each cell and for every possible cell current. Dedicated algorithms need to be employed for the determination.

The shadowing of a part of a single solar cell does not need to be taken into account in the same way as the shadowing of a solar array. For each solar cell in a solar array, it is sufficient to take the mean amount of irradiance received by it into account, even if a part of the cell is completely shadowed. However, drastic differences in the irradiance on a single solar cell can result in the formation cell damage and should therefore be minimized.

Typical issues with solar cells in space result from the exposure to the space environment. The thermal cycling of the material leads to structural problems with material and connectors. The exposure to space radiation leads to the darkening of solar cell cover glass and the degradation of the electrical characteristics. Electrostatic discharge, issues with electrical isolation and overheating of solar cells are also typical problems of solar cells in space. During solar substorms, electrostatic charge can accumulate on solar cells and cause voltages between different parts of the spacecraft in the magnitude between several 100 V and several 1000 V [97]. Potential buildups in the magnitude of 3 V/s can be observed between different parts of spacecraft that can lead to arching effects. These arching effects can alter the solar cell lamination, increase the conductive properties of the lamination material, and, in the following, cause the loss of a solar cell string. Because of this, protective measures against arching need to be taken into account during solar array design. On the lunar surface, the effect of dust accumulation on solar arrays can also result in a decrease in the power output. Because of this, the employment of dust mitigation methods is recommended. A countermeasure against this could be the implementation of a transparent conductive front layer on solar arrays. This conductive layer could be utilized in order to electrostatically ground the solar array front side and by this, reduce the adhesive property of electrostatically charged regolith.

The electrical characteristics of solar cells degrade in space, and, following this, their degradation during the mission needs to be taken into account in the generator design. The amount of radiation reaching a solar cell is dependent on the transport of radiation through front side and backside cover material. The damage that is caused by the radiation reaching the solar cell is dependent on the radiation energy, the cell technology, the density of doping atoms, and the thickness of the active region.

An interesting example for the radiation-induced degradation of solar cells occurred on board of the spacecraft INTELSAT II-F4 and ATS-1. Design or integration issues resulted in a strip of 0.38 mm at the edge of the integrated solar cells that was not covered by the cover glass [171]. Following this, this small strip was not shielded from the space environment by the cover glass. As a result, the pn-junction was damaged in orbit, which resulted in an internal shunting condition of the solar cells. Both spacecraft experienced a drop in the power generation in the magnitude of 20% during the first month in operation [171].

The expected degradation of solar cells during the mission needs to be determined during solar generator design. An accurate determination of the expected radiation damage can increase the mission duration and the mission return. In order to account for the degradation of solar cells during a mission, this type of generators is usually designed in a way that it can still generate enough power at its scheduled End of Life (EOL). This is usually done by increasing the generator size to meet the power demand requirement after the degradation during the mission. Because of this, at the beginning of a mission and during the mission, a surplus of energy can usually be harvested with the solar cells. However, whether this surplus can be transferred to the power bus depends on the employed regulator type.

Highly energetic electromagnetic radiation until the wavelength of gamma rays, as well as charged particles, can cause damages in solar cells. Protons and electrons are the main reason for solar cell damage. Particles can orientate from a natural source and they can also be generated in the spacecraft in response to the exposure to gamma radiation as secondary radiation. Heavy and fast particles, as well as energetic particles, can cause damages in solar cells. During the mission, the damages inside solar cells accumulate and alter the electrical parameters of solar cells constantly.

Typically, solar cells reach their End of Life (EOL) after they feature a degradation in the magnitude of 30% of their electrical output parameters [98]. Depending on the employed solar cell technology, solar cells typically degrade by between 0.50 $\frac{\%}{\text{year}}$ and 3.75 $\frac{\%}{\text{year}}$ [101] [20].

An example for the degradation of the output power of solar arrays during operation in Earth orbit in response to the exposure to the space environment is shown in Figure 4.26. Each of these plots belongs to one of the four spacecraft of the CLUSTER mission. The degradation of the output power of all four spacecraft experienced between 2000 and 2004 results from Solar Particle Events and the degradation experienced between 2008 and 2013. The degradation of the solar arrays of the CLUSTER 1 and CLUSTER 2 spacecraft at the end of their mission results from crossing the Van Allen belts.

Degradation can also result from micrometeorites. An example for an impact into solar cells occurred on the SAW of the Sentinel 1-A spacecraft. Images before and after the impact are shown in Figure 4.27.

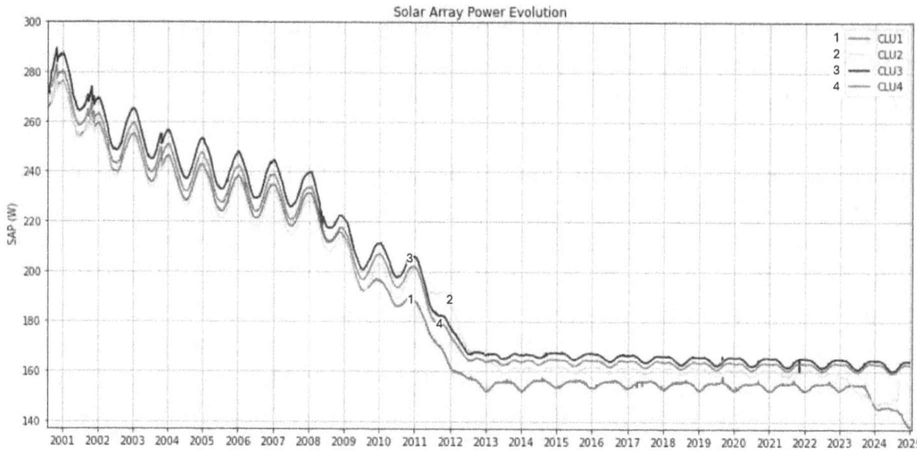

Figure 4.26 Degradation of solar arrays of the CLUSTER spacecraft during the mission (image: © ESA/CLUSTER, B. Sousa [223]

Figure 4.27 Meteorite impact into a solar array wing of the Sentinel 1-A spacecraft (image: © ESA [224])

Solar cells degrade in space, but different solar cell technologies degrade in a different way in response to the exposure of the space environment. Only a gradual degradation can be observed in silicon-based solar cells until equivalent fluences in the magnitude up to $1 \cdot 10^{16}\,cm^{-2}$ [225]. This has been related to a decrease in the minority carrier diffusion length, resulting from high energy electron and proton impacts. An increase in the short circuit current can be observed between equivalent fluences from $2 \cdot 10^{16}\,cm^{-2}$ to $5 \cdot 10^{16}\,cm^{-2}$ [225]. Greater equivalent fluences in the magnitude of $5 \cdot 10^{16}\,cm^{-2}$ and more result in a decrease in the charge carrier concentration and in the cell resistivity [225]. This results in a drop in the short circuit current and a failure of the solar cell. Solar cells based on gallium arsenide and indium phosphide technology tolerate more radiation than solar cells based on silicon technology. Especially gallium arsenide solar cells have been proven to be highly tolerant to the exposure to space radiation. In flight, significant degradation has only been observed for impacting protons featuring energies greater than 300 keV for gallium arsenide solar cells [162]. Indium phosphide solar cells have the smallest degradation rates of all available space certified solar cells and this technology also allows the annealing of radiation damage at room temperature. Dual-junction solar cells based on gallium arsenide and germanium technology degrade in response to both electron and proton radiation, but are more sensitive to electron radiation [202]. The exposure of MJ-GaAs solar cells to particles featuring energies between 0.9 MeV and 9.5 MeV results in an increased amount of damage, compared to other particle energies [202]. Dual-junction solar cells based on a combination of gallium arsenide and indium phosphide solar cells respond in a similar way to radiation as gallium arsenide solar cells, since the degradation resulting from the radiation exposure is mainly caused inside the gallium arsenide subcell. Triple-junction solar cells based on gallium indium phosphide, gallium arsenide and germanium subcells degrade in response to the exposure to electron and proton radiation. Electrons featuring energies in the range from 1 MeV to 12 MeV and protons featuring energies in the range from 50 MeV to 200 MeV cause the greatest degradation to these solar cells [202].

Different mechanisms were identified causing damage in a solar cell in response to the exposure to space radiation. Interaction between penetrating particles and the crystalline lattice of a solar cell can be subdivided into elastic and inelastic collisions. The main share of inelastic collisions in a solar cell lattice takes place between penetrating particles and solar cell electrons, but also collisions with atomic nuclei are possible. In response to an inelastic collision, a share of the energy of the penetrating particle is transferred to the nuclei or electron it is impacting. Impacted electrons can be unbound from their atomic binding and impacted nuclei can get into ionized or excited state in response to an impact. Excited nuclei can, in the following, emit secondary particles. In addition to this, the kinetic energy transfer from an inelastic collision between an impacting particle and an atomic nucleus can result in atomic displacements. Protons and electrons cause damage in solar cells by collisions with the atoms of the solar cell lattice. The damage caused by electron radiation is uniformly distributed around the path of an electron in a solar cell. The damage caused by proton radiation with energies greater than 5 MeV is also distributed uniformly along the penetration path, but the damage caused by a proton radiation with smaller energies is mainly located at the end of the penetration path [118] [162] [220]. In order to shield a solar cell from penetrating protons, a cover glass featuring a thickness of 25 μm is sufficient [162].

Ionization damage is defined as the damage caused by the removal or adding of electrons from or to atoms. Because of this, ionized atoms have less or more electrons than protons and are electrically charged. This effect is also responsible for the altering of material properties in secondary materials. Transparent materials, such as solar cell cover glass or polymers, can darken with an increase in the degree of material ionization, and also other physical or mechanical material properties can advance from ionization. A measure for the degree of ionization of a certain material is the absorbed dose.

$$Dose(\phi) = 1.6 \cdot 10^{-8} \cdot \frac{dE}{dx} \cdot \phi$$

Displacement damage is defined as the damage caused by atoms of the crystalline lattice of a solar cell that have been displaced from their initial lattice sites, and this type is the main cause for damage in space solar cells. The displaced atoms result in the formation of stable defect sites in the solar cell lattice. Since displaced atoms act as centers of recombination, the charge carrier lifetime and concentrations are also negatively influenced by the formation of stable defects. Impacting neutrons, electrons, and protons can cause displacement damage in different ways. In order to displace an atom from its lattice site, an impacting particle needs to feature a certain threshold energy and particles featuring energies less than the threshold energy cannot cause displacement damage in a solar cell. Threshold energies in the magnitude up to 145 keV have been reported [155].

$$E_d = 2 \cdot \frac{m_e \cdot E_{th}}{M \cdot m_e c^2} \cdot \left(E_{th} + m_e c^2\right)$$

In the analysis of radiation damage, only the radiation fluence has to be taken into account that has energies greater than or equal to the threshold energy. Based on this, the density of atomic displacements can be determined. It is defined as the total number of atomic displacements per unit mass.

$$N_d = n_a \cdot \sigma_{DCS} \cdot \overline{v} \cdot \phi(\geq E_d)$$

The exposure to radiation results in a decrease in the electrical output parameters of solar cells. Subsequently, a way had to be developed to determine the damage inside a solar cell and relate it to the corresponding decrease in the electrical parameters. In order to estimate the degradation of a solar cell during a space mission, the sensitivity of the solar cell to particles featuring different energies, as well as the influence of shielding materials, needs to be determined. After that, the radiation environment needs to be defined. Based on this, a relation had to be developed in order to estimate the damage resulting from the exposure to the space radiation environment. In order to utilize the results of laboratory experiments to estimate the performance degradation of a certain solar cell in space in advance of a mission, it was necessary to develop a way to relate the damage caused by the spectrum of radiation present in space to the damage caused by a certain type of reference radiation that a solar cell can be exposed to in a laboratory environment. Widely applied methods are equivalent fluence, which has been developed by NASA's Jet Propulsion Laboratory (JPL), and Non-Ionizing Energy Loss (NIEL), which has been developed by US Naval Research Laboratory (NREL). Equivalent fluence is a legacy method that requires huge efforts in laboratory experiments, but that is still omnipresent, and NIEL is a newly developed method that requires less laboratory experiments. The results of both methods are comparable. Solar cell manufacturers usually indicated the electrical parameters of their products as a function of the equivalent fluence of a certain reference radiation. The simulated current-voltage characteristic curve of a space solar cell is visualized for different values of the equivalent fluence in Figure 4.28.

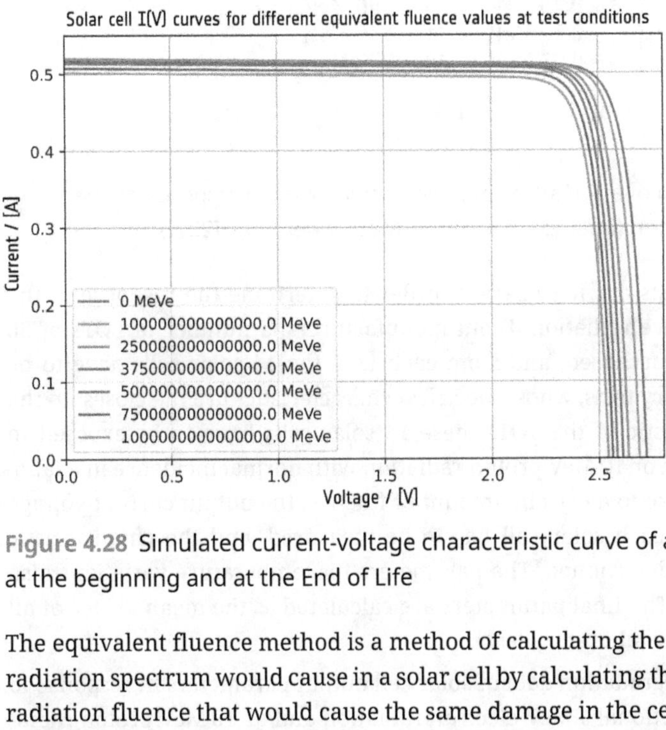

Figure 4.28 Simulated current-voltage characteristic curve of a solar cell at the beginning and at the End of Life

The equivalent fluence method is a method of calculating the damage a certain space radiation spectrum would cause in a solar cell by calculating the amount of a reference radiation fluence that would cause the same damage in the cell as the space radiation environment would cause. Utilizing the equivalent fluence method, the equivalent fluence is determined for 1 MeV electron and/or 10 MeV proton radiation, since these types of radiation are also a significant part of space radiation spectrum. In laboratories, solar cells are exposed to these types of reference radiation, and after the exposure, their output parameters are measured. For the estimation of damage inside a solar cell that will be caused by a certain omnidirectional space radiation spectrum during a mission. The expected space radiation spectrum needs to be converted to a damage equivalent fluence of the unidirectional reference radiation. The magnitude of the alteration needs to be determined for each electrical solar cell parameter separately. The response of a GaAs space solar cell to different types of radiation fluence is visualized in Figure 4.29 for the normalized Maximum Power Point power.

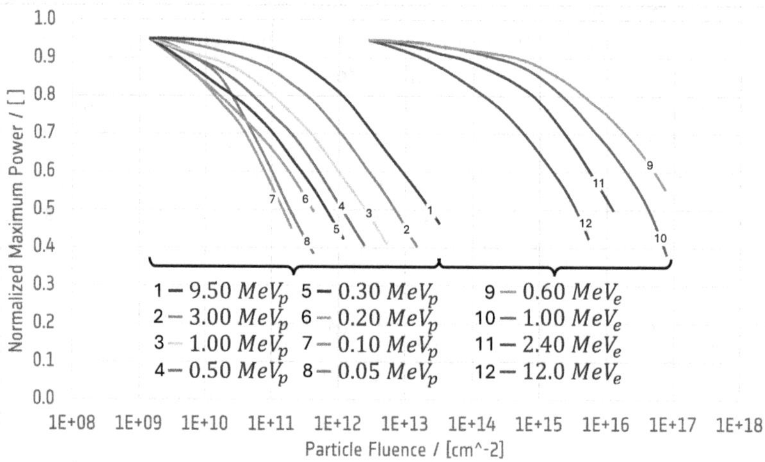

Figure 4.29 Degradation of a GaAs/Ge space solar cell power in response to the exposure to space radiation (image: data from S. Messenger et al. [227])

Extensive laboratory tests are necessary in order to determine the response of the solar cell to the reference radiation. From manufacturers, a number of LOTs of 50 solar cells have to be purchased, and from each LOT, the 14 solar cells have to be selected for the laboratory tests, whose electrical characteristics are the closest to the mean output characteristic of the LOT. These 14 solar cells have to be exposed to 1 MeV electron radiation or 10 MeV proton radiation with normal incidence in a laboratory. After the exposure to a certain amount of fluence, the output current-voltage characteristic curve of each solar cell has to be measured, and the characteristic points on it have to be determined. The parameters for temperature variations also have to be determined. The final parameters are calculated as the mean values of all solar cells involved in the laboratory tests.

With reference to the degradation data of solar cell output parameters in response to the exposure to unidirectional 1 MeV electron radiation and/or 10 MeV proton radiation, only its response to the exposure to these types of reference radiation can be estimated. However, since the space radiation is not unidirectional and since it is not monoenergetic, a relation had to be developed that relates the damage caused in a solar cell by unidirectional 1 MeV electron radiation and/or 10 MeV proton radiation in a laboratory environment to the damage caused by the actual omnidirectional radiation spectrum in space.

The parameter differential flux is a measure for the number of particles of a certain particle energy that are incident on a surface featuring a unit area in a unit time interval. The parameter integral flux is a measure for the number of particles featuring a particle energy greater than the threshold energy that are incident on a surface with reference to a unit area and a unit time interval. It is calculated from the integral of the differential flux over all particle energies greater than the value of the threshold energy. Differential fluxes are usually only available as tables containing discrete values for a number of different particle energies. Because of this, the integral flux is usually calculated as a discrete integral.

$$\phi(> E_{th}) = \int_{\infty}^{E_{th}} \phi(E)\, dE = \sum_{j=\infty}^{E_{th}} [\phi(E_j) \cdot \Delta E_{Particle,j}]$$

The parameter difference flux is defined as the number of particles featuring energies within a certain energy range that are incident on a surface featuring a unit area in a unit time interval. It can be calculated from the difference between the integral flux for the lower and the upper boundary of the energy range.

$$\Delta\phi(E_{th}) = \phi(> E_{th}) - \phi(> E_{th} + \Delta E_{Particle})$$

$$\Delta\phi(E_{th}) = \int_{\infty}^{E_{th}} \phi(E_j)\, dE_j - \int_{\infty}^{E_{th}+\Delta E} \phi(E_j)\, dE_j = \int_{E+\Delta E}^{E_{th}} \phi(E_j)\, dE_j$$

$$= \sum_{j=E_{th}+\Delta E}^{E_{th}} [\phi(E_j) \cdot \Delta E_{Particle,j}]$$

In the last step, the effective relative damage coefficients need to be determined. These damage coefficients are utilized to convert an amount of radiation fluence of a certain particle energy and type to the amount of fluence of 1 MeV electron radiation or 10 MeV proton radiation that causes the same damage in a solar cell as the type of radiation employed in laboratory experiments. Effective relative damage coefficients can be seen as conversion factors for the damage the omnidirectional space radiation spectrum causes in solar cells to the monoenergetic unidirectional 1 MeV electron reference radiation or 10 MeV proton reference radiation. In addition to this, also the transport through cover material needs to be taken into account. This is usually done by the determination of damage coefficients for different cover material thicknesses. The amount of radiation fluence that degrades the output parameters by 20% or 25% is usually defined as the parameter critical fluence [162] [171] [202] [218].

$$I_{SC}(\phi = \phi_{Critical}) = \left(1 - \frac{n_{Critical}}{100\%}\right) \cdot I_{SC}(BOL)$$

The damage coefficients are derived from laboratory experiments, where solar cells are exposed to unidirectional monoenergetic electron beams or proton beams with normal incidence of different particle energies. In order to determine the effect of omnidirectional space radiation to solar cells, the share of the space radiation fluxes that passes through the solar cell plane needs to be determined mathematically. The omnidirectional flux of space radiation passes through a unit sphere around a solar cell, but only the share of the omnidirectional flux that passes through the solar cell plane needs to be taken into account for the damage calculation. Following this, the normal component of the omnidirectional flux needs to be determined. The derivative of the normal component of the omnidirectional flux to the area of the unit sphere around the solar cell plane can be calculated by the division of the omnidirectional flux by 4π. The influence of the incidence angle can be considered by the cosines of the incidence angle.

$$\frac{d\phi_n}{dA} = \frac{\phi_0}{4\pi} \cdot \cos(\theta_{incidence})d\Omega = \frac{\phi_0}{4\pi} \cdot \cos(\theta_{incidence}) \cdot 2\pi \cdot \sin(\theta_{incidence})d\theta_{incidence}$$

In the next step, the normal component of the omnidirectional flux at the solar cell plane can be calculated by the integral over the incidence angle from zero to π. An integral from zero to $\frac{\pi}{2}$ can be utilized in order to calculate the normal component through the solar cell front side and the integral from $\frac{\pi}{2}$ to π can be utilized to calculate the normal component of the omnidirectional flux through the solar cell backside. However, due to the omnidirectional nature, it is satisfying to calculate the normal component of the omnidirectional flux at the solar cell front side and multiply it with 2 to calculate the total normal component through the solar cell front side and backside.

$$\phi_n = \int_0^{\pi} \frac{\phi_0}{4\pi} \cdot \cos(\theta_{incidence}) \cdot 2\pi \cdot \sin(\theta_{incidence})d\theta_{incidence}dA = \frac{\phi_0}{2}$$

In the case, that the back shielding of the solar cell is assumed to be sufficient, so that only the flux through the solar cell front side is relevant for the damage calculation, only the front side component needs to be considered in the calculation.

$$\phi_n = \frac{\phi_0}{4\pi} \cdot 2\pi \cdot dA \cdot \int_0^{\frac{\pi}{2}} \cos(\theta_{incidence}) \cdot \sin(\theta_{incidence})d\theta_{incidence} = \frac{\phi_0}{4}$$

Following this, the flux passing through a solar cell with an unshielded backside features a magnitude of half of the omnidirectional space radiation flux and the flux penetrating the solar cell with a perfectly shielded backside features a magnitude of one quarter of it.

It is often the case that solar cells are shielded by a number of different layers of different shielding materials. In this case, it is not satisfying to employ the sum of the different material thicknesses as shielding thickness. Instead, an equivalent shielding thickness needs to be determined that also takes into account the different densities of the different materials.

$$t_{Cover,eq} = \frac{1}{\rho_{Cover,TC}} \cdot \sum_i t_{Cover,i} \cdot \rho_{Cover,i}$$

Solar cell cover materials darken with the exposure to space radiation. The darkening of cover materials also results in a decrease in the optical transmissivity of the cover materials. A side effect from the ionization occurring inside the solar cell cover material is that particles penetrating the cover material lose a share of their energy. The energy loss can be calculated utilizing the range of a particle featuring a certain particle energy in a certain material. The range function relates the length of a possible penetration path of this particle to its energy before the cover material penetration. Particles penetrate cover materials on straight paths. Following this, the length of the penetration path is equal to the cover material thickness in normal incidence conditions. In non-normal incidence conditions, the penetration path length increases with the incidence angle.

$$l_{pen}(\theta_{incidence}, t_{Cover,eq}) = \frac{t_{Cover,eq}}{\cos(\theta_{incidence})}$$

$$l_{range}(E_{Particle,0}) = R(E_{Particle,0})$$

The energy of a particle, after its path through the cover material, can be calculated utilizing the inverse of the range function. For this, the maximum possible penetration path length of the particle is subtracted by the total penetration path length that the particle has travelled through the cover material. This difference is utilized as an input for the inverse range function. Tables are available for the range function and for the inverse range function, relating the particle energy to the maximum possible penetration path length and vice versa.

$$E_{Particle,n}(E_{Particle,0}, \theta_{incidence}, t_{Cover,eq}) = R^{-1}[l_{range}(E_{Particle,0}) - l_{pen}(\theta_{incidence}, t_{Cover,eq})]$$

$$E_{Particle,n}(E_{Particle,0}, \theta_{incidence}, t_{Cover,eq}) = R^{-1}\left[R(E_{Particle,0}) - \frac{t_{Cover,eq}}{\cos(\theta_{incidence})}\right]$$

Extensive laboratory tests are required in order to determine the damage coefficients allowing to calculate a damage equivalent fluence of a reference radiation from a space radiation spectrum. During these tests, solar cells are exposed to monoenergetic electron and proton radiation in a variety of different particle energy:

- electron radiation: 0.6 MeV, 1 MeV, 2.4 MeV, 12 MeV
- proton radiation: 0.05 MeV, 0.1 MeV, 0.2 MeV, 0.3 MeV, 0.5 MeV, 1 MeV, 3 MeV, 9.5 MeV

For each particle and energy, the degradation of solar cells is measured in 8 fluence steps [218]. During the laboratory experiments, the critical fluence is determined for the short circuit current, the open circuit voltage, and Maximum Power Point power for each combination of particle type, particle energy and cover material thickness. Based on the critical fluence, the unidirectional damage coefficients are determined by normalizing the critical fluence of electron radiation of a certain energy to the critical fluence of 1 MeV electron radiation, as well as by normalizing the critical fluence of proton radiation of a certain energy to the critical fluence of 10 MeV proton radiation in a second step [202] [218]. Only critical fluence values can be utilized for the damage coefficient determination that have been determined with equal cover material type and thickness.

$$D_{Uni,e}(E_{Particle,n}, \theta_{incidence} = 0°, t_{Cover,eq}) = \frac{\phi_{Critical,e}(E_{Particle,n}, t_{Cover,eq})}{\phi_{Critical,e}(1\ \text{MeV}, t_{Cover,eq})}$$

$$D_{Uni,p}(E_{Particle,n}, \theta_{incidence} = 0°, t_{Cover,eq}) = \frac{\phi_{Critical,p}(E_{Particle,n}, t_{Cover,eq})}{\phi_{Critical,p}(10\ \text{MeV}, t_{Cover,eq})}$$

Because of the unidirectional, monoenergetic nature of the laboratory radiation, the same geometrical aspects have to be considered for the conversion between unidirectional and omnidirectional damage coefficients, as they have been utilized during the conversion between unidirectional and omnidirectional fluxes. For this calculation, perfect back shielding of the solar cell is usually assumed. Damage coefficients have to be determined for the bare solar cell and, separately, also for solar cells with implemented cover glass in different thicknesses. When cover glass is implemented, the energy loss in the glass needs to be taken into account in the calculation.

Since the damage a single electron of a certain energy is causing in a solar cell is linearly distributed along its penetration path through the solar cell, the total damage caused by an electron is proportional to the length of the penetration path. This length increases with the incidence angle, and because of this, the incidence angle can be utilized in order to scale the unidirectional damage coefficients to omnidirectional damage coefficients by the inverse cosines of the incidence angle [162] [171] [220]. This relation can be utilized for the calculation of the omnidirectional damage coefficients in the following step.

$$D_{Uni,e}(E_{Particle,n}, \theta_{incidence}, t_{Cover,eq}) = \frac{D_{Uni,e}(E_{Particle,n}, \theta_{incidence} = 0°, t_{Cover,eq})}{\cos(\theta_{incidence})}$$

$$D_{Omni,e}(E_{Particle,0}, t_{Cover,eq}) = \frac{1}{4\pi} \cdot \int_0^{\frac{\pi}{2}} D_{Uni,e}(E_{Particle,n}, \theta_{incidence}, t_{Cover,eq}) \cdot 2\pi$$
$$\cdot \sin(\theta_{incidence}) \cdot \cos(\theta_{incidence}) d\theta_{incidence}$$

$$D_{Omni,e}(E_{Particle,0}, t_{Cover,eq}) = \frac{1}{2} \cdot D_{Uni,e}(E_{Particle,n}, \theta_{incidence} = 0°, t_{Cover,eq})$$

Gallium arsenide solar cells are manufactured very thin, and because of this, infinite back shielding cannot be assumed for this type of solar cells. In this case, the equation needs to be calculated for the complete unit sphere around the solar cell instead of for the half-sphere around the solar cell's front side.

For proton radiation, a differentiation needs to be made between high, medium and low energy protons. Medium and high energy protons pass through a solar cell and low energy protons are stopped inside it. Because of this, low energy protons cause a different kind of damage than medium and high energy protons. The damage caused by medium and high energy proton radiation is distributed uniformly around the penetration path inside the solar cell. Following this, the same relation for the conversion of unidirectional damage coefficients for normal incidents to unidirectional damage coefficients for non-normal incidence can be utilized as it has also been utilized for electron radiation.

$$D_{Uni,p,HM}(E_{Particle,n}, \theta_{incidence}, t_{Cover,eq}) = \frac{D_{Uni,p}(E_{Particle,n}, \theta_{incidence} = 0°, t_{Cover,eq})}{\cos(\theta_{incidence})}$$

For the calculation of the unidirectional damage coefficients for non-normal incidence of low energy proton radiation, two parameters need to be employed as scaling factors: the total number of atomic displacements caused by a proton featuring a certain energy before penetrating the solar cell cover glass and the total number of atomic displacements caused by a proton with a certain energy after leaving the solar cell cover glass. The total number of atomic displacements can be calculated employing a dedicated model.

$$D_{Uni,p,L}(E_{Particle,n}, \theta_{incidence}, t_{Cover,eq}) = D_{Uni,p}(E_{Particle,n}, \theta_{incidence} = 0°, t_{Cover,eq}) \cdot \frac{N_{td}(E_{Particle,0})}{N_{td}(E_{Particle,n})}$$

The omnidirectional damage coefficients for low energy proton radiation and the omnidirectional damage coefficients for medium and high energy proton radiation have to be calculated separately. Omnidirectional damage coefficients for proton radiation of the complete energy range are calculated as the sum of both in a second step. Perfect back shielding is usually assumed for this operation.

$$D_{Omni,p,HM}(E_{Particle,0}, t_{Cover,eq}) = \frac{1}{4\pi} \cdot \int_{0}^{\theta_{Incidence,C}} \frac{D_{Uni,p}(E_{Particle,n}, \theta_{incidence} = 0°, t_{Cover,eq})}{\cos(\theta_{incidence})} \cdot 2\pi \cdot \sin(\theta_{incidence}) \cdot \cos(\theta_{incidence}) d\theta_{incidence}$$

$$D_{Omni,p,L}(E_{Particle,0}, t_{Cover,eq}) = \frac{1}{4\pi} \cdot \int_{\theta_P}^{\frac{\pi}{2}} D_{Uni,p}(E_{Particle,n}, \theta_{incidence} = 0°, t_{Cover,eq})$$

$$\cdot \frac{N_{td}(E_{Particle,0})}{N_{td}(E_{Particle,n})} \cdot 2\pi \cdot \sin(\theta_{incidence}) \cdot \cos(\theta_{incidence}) d\theta_{incidence}$$

$$D_{Omni,p}(E_{Particle,0}, t_{Cover,eq}) = D_{Omni,p,L}(E_0, t) + D_{Omni,p,HM}(E_0, t)$$

Electron radiation affects the open circuit voltage, the short circuit current, as well as the Maximum Power Point voltage and current in the same way, so that the same damage coefficients can be utilized for all four parameters. The damage caused by proton radiation affects the short circuit current and the open source voltage differently. Because of this, one damage coefficient has to be calculated for the short circuit current and another has to be calculated for the open circuit voltage and the Maximum Power Point power. Besides, a perfect back shielding cannot be assumed for gallium arsenide solar cells due to the small share of the pn-junction of the total solar cell thickness. The back-shielding of gallium arsenide solar cells can still be classified as satisfying, so that also for this type of solar cells, damage coefficients for perfect back shielding can be utilized in a good approximation.

The damage equivalent fluence is calculated as the integral over the complete range of particle energies as a product of the difference flux with the radiation damage coefficients. For electron radiation, it is sufficient to simulate one damage equivalent fluence parameter and utilize this value for all solar cell output parameters. Separate damage equivalent fluence parameters also have to be simulated for the open circuit voltage and for the short circuit current of proton radiation.

$$\phi_{1MeVe} = \sum_{E_{Particle,0}=0}^{\infty} \{[\phi_e(> E_{Particle,0}) - \phi_e(> E_{Particle,0} + \Delta E_{Particle})]$$

$$\cdot D_{Omni,e}(E_{Particle,0}, t_{Cover,eq})\}$$

$$\phi_{10MeVp,Isc} = \sum_{E_{Particle,0}=0}^{\infty} \{[\phi(> E_{Particle,0}) - \phi(> E_{Particle,0} + \Delta E_{Particle})]$$

$$\cdot D_{Omni,p,Isc}(E_{Particle,0}, t_{Cover,eq})\}$$

$$\phi_{10MeVp,Voc} = \sum_{E_{Particle,0}=0}^{\infty} \{[\phi(> E_{Particle,0}) - \phi(> E_{Particle,0} + \Delta E_{Particle})]$$

$$\cdot D_{Omni,p,Voc}(E_{Particle,0}, t_{Cover,eq})\}$$

Usually, the solar cell output parameters are determined for 1 MeV electron radiation. However, in case that the environment of the mission is dominated by proton radiation, it is recommended to determine the damage equivalent fluence of 10 MeV proton radiation and use the following relation in order to convert it to the damage equivalent fluence of 1 MeV electron radiation. The mean damage caused by 1 MeV electron radiation in a silicon solar cell is greater by a factor of 3000 than the mean damage caused by 10 MeV proton radiation. For gallium arsenide solar cells, factors between 400 and 1400 can be employed.

$$\phi_{1MeVe} = D_{PE} \cdot \phi_{10MeVp}$$

A similar approach can also be utilized in order to estimate the darkening of solar cell cover materials with the exposure to the space radiation spectrum. In solar cell cover materials the main damage type is ionization damage. A measure for the degree of ionization inside solar cell cover materials is the absorbed dose. It can be calculated in a similar way as the damage equivalent fluence utilizing an integral over the difference flux, when an additional conversion function from fluence to absorbed dose is employed instead of the use of damage coefficients.

$$D_{Absorbed}(d_{Material}) = \sum_{E_{Particle}=0}^{\infty} \{\Delta\phi(E_{Particle}) \cdot I(E_{Particle}, d_{Material})\}$$

$$D_{Absorbed}(d_{Material}) = \sum_{E_{Particle}=0}^{\infty} \{[\phi(> E_{Particle}) - \phi(> E_{Particle} + \Delta E_{Particle})] \cdot dD_{Absorbed}(E_{Particle}, d_{Material})\}$$

Another method for the degradation simulation of solar cells is the Non-Ionizing Energy Loss (NIEL) method. This method requires a smaller number of laboratory measurements, compared to the equivalent fluence method. The results from the NREL model comply with the real degradation of GaAs and $CuInSe_2$ solar cells.

4.2 Solar Dynamic Generators

Solar dynamic generators, or sometimes also referred to as solar thermal generators, employ a conversion from solar irradiance to thermal energy, together with a second conversion from thermal energy to electrical energy in a dynamic system. They can be seen as an alternative to classical photovoltaic generators. Collectors focus solar energy on a receiver. In this receiver, the solar energy is absorbed and converted into thermal energy. A working medium is employed, which absorbs the thermal energy. The utilization of working media that can store thermal energy for relatively long durations of time offers the advantage that energy harvesting and conversion can be decoupled. An example for such a working medium is molten salt. In order to convert the thermal energy from a working medium into electrical energy, different processes are available. Conceivable processes are Brayton, Rankine or Stirling. Stirling generators achieve efficiencies from 25% to 30%, Rankine generators achieve efficiencies from 15% to 25% and Brayton generators achieve efficiencies from 20% to 35% [97] [98]. In order to release thermal energy that cannot be converted into electrical energy, a solar dynamic system also requires additional radiator systems. The power density of solar dynamic generators is usually higher than the power density of photovoltaic generators. Because of this, the required collector size can be reduced while meeting the same power demand requirements. In the 1980s, solar dynamic generators have even been considered for the International Space Station (ISS) for several months, but these have been shifted towards the utilization of photovoltaic generators for the final system.

4.3 Solar Power Regulation

Power regulation methods for solar generators can be differentiated into the methods Direct Energy Transfer (DET) and Maximum Power Point Tracking (MPPT). The DET method utilizes regulators, like the Sequential Switching Shunt Regulators (S3R) in order to connect solar array strings in parallel to the bus. The MPPT method utilizes DC/DC converters controlled by a Maximum Power Point Tracking algorithm, in order to transfer electrical power from the generator to the power bus. In principle, it is possible to utilize both methods with all of the general bus topologies. For the final EPS design, usually the method is chosen that provides a good balance of a minimum amount of power dissipation, a maximum power yield, and a satisfying voltage regulation.

When a Direct Energy Transfer (DET) regulator is chosen in order to regulate solar arrays, then these are directly connected in parallel to the main bus. Solar cells are connected in series to so-called strings in order to increase the voltage. Employing the DET method, each solar cell string is connected to one channel of the DET regulator, which utilizes solid-state switches and diodes in order to connect or disconnect strings to or from the main bus. Subsequently, the main bus voltage determines the operational point on the current-voltage characteristic curve of the solar cell strings. As the solar cell current is dependent on the solar cell voltage, the solar cell string current is determined from the main bus voltage. The total amount of current transferred from a solar array to the main bus is regulated by adjusting the number of solar cell strings connected to the main bus.

The power generation has to be matched to the power demand in an EPS. The voltage of a DC bus will feature a constant value in steady state conditions. This is different in the case that power generation and load demand are not balanced. If the load power demand is greater than the power generated by the generative systems and transferred to the bus, then the bus voltage will decrease, and if the load power demand is smaller than the power generated by the generative systems and transferred to the bus, then the bus voltage will increase.

A DET solar regulator adjusts the number of solar cell strings directly connected to the bus with reference to bus voltage measurements in order to keep the bus voltage constant. In static load conditions, a static number of strings will be connected to the bus constantly, another static number of strings will be disconnected from the bus constantly, and only one string will be connected and disconnected from the bus regularly in a Pulse-Width Modulation (PWM) mode. The PWM operation of one S3R channel results directly from the S3R control method.

Different topologies of DET regulators are available. DET topologies can be differentiated into those topologies, where a solid-state switch is implemented in series and a diode is implemented in parallel in order to connect or disconnect a solar cell string actively to and from the bus, and into those topologies, where a parallel solid-state switch and a series diode are implemented in order to shunt a solar cell string actively. The most popular DET topology is the Sequential Switching Shunt Regulator (S3R) topology that employs a parallel solid-state switch and a series diode. As solid-state switches, MOSFETs are usually utilized. When a solar cell string is shunted by a DET regulator, the output power of the string is dissipated. The power dissipation occurs in this case in the internal parasitic resistances of the solar cells, in the solid-state switch, and, when the utilized regulator implementation utilizes shunt resistor banks, also in these resistors. Because excess power is dissipated in DET implementations, DET systems are defined as dissipative systems. In order to prevent back currents from the bus to the solar cell strings, diodes have to be implemented in series with the solar cell strings. These diodes are the only elements that are implemented in series in between a solar cell string and the main bus in an DET topology. When a solar cell string is connected to the bus, power dissipation does only occur inside these series diodes. Because of this, typical efficiencies of DET solar regulators have a magnitude between 98% and 99% [20]. Direct energy transfer regulators are available for bus voltages between 28 V (DC) and 120 V (DC) [148].

In space, solar cells will degrade over time resulting from the exposure to the space environment. As a result of the degradation, the solar cell output parameters will decrease over time. In addition to the radiation-induced degradation, the solar array output parameters can also be altered induced by variation in the cell temperature or in the irradiance conditions. A DET topology does not feature a voltage conversion for the generator voltage in order to compensate for these variations. Following this, solar generators have to be designed in a way that they are operated at their Maximum Power Point at the end of their operational lifetime. This can be achieved, when the solar generator is designed in a way that the Maximum Power Point voltage of the generator at its expected End of Life has a value in the magnitude of the bus voltage. Because of this, solar generators are usually not operated at their Maximum Power Point during the mission in DET implementations. An example of a DET solar regulator implementation is shown in Figure 4.30.

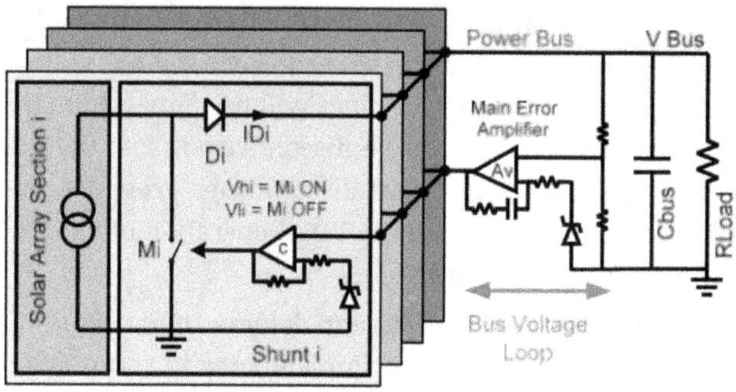

Figure 4.30 Example of solar array regulation utilizing a DET-type solar array regulator (image: © Mourra et al. [217])

In the case that it shall be always possible to operate the solar array at the Maximum Power Point and transfer power in the magnitude of the Maximum Power Point power to the bus, an MPPT regulator instead of a DET regulator needs to be implemented. This offers advantages when the load demand is not steady. However, in common space missions, like Earth observation missions, the load conditions are usually relatively static. In this case, the excess energy that would be available when an MPPT regulator would be implemented instead of a DET regulator, could not be consumed by the space systems. This is different for missions featuring variable power demand like in a lunar energy grid or in the EPS of a space station. During the initial phase of a lunar base, there will already be a high power demand, but only a part of the final generation systems will be implemented. In this case, MPPT regulators could be utilized in order to extract as much power as possible from solar generators, whereas this would not be possible when utilizing DET regulators.

Maximum Power Point Tracking (MPPT) is another method for the regulation and conditioning of the solar array output power. It employs a DC/DC converter that is controlled by a Maximum Power Point tracking control algorithm in order to transfer the solar generator output power to the main bus. Employing this method, the voltage at the solar generator is decoupled from the main bus voltage, so that the operational point of the solar generator can actively be controlled. Contrary to the DET approach, where the individual solar cell strings have been connected to individual input channels of the DET regulator, an MPPT regulator usually only has one input channel. Because of this, all solar array strings are connected in parallel to the MPPT regulator input channel. String diodes in series with solar array strings can be implemented in order to protect the individual strings, but these have internal power dissipation. MPPT algorithms actively control the power flow from the solar generator through the DC/DC converter to the main bus. They control the operational point of the solar generator by controlling the voltage at the input of the DC/DC converter, where the solar generator is connected. In case that no power needs to be transferred from the solar generator to the bus, the algorithm will control the input voltage of the converter in the direction of the open circuit voltage of the solar generator in order to decrease the solar generator output current and power. With increasing power demand, the control algorithm will dynamically decrease the voltage at the converter input in response to increase the solar generator output current. The MPPT algorithm will decrease the solar generator terminal voltage until the generator output power is sufficient to supply the load power demand. In case that the combined power demand of energy storages and electrical loads has a value in the magnitude of the maximum generator output power, the algorithm will operate the solar generator at the operational point, where the output power reaches its maximum value. This operational point is usually referred to as Maximum Power Point (MPP). During the operation, the input voltage of the DC/DC converter will oscillate around the Maximum Power Point Voltage of the solar generator. Within the boundary conditions of the chosen converter topology, the converter input voltage, which is equal to the solar generator terminal voltage, can be controlled independently from the bus voltage. When a step-down type of converter topology is chosen for the MPPT regulator, the solar array Maximum Power Point Voltage needs to be higher than the bus voltage and when a step-up type of converter topology is chosen, the solar array Maximum Power Point Voltage needs to be lower than the bus voltage. In these cases, a common margin, between the solar generator Maximum Power Point Voltage and the bus voltage is 10% below or above the bus voltage. Combined step-down and step-up type topologies, like the B2R topology that was certified for space by ESA/ESTEC, enable a more flexible solar generator design. When a such a topology is utilized, the solar generator can be designed in a way, that its Maximum Power Point Voltage has a value in the magnitude of the bus voltage. This allows an increase in the regulator efficiency because the regulator efficiency increases with a decreasing voltage difference between regulator input and output. On board of European space missions, both step-down and step-up type regulator topologies have been implemented.

It is always possible to transfer the maximum possible amount of power from the solar generator to the main bus during all mission phases when MPPT regulators are implemented because the operational point of the solar generators can actively be controlled, while with DET-type regulator topologies, the bus voltage has to have a value in the magnitude of the solar generator's Maximum Power Point voltage in order to allow an operation at the Maximum Power Point. MPPT regulators are defined as non-dissipative systems because they usually do not feature dissipative elements. Within an MPPT topology, power is dissipated in the regularly switching MOSFETs, in diodes, and also in parasitic resistances of inductors. Internally dissipated power accounts for the internal losses of the regulator. Power dissipation inside an MPPT regulator can be accounted in a magnitude between 4% to 7% of the output power [98] [217]. An example of a Maximum Power Point Tracking solar energy regulator implementation into an unregulated power bus is shown in Figure 4.31.

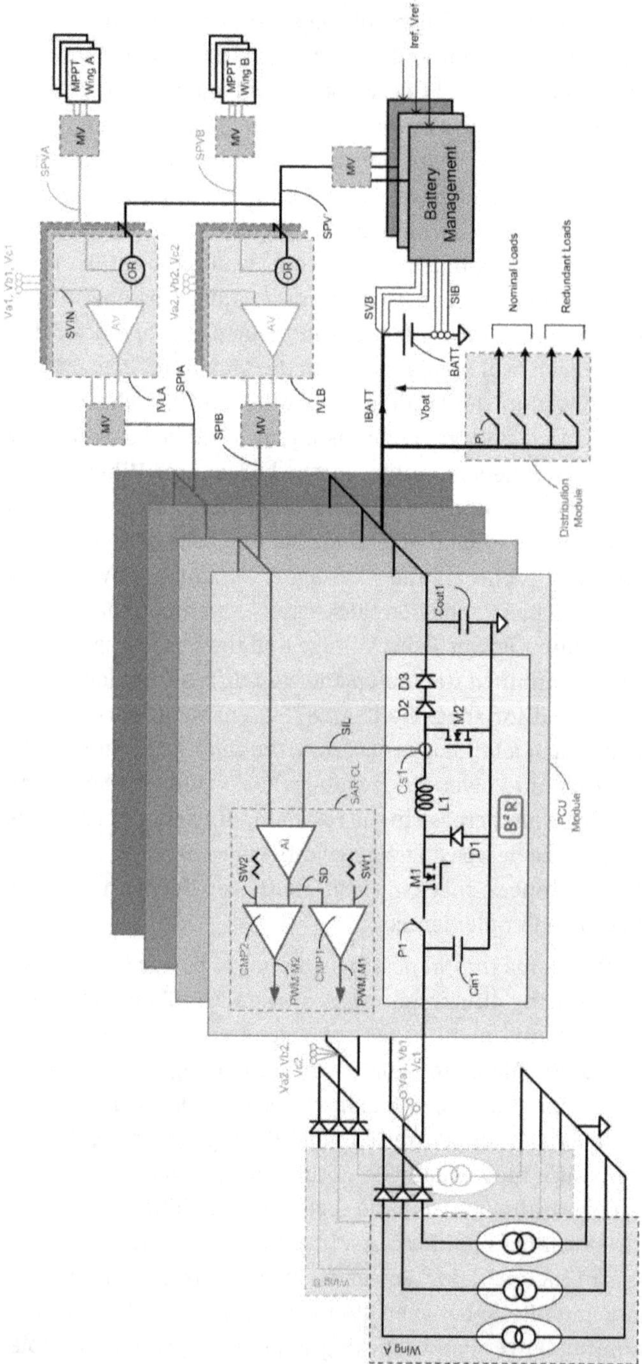

Figure 4.31 Example of an unregulated spacecraft power bus, utilizing SAWs as power source, a buck-boost type SAW regulator, batteries as energy storages, a main bus capacitor and a power distribution unit PDU (image: © Mourra et al. [217])

A common assumption is that the Maximum Power Point Tracking topology would be superior to the Direct Energy Transfer (DET) topology because MPPT regulators are able to transfer the maximum possible amount of power to the bus at any point. However, in reality the choice of the regulator topology is based on the balance of a number of variables. DET regulators are relatively simple and consist of less parts then MPPT regulators. Following this, DET regulators enable mass and cost savings. Since the only losses inside a DET regulator occur in the series semiconductors, DET regulators are usually more efficient than MPPT regulators in relatively stable conditions. The power dissipated in the dissipative elements of a DET implementation resulting from the disconnection of a solar array string from the bus is usually not accounted as power losses in the review of DET efficiency. Usually, the amount of additional power resulting from the utilization of an MPPT regulator is smaller than the additional losses occurring inside the MPPT topology. MPPT regulators are beneficial when, during the operation, a high variability of the environmental conditions or a variability in the load demand can be expected. Solid-state switches inside DET regulators are switched less often. Because of this, DET regulators are more reliable than MPPT regulators and can be in operation for longer periods of time. DET regulators also feature less output voltage noise. Utilizing MPPT regulators, the solar generator size can be reduced in a magnitude between 8% and 25%, while meeting the same output power requirement [97]. MPPT regulators allow to design solar generators more flexible, but additional efforts in the control system design have to be taken.

4.4 Thermal to Electrical Power Conversion

Thermal to electrical power conversion methods require a certain temperature gradient. In order to achieve this, thermal energy has to be supplied to the hot part of the generator by the heat source and the cold part has to be cooled by the application of an active or passive thermal management method. Thermoelectric generators or thermionic generators can be utilized as static generators for the conversion from thermal power to electrical power and Stirling, Rankine or Brayton generators can be utilized as dynamic generators.

Thermoelectric generators are devices enabling the conversion from thermal to electrical power. Inside thermoelectric generators, the Seebeck effect is utilized. In order to utilize this effect, a suitable material combination needs to be implemented in the device. The Seebeck effect allows to convert an applied temperature gradient at a pn-junction into a voltage at the terminals of a Seebeck type generator. The electrical field results in a current flow when an electrical circuit is connected to the Seebeck generator. Seebeck generators achieve efficiencies between 5% and 20% [97] [98]. They can be connected in series and in parallel to increase the generator voltage and current, in a similar way as it is also possible with solar cells.

Thermionic generators utilize two electrodes inside a sealed container that is filled with an ionized gas [98]. The emitter electrode needs to be hotter than the collector electrode. Electrons are emitted from the emitter electrode, and after that, these condense at the collector electrode. It is recommended to use thermionic generators in hot temperatures. Thermionic generators achieve efficiencies between 5% and 20% [97] [98].

Dynamic power conversion processes employ a conversion from thermal to mechanical power in an interim step, together with a conversion from mechanical power to electrical power in a second step. Thermodynamic cycles employ the two isothermal processes compression and expansion and the two constant volume processes heating and cooling. Stirling generators achieve efficiencies between 25% and 30%, Rankine generators achieve efficiencies between 15% and 25% and Brayton generators achieve efficiencies between 20% and 35% [97] [98].

4.5 Nuclear Energy

Nuclear generators are an interesting systematic option for the exploration of space. These generators can either employ the by-product thermal energy resulting from the decay of a nuclear isotope, or they can employ a nuclear chain reaction in a nuclear reactor system [148]. For both types of approaches, static power conversion systems and dynamic power conversion systems can be employed in order to convert thermal energy to electrical energy. Nuclear heat sources are implemented inside nuclear generators and supply a generator system with thermal energy. Because of this, this type of generative systems is independent from environmental conditions. Nuclear generators have proven a high reliability, a long life duration and are maintenance-free, even under challenging environmental conditions. Risks resulting from the employment of nuclear generators in space systems have been classified to be lower than those risks imposed by the majority of different available propellants. However, nuclear generators in the kilowatt class feature significant risks during launch and landing. Failures during these mission phases have the potential of contaminating large areas around the launch or landing site. Because of this, these systems need to be designed in a way that even in the case of failures, these systems do not expose the internal radionuclides to the environment.

Nuclear generators have been developed for various power demands, but recent applications have been limited to low power demand applications. For a number of space missions, nuclear-based generators are of potential interest. Examples are missions to the outer solar system, where solar energy is not available in a satisfying magnitude, and military missions in Low Earth Orbit, in order to limit the visibility of a spacecraft by minimizing its cross-sectional area compared to satellites utilizing Solar Array Wings.

Nuclear generators generally consist of a nuclear heat source, a generator for the conversion from thermal power to electrical power, and a radiator to dump excess thermal power from the heat source that could not be converted into electrical power. Excess heat can also be utilized inside the spacecraft in order to keep its subsystems within a desired temperature range. In addition to nuclear-based electrical generators, radioisotope heater units are also available that supply space systems with thermal power instead of electrical power. An overview schematic of a nuclear generator is shown in Figure 4.32.

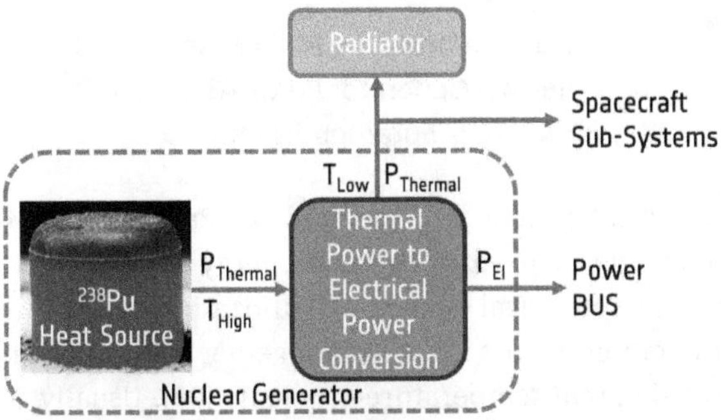

Figure 4.32 Overview schematic of a nuclear generator (image: © US department of energy (^{238}PU) [233])

As nuclear heat sources, plutonium 238, polonium 210, curium 242, uranium 235, and enriched uranium are available. In RTGs, the decay of plutonium 238 is often employed, which decays to uranium 234. As a by-product, an alpha-particle and 5.6 MeV of energy are released [148].

$$^{238}\text{Pu} \rightarrow {}^{234}\text{U} + \alpha + 5.6\ \text{MeV}$$

For the integration into an RTG, plutonium 238 is pressed into pallets and integrated into a casing. The cased plutonium 238 pallet is referred to as General Purpose Heat Source (GPHS). Each gram of plutonium 238 delivers 0.5 W of thermal power and each General Purpose Heat Source delivers 62 W [48]. An image of such a plutonium 238 pallet is shown in Figure 4.32 and 8 General Purpose Heat Sources integrated into the RTG system Multi Mission Radioisotope Thermoelectric Generator (MMRTG) can be found in the systematic overview visualization in Figure 4.33.

In nuclear reactors, such as the reactor onboard of the COSMOS-954 satellite, uranium 235 has been utilized as source of energy [48]. The nuclear reactor system TOPAZ-1 was able to deliver 5 kW of electrical power from 12 kg of nuclear isotopes during a period between 3 and 5 years [48]. The total mass of a TOPAZ-1 reactor had a value of 320 kg [48].

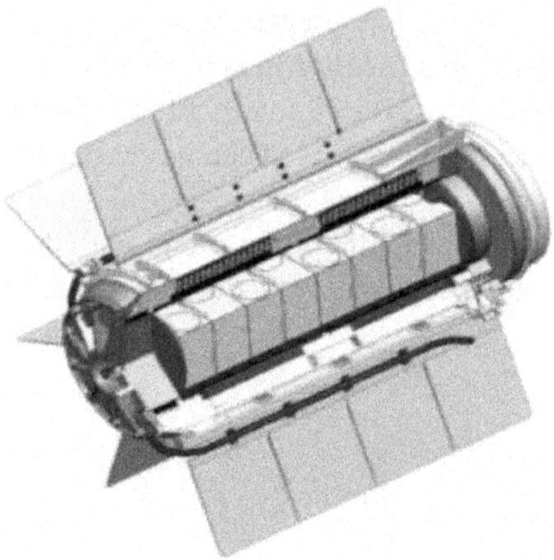

Figure 4.33
Visualization of the Multi Mission Radio Thermal Generator (MMRTG) (image: © NASA, S. Surampudi [148])

Usually, thermoelectric generators are integrated into the walls of Radio Thermal Generators (RTGs) in order to convert the by-product thermal energy from the nuclear decay of a nuclear heat source into electrical energy. The utilization of thermionic generators is also a systematic option for the utilization together with nuclear generators, but these require relatively great temperature gradients [98]. Usually, a nuclear chain reaction needs to be employed for this application. Besides static power conversion methods, also dynamic power conversion methods can be utilized. Possible power conversion methods are Stirling generators, Rankine generators, and Brayton generators.

An example for a nuclear generator system utilizing static energy conversion is the Radioisotope Thermal Generator (RTG) and an example for a nuclear generator system utilizing dynamic energy conversion technology is the Dynamic Isotope Power System (DIPS). Usually, RTGs have an output power in the domain of less than 1 kW and DIPS have an output power in the domain between 1 kW and 10 kW [97].

In total, a number of 25 spacecraft utilized RTGs [48] [148]. These missions include the Apollo missions, the Viking missions, the Voyager missions, the Cassini mission and the Mars science laboratory Curiosity.

Nuclear fission reactors had only been utilized in some missions until the 1980s, but these systems are taken into account as possible systematic options for missions to the Moon and Mars. The American SNAP 10-A nuclear reactor was utilized in a 43–day-long mission in 1965 [48] [97]. In addition to this, the RORSAR and COSMOS-954 missions carried out by the U.S.S.R. utilized the nuclear reactor models TOPAZ and RORSAT. While the TOPAZ reactor utilized a thermionic generator, the RORSAT and SNAP 10-A reactors utilized thermoelectric generators. The reactor SP-100 was specifically designed for the application on the Moon and Mars.

4.6 Topologies of Space Electrical Power Systems

A number of different topologies are available for the design of spacecraft Electrical Power Systems (EPSs). These basic topologies can be differentiated by the method utilized for power regulation, conditioning, and connecting the generative system to the EPS, and they can also be differentiated by the degree of control they offer for the bus voltage. Solar arrays are commonly assumed as generative systems in space EPS topologies because by far the majority of spacecraft utilizes photovoltaic generators. Because of this, standard space EPS topologies are usually visualized featuring solar power generators and solar regulators in literature. However, in principle it would also be possible to utilize other generative systems like nuclear-based generators. A number of standard power bus topologies is visualized in Figure 4.34. An example for a fully regulated power bus is shown in Figure 4.30 and an example for an unregulated power bus is shown in Figure 4.31. An overview over the technical parameters of different topologies is given in Table 4.2.

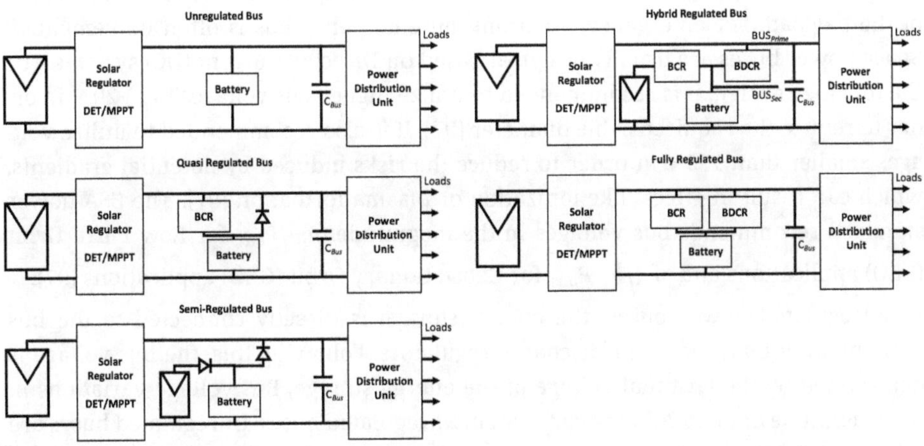

Figure 4.34 Standard spacecraft power bus topologies

Table 4.2 Comparison of different power bus topologies (in empty fields, data was not available) [112] [119] [147]

EPS topology	Bus voltage	Voltage variation	Power dissipation
fully regulated bus	24 V_{DC}, $P < 2.5$ kW 50 V_{DC}, $P < 8$ kW 100 V_{DC}, $P > 8$ kW 124 V_{DC}, $P > 8$ kW	± 0.5% in steady state ± 1% at 50% load step ± 5% during power up ± 10% when fuse is triggered	3–13% BDR 3–10% BCR

Table 4.2 Comparison of different power bus topologies (in empty fields, data was not available) [112] [119] [147] *(continued)*

EPS topology	Bus voltage	Voltage variation	Power dissipation
unregulated bus	28 ± 6 V_{DC}, $P < 2$ kW 35 ± 7 V_{DC}, $P < 3.5$ kW 42 ± 8 V_{DC}, $P < 5$ kW	± 20%	
high voltage bus	300–1500 V_{DC}, 0.3–100 kW		

Power buses are utilized in spacecrafts in order to distribute electrical energy. The bus voltage has to be kept within a predefined voltage domain by a dedicated control system. Different power bus topologies are available for the spacecraft EPS design. The choice of the bus topology results in additional implications for EPS design. Multiple power buses can be implemented into a space system. As a minimum, one main bus and one additional backup main bus need to be implemented in a space system. In addition to this, space systems often also feature a dedicated power bus for the supply of the payload. For emergency situations, an emergency bus is often implemented. Space power buses are usually designed based on Directed Current (DC) systems. For the topology design, it is recommended to utilize higher bus voltages than 20 V in order to reduce the required cable diameter [97]. It is also recommended to utilize voltages smaller than 125 V in order to reduce the risks induced by potential gradients, which can result in effects like ionization or plasma formation [97]. The ESA power standard recommends bus voltages in the magnitude of $\sqrt{P_{\text{bus}}}$ for Low Earth Orbit (LEO) applications and of $\sqrt{\frac{1}{2} \cdot P_{\text{bus}}}$ for Geostationary Orbit (GEO) applications [97].

In unregulated power buses, the energy storage is directly connected to the bus without additional charge or discharge regulators. Following this, the bus voltage is determined by the terminal voltage of the energy storage. Bus voltage variations in the magnitude of up to 20% are common in unregulated buses. Unregulated buses are relatively simple and have a relatively small number of required parts. Because of this, unregulated buses offer significant mass reductions. Typical bus voltages of unregulated buses are 28 V (DC) with a variation of ±6 V for buses with a power generation capability below 2 kW, 35 V (DC) with a variation of ±7 V for buses with a power generation capability between 2 kW and 3.5 kW, as well as 42 V (DC) with a variation of ±8 V for buses with a power generation capability between 3.5 kW and 5 kW [97]. The unregulated bus topology can be defined as more efficient than other topologies because battery charge or discharge regulators that have additional power dissipation are not part of this topology. It can also be defined as more reliable than other topologies because it features a smaller number of units that may principally fail. However, since not all loads are tolerant to the comparable high voltage variations, additional DC/DC converters often need to be implemented in order to supply the individual loads. In each of these load converters, power dissipation in a magnitude between 1% and 4% can be expected [97]. In addition to this, the required efforts that have to be taken into the design of the main bus filter in order to meet the EMV requirements of the space system are usually increased with unregulated buses compared to other topologies. Due to the high voltage variation, also up to 50% more losses in the cable harness or an increase in the cable diameter by up to 25% needs to be accounted with this topology [97]. In order to account for these issues, it is highly important to analyze the bus voltage variation during the design phase of an unregulated bus. Not all energy storage technologies are tolerant to be directly connected in parallel to a power bus and because of this, the available energy storage technologies are reduced with unregulated buses to those that are tolerant to this implementation. In addition to this, energy storages can only be connected to unregulated buses when these are tolerant to deep discharge or overcharge conditions. With unregulated buses, deep discharging of the energy storage cannot be prohibited in any case, so this topology shall only be utilized when the energy storage is tolerant to deep discharging or when deep discharging can be prohibited by other countermeasures. A recommendation for the design of unregulated buses utilizing a DET solar regulator topology is to match solar generator and battery voltages in a way that during battery charging and discharging, the bus voltage varies around the Maximum Power Point of the solar generator.

When high resistive current sources and low resistive voltage sources are connected directly together, the operational condition power lock-up can occur. This could for example be the case when solar cells and batteries are implemented together in an unregulated bus utilizing a DET regulator topology. Power lock-up occurs during high-demand conditions, where the generative systems are not capable of supplying the connected loads themselves. As a result, the bus voltage will decrease and the directly connected energy storages will discharge in response. The energy storage terminal voltage, and, following this, the bus voltage will decrease resulting from the discharge. In the next step, a decreased bus voltage will alter the operational point of the generative systems, so that their output power decreases and the energy storage output power will increase in order to satisfy the load demand. The power lock-up condition can lead to a complete energy storage discharge. Unfortunately, in relatively low states of charge or complete discharge state, the energy storage voltage will feature a minimum value. The operational point of the solar generator will be determined by the bus voltage, and in common designs, the generative system will not be able to transfer the maximum possible amount of power to the bus in low voltage conditions. If load shedding systems have been implemented into the Power Conditioning Unit (PDU), the PDU has to disconnect loads in low bus voltage conditions in order to enable the solar generators to recharge the energy storages again to recover the bus voltage. It is also possible that load shedding systems have not been integrated into the PDU or that the load shedding fails. In this case, it is possible that the connected loads will consume all the power transferred by generators, and a recovery of the bus voltage will not necessarily be possible. Another possibility is that malfunctioning in the energy storages occurs as a result from the bus voltage collapse.

The fully regulated power bus topology is the most complex spacecraft EPS topology. When a fully regulated bus topology has been implemented, the bus voltage is decoupled from the energy storage voltage. This results from the employment of a charge regulator together with a discharge regulator in order to connect the energy storage to the power bus. The employment of a combined charge/discharge regulator is also conceivable. The fully regulated bus topology is less efficient than other bus topologies. However, fully regulated buses are implemented in spacecraft requiring a highly regulated voltage. The bus voltage is controlled by a complex control system. Because of this, more efforts have to be taken into the design of control systems for the bus voltage, as it would be the case for other bus topologies. In nominal operation, the requirement is that bus voltage variations are smaller than 0.5% of the reference voltage [97]. During variations in the bus load in the magnitude of 50% of the load, the requirement is that bus voltage variations shall be smaller than 1% of the reference voltage [97] [147]. During power-up, the requirement is that the voltage ripple is smaller than 10% of the reference voltage [97]. Typical bus reference voltages of fully regulated buses are 24 V (DC) when the power capacity of the bus is smaller than 2.5 kW, 50 V (DC) when the power capacity of the bus is smaller than 8 kW, and 100 V (DC) or 124 V (DC) when the power capacity of the bus is greater than 8 kW [97]. It is recommended to design battery storages in a way that their voltage is lower than the bus voltage in a magnitude of 20% [20]. In this case, a step-down converter topology needs to be utilized as a battery charge regulator and a step-up converter topology needs to be utilized as a battery discharge regulator. An infinite number of batteries and solar generators can be connected in parallel to a fully regulated bus because these systems are connected to the bus utilizing dedicated regulators. Subsequently, fully regulated buses can be expanded incrementally by implementing additional power generators or energy storages in parallel to the bus. This type of buses also features the advantage that the highly stable bus voltage allows to connect a great share of the loads directly to the power bus. Additional regulators are only required when the required input voltage of a load is different from the bus voltage. The highly stable bus voltage allows to design more efficient load converters. Increases in the load converter efficiency in a magnitude from 1% to 4% can be achieved [97]. Another advantage of fully regulated buses is that the spacecraft thermal control system design is simpler. Thermal control systems are usually implemented based on electrical heater units that are operated by a certain control logic. These heaters produce a constant heating power at constant input voltages. In case that the input voltage is already constant, compensations for input voltage fluctuations do not need to be taken into account in the design. Fully regulated buses feature better EMV characteristics than other bus topologies, which results from a highly constant bus voltage. During the bus design, the bus voltage does not necessarily need to be analyzed to the same extend as it would be required with other bus topologies. Due to the decoupling of the energy storage, power lock-up conditions cannot occur in a fully regulated bus topology. However, between 3% to 13% of the charge power and between 3% and 10% of the discharge power are usually dissipated inside the dedicated regulators and account to the internal power losses [97]. In addition to this, the discharge regulators need to be designed to supply peak loads. This is an issue because peak loads are usually much greater than nominal loads. Because of this, the implemented power transfer capacity of the discharge converter will not be utilized completely during most operational phases. Resulting from the increased number of DC/DC converters, fully regulated power buses usually feature higher system masses than other bus topologies.

The quasi-regulated bus topology is relatively similar to a fully regulated bus topology. Here, the energy storage is connected to the main bus utilizing a charge regulator and utilizing a diode for discharging. During operational periods when primary power is transferred to the bus, the bus voltage will be regulated by the regulator of the energy source. The energy storage can be charged or it can also be not in operation during these periods. In order to discharge the energy storage with this topology, the bus voltage needs to decrease to a value that allows the diode to transition into conduction state. The diode has to return into blocking state again in order to terminate the battery discharge operation. This is the case when the difference between bus voltage and terminal voltage of the energy storage turns negative. This topology can be seen as a good balance between the fully regulated bus and the unregulated bus. Since a discharge regulator is not required by this topology, this topology features a smaller level of complexity than the fully regulated bus. In addition to this, cost and mass savings can be achieved because this topology utilizes one regulator less than the fully regulated bus topology. The energy storage is charged using a charge regulator, so that this operation can actively be controlled. However, this type of buses has relatively low efficiencies and high electromagnetic disturbances. In case that loads that are requiring a certain level of voltage stability are connected to a quasi-regulated bus, these loads usually need to be connected employing an additional load converter.

The energy storage is charged by a dedicated generator that is directly connected to the energy storage over an additional diode in a semi-regulated bus. It is connected to the main bus utilizing an additional diode for the discharge operation. The bus voltage of a semi-regulated bus is regulated by the regulator of the generative system during generative periods. During energy storage discharge periods, the bus voltage is determined by the energy storage terminal voltage. The discharge of the energy storage will start when the voltage difference between the power bus and the energy storage terminal voltage is greater than the threshold voltage of the diode, and the discharge will continue until this difference turns negative. Loads requiring a more stable voltage supply than the variation caused by the energy storage have to be connected to a semi-regulated bus via an additional load converter. This topology utilizes fewer regulators than the fully regulated bus topology, which leads to savings in costs and mass as well as a smaller level of complexity than the fully regulated bus topology.

A hybrid bus topology utilizes a discharge regulator to transfer power from the energy storage to the power bus. The energy storage is charged by a charge regulator that is directly supplied by a dedicated generator that is directly connected to the energy storage but not to the power bus. During periods when the bus is supplied by the generative systems, the bus voltage is regulated by the regulator of the generative systems. During energy storage discharge periods, the bus voltage is regulated by the discharge regulator of the energy storage. Because of this, hybrid buses also feature highly stable bus voltages that allow to connect loads directly to the bus. It is also possible to implement a second power bus directly connected in parallel to the energy storage because it is decoupled from the main bus. This second power bus can be seen as an unregulated bus. This offers advantages when pulsed loads need to be supplied. Loads demanding pulsed power can be connected directly to the unregulated voltage power bus, so that the discharge regulator does not necessarily need to be designed for the transfer of pulse power to the regulated voltage power bus. The design of hybrid buses is complex because multiple power buses, dedicated generators for battery charging, a number of different converters and regulators, as well as different bus interconnection possibilities, need to be taken into account.

In addition to the bus topologies discussed in this chapter, high voltage bus topologies are also available. These offer the advantage of high voltages supply for electric propulsion systems [97]. Their voltages have values in the domain between 300 V (DC) and 1500 V (DC) [97]. These buses have a power capacity in the domain between 30 kW and 100 kW [97].

Space Electrical Power Systems have to be single-point-of-failure-free as a requirement. This can be achieved by connecting a redundant set of two converter or regulator units in parallel to provide the functionality of one converter or regulator. Inside a converter or regulator, series diodes have to be implemented as a set of two in series connected diodes. For the case of a regulator failure, power switches have to be implemented in series with converter or regulator units in order to connect or disconnect them from the power bus. An example for a single point of failure free implementation can be found in Figure 4.31. In order to stabilize the bus voltage, a central main bus capacitor is usually connected in parallel to main buses.

4.7 Power Distribution System

The Power Distribution System (PDS) distributes electrical power in a space system or spacecraft and connects electrical loads to the power bus in order to supply these with electrical power. It consists of the cabling, dedicated fault protection systems, and switches in order to connect or disconnect loads to and from the power bus. It is usually implemented by a physical Power Distribution Unit (PDU) in which the functionalities are integrated together with the cable harness. The PDS is usually based on Direct Current (DC) systems in space implementations.

Centralized Power Distribution Systems and decentralized Power Distribution Systems can be differentiated. Centralized DC PDSs employ regulated power buses. Loads are usually connected directly to the bus and only a minority of loads is connected to the bus utilizing additional converters. Decentralized PDS usually employ unregulated buses [98]. Loads are usually connected to the bus with individual load converters at each load that are distributed throughout the space system or spacecraft.

A Power Distribution Unit (PDU) needs to be capable of keeping power quality, EPS reliability and survivability, as well as the EPS costs within the requirements. It needs to be designed with focus on minimizing EPS mass and the internal power dissipation while meeting the system requirements. Following this, the PDU design depends on the characteristics of the power bus, the characteristics of the systems connected to it, as well as the requirements of the different loads. The PDU is utilized in order to connect and disconnect different loads to and from the power bus by employing a set of dedicated switches inside the unit. These switches are usually implemented utilizing a set of two or more parallel switches for redundancy reasons. They can be implemented utilizing mechanical or solid-state relays. In case that mechanical relays are employed, currents are only allowed to rise until 50% of the nominal relay current, and because of this, current limitation capabilities have to be integrated into the PDU [97]. MOSFETs are usually implemented in PDUs as solid-state switches.

The cabling connecting different loads to the PDU is usually referred to as cable harness. Each load needs to be supplied by an individual cable and an additional back cable. Backup systems also need to be connected, employing individual cables.

4.8 Power Transmission

A future lunar base EPS will feature transmission lines with a total length of multiple kilometers. The power transmission is usually based on cables and Direct Current (DC) systems in space implementations. The system of cables implemented in a spacecraft is usually referred to using the term cable harness. Taking into account the effects of long cables is of particular importance in the case of distributed Electrical Power Systems (EPSs) featuring long transmission distances. Subsequently, the cable harness needs to be designed with focus on minimizing power dissipation. Within the cable harness, power dissipation mainly occurs resulting from the internal resistance of the employed cables.

The internal resistance of a cable can be defined with reference to the cable material. Each cable material features the intrinsic property specific resistance. The value of the specific resistance varies with the material temperature and its value is usually calculated for a certain temperature employing one or two temperature coefficients.

$$\rho_{Electric}(T) = \rho^0_{Electric} \cdot (1 + \alpha_{\rho_{Electric}} \cdot (T - T_{TC}))$$

$$\rho_{Electric}(T) = \rho^0 \cdot (1 + \alpha_{\rho_{Electric}} \cdot (T - T_{TC}) + \beta_{\rho_{Electric}} \cdot (T - T_{TC})^2)$$

The specific resistance of the cable material can be utilized in order to calculate the electrical resistance of a cable. It can be calculated as product of the specific resistance of the cable material with the division of the cable length by the cable diameter.

$$R_{Wire} = \rho_{Electric}(T) \cdot \frac{l_{Wire}}{A_{Wire}}$$

Electrical power transmission requires a positive and a negative cable. The series resistances of both cables need to be taken into account in the analysis of the internal power dissipation. A visualization of the resistances in this simplified model of power transmission lines is shown in Figure 4.35.

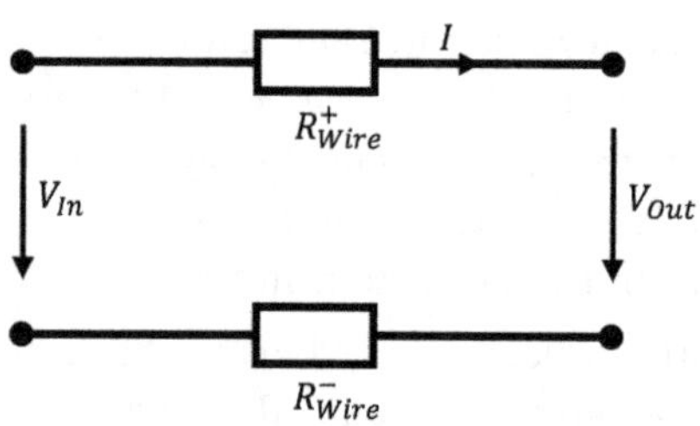

Figure 4.35 Electrical resistances in power transmission lines

The series resistances of the positive and of the negative cable can be seen as a series connection of two resistances. Subsequently, the series resistance of a transmission line can be calculated as the sum of both series resistances. The equation can be simplified even further when equal temperatures, materials, lengths, and diameters can be assumed for both transmission lines.

$$R_{Wire}(T) = R^+_{Wire}(T) + R^-_{Wire}(T) = \rho(T) \cdot \frac{l^+_{Wire} + l^-_{Wire}}{A_{Wire}}$$

The voltages occurring at a transmission line can be calculated employing Ohm's law. These voltages reduce the value of the output voltage that can be measured at the end of the transmission line. The remaining voltage after the transmission can be calculated employing Kirchhoff's law.

$$V^+_{Wire} = R^+_{Wire}(T) \cdot I_{Wire}$$

$$V^-_{Wire} = R^-_{Wire}(T) \cdot I_{Wire}$$

$$V_{Wire} = V^+_{Wire} + V^-_{Wire} = \left[R^+_{Wire}(T) + R^-_{Wire}(T)\right] \cdot I_{Wire} = R_{Wire}(T) \cdot I_{Wire}$$
$$= \rho(T) \cdot \frac{l^+_{Wire} + l^-_{Wire}}{A_{Wire}} \cdot I_{Wire}$$

$$V_{Out} = V_{In} - V^+_{Wire} - V^-_{Wire} = V_{Input} - V_{Wire} = V_{In} - R_{Wire}(T) \cdot I_{Wire}$$
$$= V_{In} - \rho(T) \cdot \frac{l^+_{Wire} + l^-_{Wire}}{A_{Wire}} \cdot I_{Wire}$$

The power dissipation occurring inside a cable resulting from the internal resistance can be calculated with reference to the currents inside a cable and the cable resistance.

$$P_{Dissipated,Wire} = V_{Wire} \cdot I_{Wire} = (R_{Wire} \cdot I_{Wire}) \cdot I_{Wire} = I^2_{Wire} \cdot \rho_{Electric}(T) \cdot \frac{l_{Wire}}{A_{Wire}}$$

The power dissipated in a transmission line is equal to the sum of the dissipated power in the positive and the negative cable.

$$P^+_{Dissipated,Wire} = V^+_{Wire} \cdot I = R^+_{Wire}(T) \cdot I^2_{Wire}$$

$$P^-_{Dissipated,Wire} = V^-_{Wire} \cdot I = R^-_{Wire}(T) \cdot I^2_{Wire}$$

$$P_{Dissipated,Wire} = P^+_{Dissipated,Wire} + P^-_{Dissipated,Wire}$$

$$P_{Dissipated,Wire} = V_{Line} \cdot I_{Wire} = \left(R^+_{Wire}(T) + R^-_{Wire}(T)\right) \cdot I^2_{Wire}$$
$$= \rho_{Electric}(T) \cdot \frac{l^+_{Wire} + l^-_{Wire}}{A_{Wire}} \cdot I^2_{Wire}$$

The total power dissipation occurring inside the complete cable harness can be calculated as the sum of the individual power dissipation in all cables.

$$P_{Dissipated,Harness} = \sum_{i=1}^{N_{Loads}} P_{Dissipated,Wire,i}$$

$$P_{Dissipated,Harness} = \sum_{i=1}^{N_{Loads}} I^2_{Wire,i} \cdot R_{Line,i}$$

$$P_{Dissipated,Harness} = \sum_{i=1}^{N_{Loads}} I^2_{Wire,i} \cdot \rho_{Electric}(T) \cdot \frac{l^+_{Wire} + l^-_{WIre}}{A_{Wire}}$$

Cables have to be designed as short as possible in order to minimize the internal power dissipation. The power dissipation can be decreased further by increasing the cable diameter. It is recommended to choose the cable length and diameter in a way that the power dissipation is limited to between 1% and 3% of the input power of the transmission line. The voltage dissipation in the cables utilized for battery charging should be smaller than 3% of the voltage and the voltage dissipation in the cables utilized for battery discharging should be smaller than 7% of the voltage [151]. Cables also require electrical isolation from the environment, which adds additional mass and volume to a transmission line. The isolation needs to be designed to withstand the lunar environment. Especially the temperature variation, radiation degradation of the isolation materials, and the exposure of the isolation materials to abrasive lunar regolith need to be taken into account.

The cable harness usually accounts for between 10% and 25% of a space system's mass [98]. Because of this, a balance needs to be found between the cable harness mass and meeting the power dissipation requirements.

4.9 Power Converters and Regulators

Power electronics are utilized in space Electrical Power Systems (EPSs) in a wide range of different applications such as the transformation between different voltage levels or the control of the power flow. Usually, Direct Current (DC) systems are utilized in space EPSs. Typical applications are solar power regulators, battery charge regulators, battery discharge regulators, or load converters. For example, on board of the ROSETTA spacecraft, a 2-Inductor step-down topology has been utilized as a solar array regulator.

Power converters and regulators consist of a number of different electrical components. The utilization of components featuring a high power dissipation is usually minimized in order to maximize the efficiency. Because of this, mainly diodes, transistors, capacitors, inductors, and transformers are utilized in power electronics. Diodes are operated as passive electrical switches, and transistors are operated as active electrical switches. Capacitors are usually implemented in parallel to the input and output in order to stabilize the input and output voltages. Inductors are usually implemented in order to stabilize the current and to store energy in between switching subintervals. The utilization of transformers allows to galvanically isolate the converter or regulator input from the output. In addition to this, they can also be utilized to store energy between different switching intervals.

The incremental property of an inductor is the inductance, which is defined as the ratio between the magnetic flux linkage and the inductor current. The magnetic flux linkage can be defined as the product of the number of inductor turns and the magnetic flux inside the inductor, or as the product of the inductance and the inductor current. The derivative of the magnetic flux linkage is equal to the inductor voltage. The voltage at an inductor can vary discontinuously, but the inductor current can only feature a steady progression. The value of the inductor current can be calculated as the time integral over the division of the inductor voltage by the inductance. The energy stored inside the magnetic field of the inductor can be calculated as the product of 0.5, the inductance and the squared inductor current. The schematic of an inductor is shown in Figure 4.36.

$$L = \frac{\psi_L}{I_L} = \frac{N_L \cdot \phi_L}{I_L}$$

$$\psi_L = L \cdot I_L$$

$$V_L = \frac{d}{dt}\psi_L = \frac{d}{dt}L \cdot I_L = L \cdot \frac{d}{dt}I_L$$

$$I_L = \int_0^t \frac{1}{L} \cdot V_L(\tau)\, d\tau$$

$$E_L = \frac{1}{2} \cdot L \cdot I_L^2$$

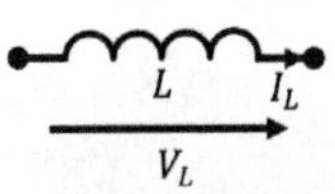

Figure 4.36
Inductor schematic

Capacitors feature the property to store energy inside the internal electric field. The capacitor current is defined as the deviation of the capacitor charge over time. The incremental property of a capacitor is the capacity. The capacity can be defined as the ratio between the capacitor charge and voltage.

The capacitor current can vary discontinuously, but the capacitor voltage needs to have a steady progression. The capacitor voltage can be determined as the time integral of the division of the capacitor current by the capacitance. The energy stored inside a capacitor is equal to the product of 0.5, the capacitance and the squared capacitor voltage. The schematic of a capacitor is shown in Figure 4.37.

$$I_C = \frac{dQ}{dt}$$

$$Q_C = C \cdot V_C$$

$$C = \frac{Q_C}{V_C}$$

$$I_C = \frac{dQ_C}{dt} = C \cdot \frac{dV_C}{dt}$$

$$V_C = \int_0^t \frac{1}{C} \cdot I_C(\tau)\, d\tau$$

$$E_C = \frac{1}{2} \cdot C \cdot V_C^2$$

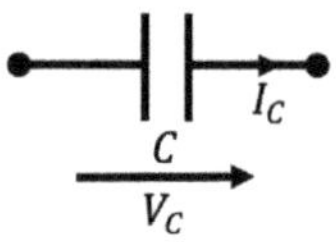

Figure 4.37
Capacitor schematic

Resistors are not commonly utilized in power electronics. However, other components like capacitors or inductors can have parasitic internal resistances. In addition to this, there are implementations of solar power regulators utilizing resistor banks. The relation between the resistor voltage and the current is defined by Ohm's law. Both the voltage and the current can have a discontinuous progression. The power at a resistor can be defined as the product of resistor voltage and current. Resistors do not store energy. The schematic of a resistor is shown in Figure 4.38.

$$V_R = R \cdot I_R$$

$$P_R = V_R \cdot I_R = R \cdot I_R^2$$

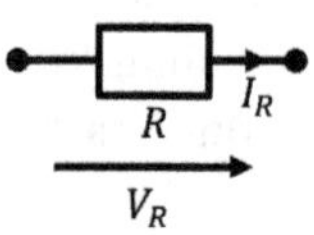

Figure 4.38
Resistor schematic

Diodes feature a complex relation between diode voltage and current. This behavior can be described utilizing the Shockley equation. In power electronics, diodes are usually operated as passive switches, so that their behavior in power electronics can be described as a switch in a good approximation. The schematic of a diode has been visualized in Figure 4.39.

When a certain positive voltage is applied to a diode, it gets into conductive state. In conductive state, the diode current features a positive value, while only a small voltage occurs at the diode terminals. When negative voltages are applied to a diode, the diode transitions into blocking state. In blocking state, only very small negative currents can flow through the diode. In the case of high negative voltages, breakthrough effects occur, in which high negative currents also flow. Subsequently, highly negative voltages need to be prevented.

$$V_D = \begin{cases} V_{D,F} & \text{conducting} \\ V_D & \text{blocking} \end{cases}$$

$$I_D = \begin{cases} I_D & \text{conducting} \\ 0\,\text{A} & \text{blocking} \end{cases}$$

$$P_D = V_D \cdot I_D$$

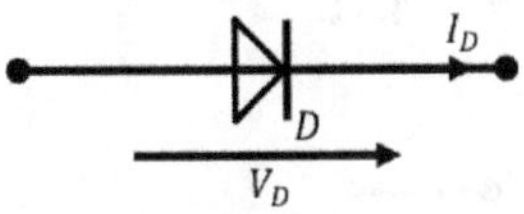

Figure 4.39
Diode schematic

Transistors can be controlled actively and are utilized in power converters and regulators as active switches. These are employed to switch between different converter or regulator states. By applying a voltage between the basis and the emitter contacts, the operational state of a transistor and the current flow between the collector and the emitter contacts can be controlled. Transistors feature three operational states. In the blocking state, transistors do not conduct current, and in the conducting state, transistors conduct current between the emitter and the collector. The conducting state can be subdivided into the linear domain and into the saturation domain. In order to minimize power dissipation, transistors are only operated either in the blocking or in the saturation domain in power electronics. The schematic of a transistor has been visualized in Figure 4.40.

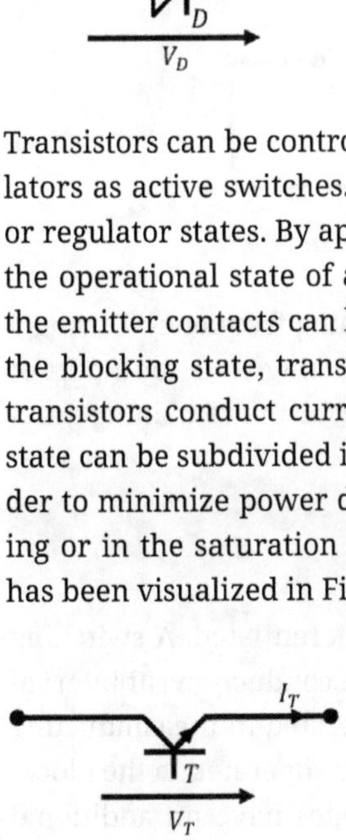

Figure 4.40
Transistor schematic

For simplification reasons, transistors are usually considered as switches in power electronics. In the blocking state, transistors usually do not conduct currents, but can feature any positive voltage. In the conducting state, usually only small voltages occur at transistors, but any positive current can be conducted. The power at a transistor can be defined by the product of voltage and current.

$$V_T = \begin{cases} V_{Forward} & C = 1 \\ V_T & C = 0 \end{cases}$$

$$I_D = \begin{cases} I_T & C = 1 \\ 0\,\text{A} & C = 0 \end{cases}$$

$$P_T = V_T \cdot I_T$$

MOSFETs are often utilized in power electronics. The utilization of snubber circuits is highly recommended. These are implemented in parallel with switching components in a converter topology.

The operation of power converters and regulators is based on the switching between different converter states utilizing transistors. The state of a transistor is controlled by the value of a control signal, which has to be generated by a control logic. The amplitude of such a control signal needs to rise to a value that is high enough to bring the transistor into conducting state. The different operational states of a transistor operated as a switch in power electronics are visualized in Figure 4.41.

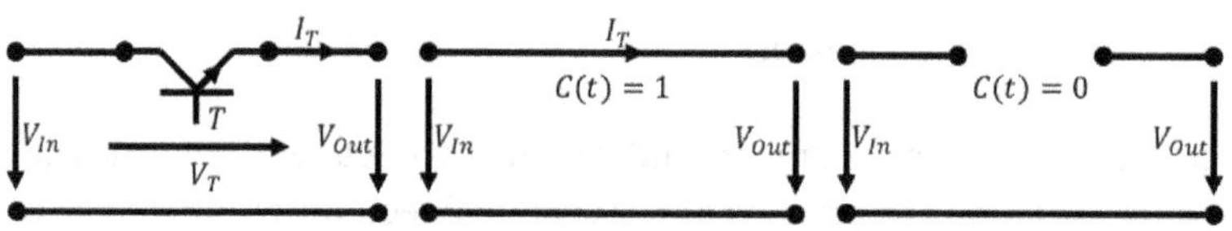

Figure 4.41 Transistor states utilized in power electronics

The duration of a switching interval is defined by the switching frequency. During a switching interval, a transistor switch is operated in a conducting state for a part of the interval, and, after this, it transitions into a blocking state for the remaining duration of the switching interval.

$$T_S = \frac{1}{f_S}$$

Single-pulse topologies and dual-pulse topologies can be differentiated. A switching interval of a single-pulse topology can be subdivided into one conduction subinterval and one blocking subinterval. All transistor switches are operated in the conducting state during the first subinterval and all transistor switches are operated in the blocking state during the second subinterval. Dual-pulse topologies have an additional conduction state. They have to have a minimum number of two transistor switches, which are grouped into two groups and have to be operated in alternating switching mode. It is important that the switching in dual-pulse topologies happens uniformly. A switching interval of a dual-pulse topology can be subdivided into 4 subintervals. The first two subintervals and the last two subintervals each have a cumulated duration of half of a switching interval. During the first subinterval, the first group of transistor switches is operated in the conducting state, and the second group of transistor switches is operated in the blocking state. During the second subinterval, all transistor switches are operated in the blocking state. The first group of transistor switches is operated in the blocking state, and the second group of transistors is operated in the conduction state during the third subinterval. Both groups of transistors are operated in the blocking state during the last subinterval. Effectively, dual-pulse topologies feature a doubled switching frequency compared to a single-pulse topology, which results in a smoother output voltage. The progression of the control signal during a switching interval is visualized for single-pulse and dual-pulse topologies in Figure 4.42.

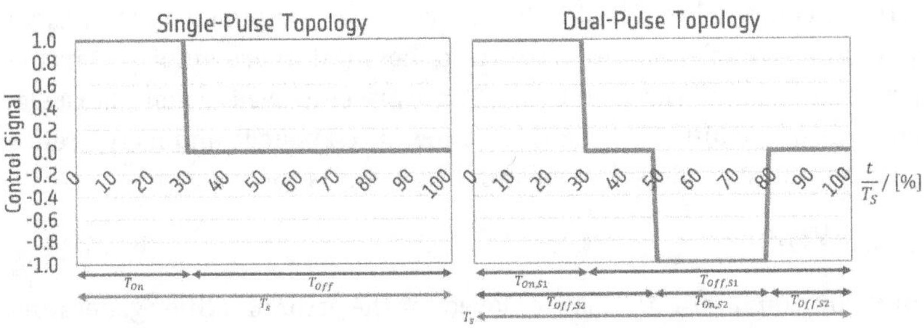

Figure 4.42 Progression of the control signal for a switching subinterval in single-pulse and dual-pulse topologies for a duty cycle of 30%

The parameter duty cycle can be defined as the ratio between the duration of one conduction subinterval and the duration of a total switching interval. It can feature a value between 0 and 1 for single-pulse topologies and between 0 and 0.5 for dual-pulse topologies. The duration of one conduction subinterval can be calculated as the product of the duty cycle and the duration of the switching interval. In single-pulse-topologies, the duration of the blocking subinterval can be calculated as difference between the duration of one switching interval and one conduction subinterval. The duration of the blocking subintervals of dual-pulse topologies can be calculated as difference between half the duration of one switching interval and the duration of one conduction subinterval.

$$D_{PE} = \frac{T_{On}}{T_s}$$

$$T_{On} = D_{PE} \cdot T_s$$

$$T_{Off,Single-Pulse} = T_s - T_{On} = T_S - D_{PE} \cdot T_s = T_S \cdot (1 - D_{PE})$$

$$T_{Off,Dual-Pulse} = \frac{1}{2} \cdot T_s - T_{On} = \frac{1}{2} \cdot T_S - D_{PE} \cdot T_s = T_S \cdot \left(\frac{1}{2} - D_{PE}\right)$$

In the operation of DC/DC converters, three conduction modes can be differentiated. In the operation at low duty cycles, it is possible that the converter output current runs out during the transistor blocking subinterval, before the beginning of a new transistor conduction subinterval. The operation at a duty cycle, where the output current features a value greater than zero ampere at the end of a transistor blocking subinterval, is referred to as Continuous Conduction Mode (CCM). The operation at a duty cycle, where the output current exactly reaches a value of zero ampere at the end of a transistor blocking subinterval, is referred to as the Boundary Conduction

Mode (BCM), and the operation at a duty cycle, where the output current reaches a value of zero ampere before the beginning of the new transistor conduction subinterval, is referred to as Discontinuous Conduction Mode (DCM).

Error amplifiers are commonly utilized in the converter and regulator control as an incremental part. For example, (Main) Error Amplifiers (MEA) are employed in the control of solar regulators and DC Main buses. These are employed to generate a signal in the magnitude of the difference between a measured voltage and its reference value. This value is usually employed as the input signal for a control logic.

$$V_{Error} = V_{Meas.} - V_{Ref.}$$

The term Main Error Amplifier (MEA) is employed for the error amplifier generating the difference signal for the main bus voltage. A differential amplifier topology has to be employed for the implementation of an error amplifier. A simplified visualization of an error amplifier is visualized in Figure 4.43. For the real implementation, additional parts, such as resistors, are required in order to tune the error amplifier.

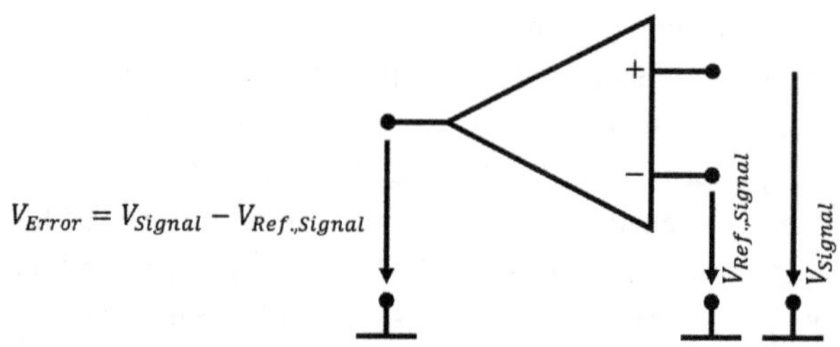

Figure 4.43 Working principle of error amplifiers in DC/DC converters

The buck converter is a common single-pulse DC/DC converter topology that features the property to step down the voltage at the converter input terminals to a smaller voltage at the converter output terminals. The output voltage of a buck converter can feature values between zero volts at a duty cycle of zero and the value of the converter input voltage at a duty cycle of one. This converter type consists of a transistor, a diode, and an inductor. The schematic of a buck converter is visualized in Figure 4.44.

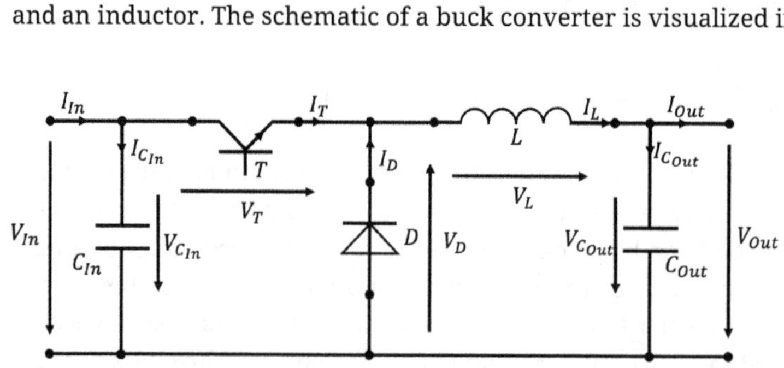

Figure 4.44 Buck converter topology schematic

During the transistor conduction subinterval, the diode is operated in the blocking state, and the energy is transferred from the converter input, through the inductor, to the converter output. The inductor voltage approximately has a value of the difference between the converter input and the output voltage during this subinterval. During the transistor blocking subinterval, the diode is operated in conducting state, and the energy stored inside the magnetic field in the inductor is dissipated in order to drive the inductor current. Energy is not transferred from the converter input to the converter output during this subinterval. The inductor voltage can be determined employing the voltages in the converter output mesh during this subinterval. The inductor current results from the inductor voltage and can be calculated as the time integral of the division of the inductor voltage by the inductance. During the transistor conduction subinterval, the inductor current increases, and during the transistor blocking subinterval, it decreases. In steady state, the duty cycle of a buck converter is equal to the division of the output voltage by the input voltage, and it is also equal to the division of the mean input current by the mean output current.

$$V_L \approx \begin{cases} V_{In,Converter} - V_{Out,Converter} & \text{transistor conduction} \\ -V_{Out,Converter} & \text{transistor blocking} \end{cases}$$

$$I_L(t_0 + \Delta t) = \frac{1}{L} \cdot \int_{t_0}^{t_0+\Delta t} V_L(\tau)d\tau + I_L(t_0)$$

$$D_{Buck} = \frac{V_{Out}}{V_{In}} = \frac{\overline{I_{In}}}{\overline{I_{Out}}}$$

The boost converter topology is a common single-pulse DC/DC converter topology that has the property to step up the voltage at the converter input terminals to a greater voltage at the converter output terminals. This converter type can feature output voltages equal to the input voltage at a duty cycle of zero or greater than the input voltage at higher values of the duty cycle. It consists of a transistor, a diode, and an inductor. The boost-converter schematic is visualized in Figure 4.45.

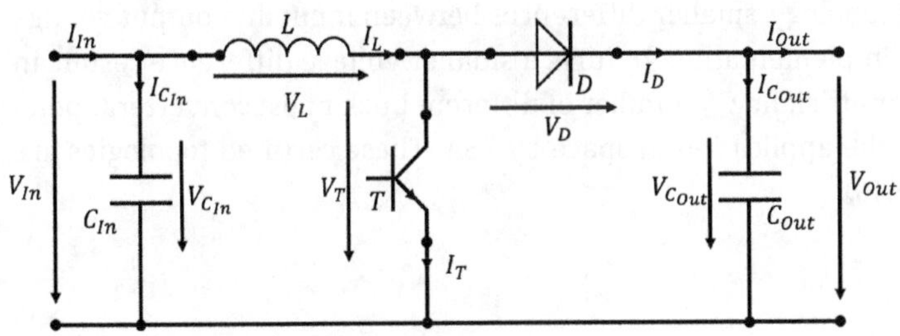

Figure 4.45 Boost converter topology schematic

During the conduction subinterval of the transistor, the diode is operated in the blocking state and the inductor is shortened to the low potential of the converter topology. The inductor voltage can be calculated employing Kirchhoff's law for the voltages in the converter input mesh and approximately has a value equal to the converter input voltage during this subinterval. During the second subinterval, the transistor is operated in blocking state and the diode is operated in conducting state. The inductor voltage approximately has a value equal to the difference between the converter input and output voltage during this subinterval. The inductor current can be calculated employing the time integral over the division of the inductor current by the inductance. It increases during the first subinterval and decreases during the second subinterval. Electrical power is only transferred from the converter input to the converter output during the second subinterval. In steady state, the duty cycle of a boost converter is equal to 1 subtracted by the division of the input voltage by the output voltage, and it is also equal to 1 subtracted by the division of the mean output current by the mean input current.

$$V_L \approx \begin{cases} V_{In,Converter} & \text{transistor conduction} \\ V_{In,Converter} - V_{Out,Converter} & \text{transistor blocking} \end{cases}$$

$$I_L(t_0 + \Delta t) = \frac{1}{L} \cdot \int_{t_0}^{t_0+\Delta t} V_L(\tau) d\tau + I_L(t_0)$$

$$D_{Boost} = 1 - \frac{V_{In}}{V_{Out}} = 1 - \frac{\overline{I_{Out}}}{\overline{I_{In}}}$$

The buck-boost converter topology is a combination of the buck and the boost converter topologies. Depending on the operational duty cycle, buck-boost converters can act as buck converters or as boost converters, featuring output voltages from zero volts to voltages greater than the converter input voltage. Buck converter topologies and boost converter topologies are typically designed in a way that they have a voltage difference in the magnitude of 20% between input and output [97]. Employing a buck-boost converter topology, smaller differences between input and output voltage can be utilized in the implementation. In turn, a smaller voltage difference results in an increased converter efficiency. A number of different buck-boost converter topologies was certified for the application in space by ESA. These certified topologies are visualized in Figure 4.46.

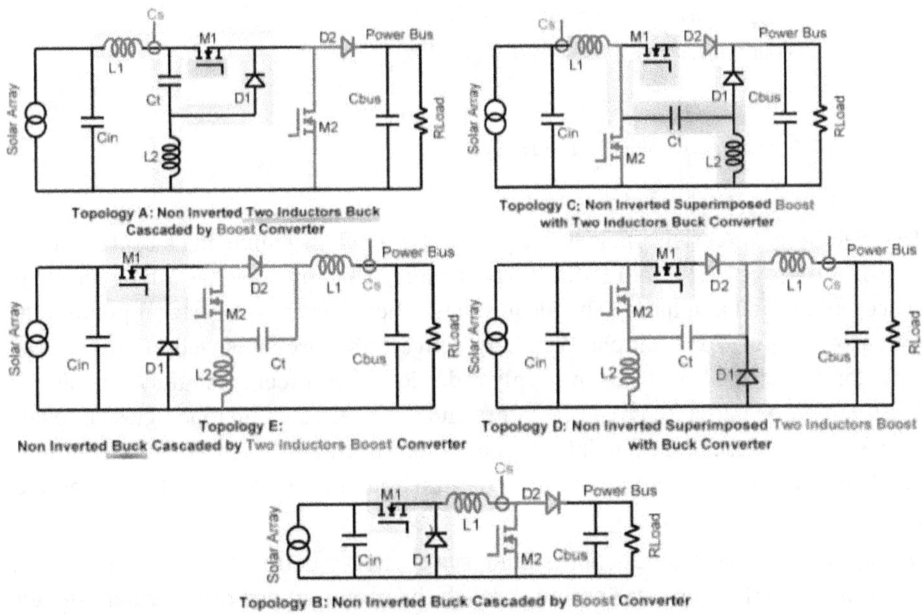

Figure 4.46 Space certified buck-boost topologies (image: © Mourra et al. [217])

Buck-boost converter topologies are of potential interest as solar array regulators. The output voltage of solar arrays varies resulting from radiation induced degradation, from temperature effects, and from irradiance variations. The utilization of a buck-boost converter topology allows to track the Maximum Power Point over a wider voltage range than it would be possible with a buck or a boost converter topology.

In the following, the working principle of the buck-boost converter topology B (Figure 4.46) will be discussed exemplary. The buck-boost converter topology B is a single-pulse converter topology employing two simultaneously switched transistors, two diodes, and one inductor. During the first subinterval, both transistors are operated in the conducting state and both diodes are operated in the blocking state. Subsequently, the inductor is shortened to the converter's low potential, and the electrical power is not transferred from the converter input to the output during this subinterval. Because of this, the inductor voltage approximately has a value equal to the converter input voltage during this subinterval. Energy is stored inside the inductor, but not transferred to the converter output during this subinterval. During the second subinterval, both transistors are operated in blocking state and both diodes are operated in conducting state. As a result, the inductor voltage has a value equal to the negative output voltage during this subinterval and electrical power is transferred from the inductor to the converter output. The value of the inductor current can be determined employing the time integral of the division of the inductor voltage by the inductance. During the first subinterval, the inductor current increases, and during the second sub-interval it decreases.

$$V_L \approx \begin{cases} V_{In,Converter} & \text{transistor conduction} \\ -V_{Out,Converter} & \text{transistor blocking} \end{cases}$$

$$I_L(t_0 + \Delta t) = \frac{1}{L} \cdot \int_{t_0}^{t_0+\Delta t} V_L(\tau) d\tau + I_L(t_0)$$

The Weinberg converter topology is a dual-pulse topology employing a center-tapped transformer together with a flyback transformer. This topology has a long legacy in space exploration and has a galvanical isolation between the converter input and the converter output. For example, Weinberg converters were employed to operate battery storages of the ISS EPS or to supply radar loads. Weinberg step-down topologies are available as battery charge regulators and Weinberg step-up topologies are available as battery discharge regulators. In addition to this, also bidirectional Weinberg topologies are available that combine the functionalities of battery charge and discharge regulators.

During the transistor conduction subintervals, energy is transferred from the converter input to the converter output through the center-tapped transformer, and energy is also stored inside the magnetic field of the flyback transformer. The transistor T_0 and the diode D_1 are operated in the conducting state, while the transistor T_1 and the diodes D_0 and D_3 are operated in the blocking state during the first subinterval. During the third subinterval, the transistor T_1 and the diode D_0 are operated in the conducting state, while the transistor T_1 and the diodes D_1 and D_3 are operated in the blocking state. During the subintervals two and four, both transistors, as well as the diodes D_0 and D_1, are operated in the blocking state, while the diode D_3 is operated in the conducting state. Energy is only transferred from the magnetic field inside the flyback transformer to the converter output during these subintervals, but not from the converter input. Inside the center-tapped transformer, no energy is stored during the operation of a Weinberg converter. For the operation of a Weinberg converter, it is important that the switching happens uniformly. Subsequently, both transistor conduction intervals need to have equal durations, and also both transistor blocking subintervals need to have equal durations.

In literature, a number of converter schematics are referred to as Weinberg converters that have slight differences. The Weinberg converter schematic from the power topologies handbook from Texas Instruments is visualized in Figure 4.47. There are also more advanced topologies available that have additional advantages like bidirectional power flow.

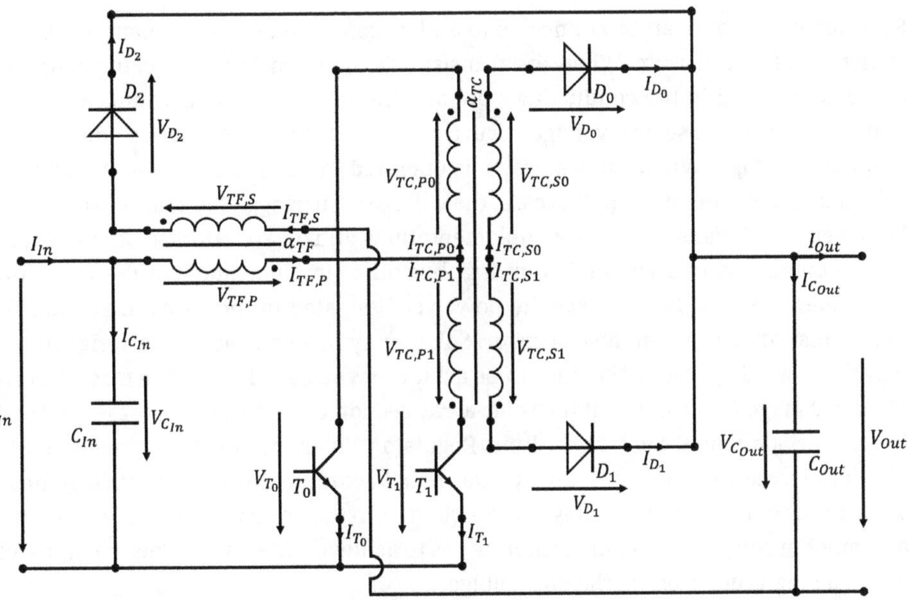

Figure 4.47 Weinberg converter topology schematic

The Sequential Switching Shunt Regulator (S3R) topology is the most common regulator type for spacecraft solar power. It features a high reliability, high efficiency low mass, and a simple schematic, but it also has the disadvantage that Maximum Power Point Tracking (MPPT) is not possible. In the implementation with S3R, solar arrays are subdivided into sections, and each section is connected to one of multiple S3R input channels. The schematic of one S3R channel is visualized in Figure 4.48.

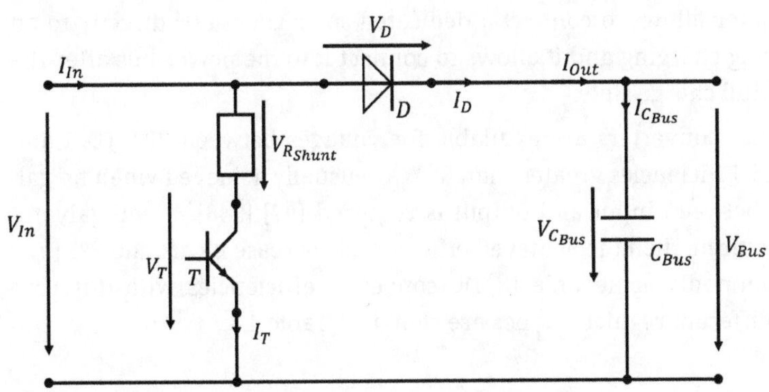

Figure 4.48 Single channel Sequential Switching Shunt Regulator schematic

Solar array sections can be connected to and disconnected from the main bus by the switching of a transistor. When the transistor is operated in the blocking state, the diode is operated in the conducting state and the input channel is connected to the main bus. In this case, the voltage occurring at a solar array section is equal to the main bus voltage. When the transistor is operated in conducting state, the diode is operated in blocking state. In this case, the voltage occurring at the solar array section has a very small value. There are implementations utilizing resistors to dissipate electrical power during a channel shunting, but other implementations do not utilize additional resistors. In this case, the power is dissipated in the parasitic resistances of the solar arrays and in the transistor. Solar array sections have to be designed in a way that a good operational point can be achieved when a voltage in the magnitude of the bus voltage is occurring at the solar array section terminals when these are regulated by a Sequential Switching Shunt Regulator. Solar array sections regulated by S3Rs are typically designed in a way that their Maximum Power Point voltage has a value around the main bus voltage. Subsequently, in the design of solar array sections, a number of solar cells is connected in series to achieve a Maximum Power Point voltage in a magnitude close to the bus voltage.

Sequential Switching Shunt Regulators usually have multiple input channels. Each channel can connect or disconnect a solar array section in parallel to the main bus. In case that multiple S3R channels are connecting solar array sections to the main bus, these sections are connected in parallel to the bus. The S3R output current is equal to the sum of the output currents of the individual channels in this case.

DC/DC converter-based charge regulators are usually implemented employing a step-down topology in space EPSs. Battery discharge regulators based on DC/DC converters are usually implemented employing a step-up topology in space EPSs. Energy storages can also be connected to a power bus employing line regulators such as the S4R regulator. An S4R regulator allows to connect a dedicated solar generator directly to an energy storage during charging and it allows to connect it to the power bus after the storage reached its full charge state.

Space certified DC/DC converters are available for voltages between 22 V (DC) and 120 V (DC) [97] [148]. Efficiencies greater than 85% are usually achieved when no galvanically isolation between input and output is required [97] [148]. When galvanically isolation is required, then the achieved efficiencies decrease by around 3% [97]. An overview of commonly achievable DC/DC converter efficiencies with different voltage levels and different regulator types are shown in Table 4.3.

Table 4.3 Efficiency of different DC/DC converter topologies at different voltage levels [97]

Voltage			20 V	30 V	40 V	50 V	70 V	100 V
Solar array regulator	**Direct Energy Transfer (DET)**	**Serial regulator**	94.0%	95.5%	96.5%	97.0%	97.5%	98.0%
		Shunt regulator	94.0%	96.0%	97.0%	97.7%	98.2%	98.8%
	DC/DC converter with MPPT	**Buck Converter**	93.0%	94.1%	95.2%	96.0%	96.6%	97.0%
		B2R converter	93.0%	94.0%	95.0%	95.5%	96.0%	96.5%
Battery charge/dis-charge regulator		**Buck converter**	90.0%	92.0%	93.5%	95.0%	96.0%	97.0%
		Boost converter	87.0%	91.0%	93.0%	94.5%	96.0%	97.0%
Battery charge regulator		**Shunt regulator**	92.0%	94.0%	95.5%	96.5%	97.5%	98.0%

Power dissipation inside the internal elements of the converter topology results from the conversion of a certain voltage level to another voltage level. The internal power dissipation accounts for the converter losses and reduces the converter efficiency. Following this, the converter output power usually features a smaller value than the converter input power.

$$P_{Converter,Output} = P_{Converter,Input} \cdot \eta_{Converter}$$

When a certain amount of power has to be supplied to a system at the converter output, the converter input power needs to be increased by the magnitude of the power dissipation in order to meet the load demand requirement.

$$P_{Converter,Input} = \frac{P_{Converter,Output}}{\eta_{Converter}}$$

The power supplied to a DC/DC converter is equal to the sum of the internal dissipated power and the converter output power. Following this, the amount of internal dissipated power can be calculated as the difference between the converter input and the output power.

$$P_{Converter,Dissipation} = P_{Converter,Input} - P_{Converter,Output}$$

$$P_{Converter,Dissipation} = P_{Converter,Input} - P_{Converter,Input} \cdot \eta_{Converter}$$

$$P_{Converter,Dissipation} = P_{Converter,Input} \cdot (1 - \eta_{Converter})$$

DC/DC converters can usually be operated within a temperature range between −55 °C and +75 °C in space applications [148]. Another requirement for the development of DC/DC converters for the application in space is to minimize volume and mass of these systems. This can be achieved by an optimization of the individual regulators or by combining the functionalities of different regulators, such as charging and discharging, into a single unit. Combined converter units allow reductions in the overall converter volume and mass in an EPS.

Multi-junction solar cells have a greater parasitic capacity than other solar cell types. Resulting from this, current spikes can occur when the solar cell strings consisting of multi-junction solar cells are connected to the main bus by a solar array regulator. A possible countermeasure for this is the characterization of the solar cell string capacity. Solar cell strings featuring a low capacity can be employed for low MEA error voltages, and, following this, would be switched more often. In turn, solar cell strings featuring a high capacity would be disconnected and connected to the main bus less often. A higher voltage stabilization can be achieved by employing high and low power solar array sections. High-power solar array sections would be switched on and off less often and low power sections would in turn be switched on and off more often.

4.10 Control Systems

Control systems are crucial subsystems in the operation of space EPSs and their subsystems. A control system needs to be specifically designed for a certain hardware topology in order to meet operational goals and requirements. Requirements should be met in and outside the nominal operational state. In space EPSs, different types of control systems are employed. Common use cases are the control of DC/DC converters, power regulators, power buses, or Maximum Power Point Tracking.

DC/DC converter control can be implemented employing analog or digital systems. Common methods are Feedforward-Proportional-Integral-Derivative (FPID) control or Fuzzy Logic Control (FLC). As control parameter, a voltage or current value inside a converter topology can be employed. DC/DC converters are usually controlled with reference to the converter output voltage or current.

Current control has been employed in the DDCU converter units on board of the International Space Station (ISS) and in the solar array regulators on board of the ROSETTA spacecraft. It is also recommended to use current control methods in battery regulators because too high battery currents can significantly degrade the battery performance and cycle life.

Converter control loops consists of multiple steps. In the first step, the difference between the measured control parameter and its reference value has to be calculated, which is usually also referred to as the control error. In the next step, the new value of the control parameter reference value is calculated employing a dedicated control method. The resulting reference value is employed in the last step in the determination of the operational duty cycle value.

The simplest control method for DC/DC converters is to operate them at a constant duty cycle. Based on the duty cycle value, the control signal is generated that is in the next step employed to control the switching of semiconductor switches. However, this control method does not feature compensations against fluctuations in the input voltage or in the load demand. Subsequently, good control results can only be achieved in optimal cases. The topology of such a control system is visualized in Figure 4.49.

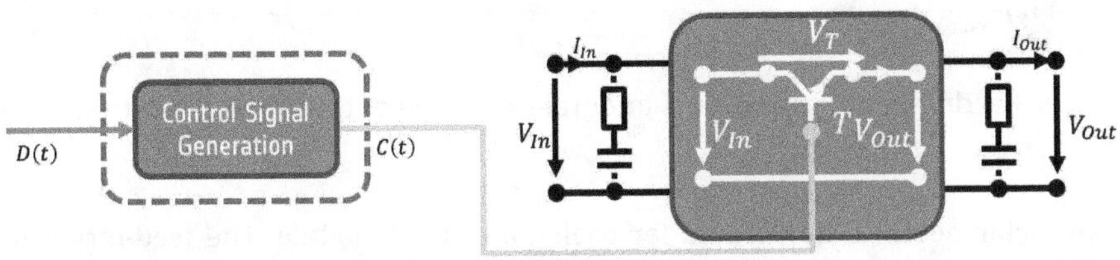

Figure 4.49 Basic DC/DC converter control system

More complex converter control methods control the operation of a DC/DC converter with reference to a reference value for the converter output voltage or current that the converter should achieve in steady state. In this case, the duty cycle value is not directly defined. Instead, it is calculated in a preceding step with reference to the reference voltage or current and usually also with reference to measurements of the converter state.

Transfer functions can be utilized for this purpose. The converter transfer function can be defined as the relation between the duty cycle and the converter input and output voltages or currents. It is converter topology specific and is usually determined from a mean value analysis of the converter output parameters. In control engineering, the transfer function can be utilized in order to determine the value of the duty cycle for a measured value of a converter input parameter and the reference value of a converter output parameter.

$$D_{PE} = f_{Transfer,V}(V_{Ref.,Out}, V_{In,Converter}) = f_{Transfer,I}(I_{Ref.,Out}, I_{In,Converter})$$

A common method for the converter control is a combination of Feedforward-Proportional-Integral-Derivative (FPID) control. These types of control algorithms are widely applied in the operation of DC/DC converters for their good stability, satisfying and stable voltage regulation, as well as fair implementation complexity properties. The structure of an FPID DC/DC converter voltage control system is visualized in Figure 4.50.

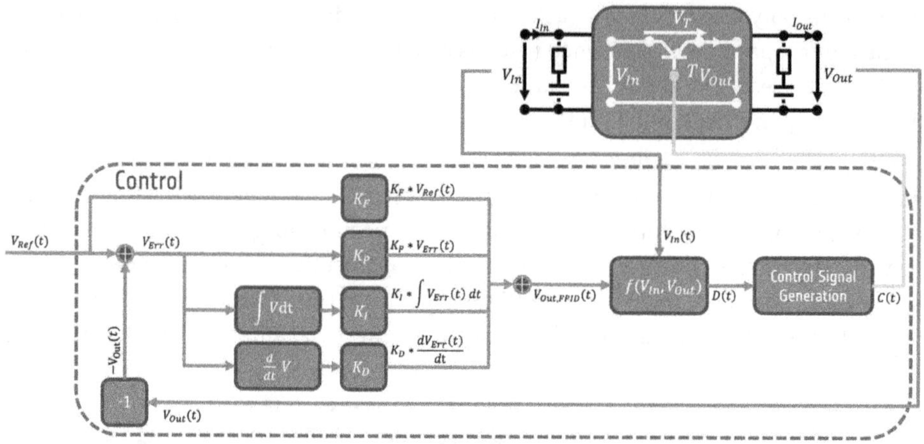

Figure 4.50 Feedforward-Proportional-Integral-Derivative output voltage control system

One gain factor needs to be defined for each control submodule. The feed-forward submodule takes the value of the converter output reference voltage as the input value. Subsequently, no measurements from the converter implementation are required for the utilization of a feed-forward submodule. The feed-forward submodule is utilized in order to define the mean value of the converter output voltage, which the converter should achieve in steady state. The output value of the feed-forward submodule is calculated as the product of the feed-forward gain factor and the reference voltage. Usually, a value of one is utilized for the feed-forward gain factor because other values would result in a steady state error.

$$V_{Out,F}(t) = K_F \cdot V_{Ref.}(t)$$

The proportional, integral and derivative control submodules take the voltage error between the converter output reference voltage and the measured converter output voltage as the input value. The error value is employed to compensate control errors. It is calculated as the difference between the output reference voltage and the measured output voltage.

$$V_{Error}(t) = V_{Ref.,Out}(t) - V_{Out}(t)$$

The output of the proportional control module can be calculated as the product of the voltage error and the proportional gain factor. An increase in the proportional gain factor results in a decrease of the steady state error, but it does not improve the dynamic behavior.

$$V_{Out,P}(t) = K_P \cdot V_{Error}(t)$$

The integral control submodule takes the voltage error as input and calculates a time integral of the voltage error. The integral gain factor is employed in order to amplify the output value of the integral submodule. Its output is equal to the product of the integral gain factor with the time integral over the voltage error. The integral submodule allows to reduce the noise in the converter output voltage, but it deteriorates the dynamic behavior of the converter.

$$V_{Out,I}(t) = K_I \cdot \int_0^t V_{Error}(\tau)\, d\tau$$

The derivative control submodule also takes the voltages error as input. The derivative gain factor is employed in order to amplify the time derivative of the voltage error. The output of a derivative control module is equal to product of the derivative gain factor with the time derivative of the error voltage. The application of derivative control submodule allows to decrease overshoot.

$$V_{Out,D}(t) = K_D \cdot \frac{dV_{Error}(t)}{dt}$$

The additional implementation of a derivative submodule in a Feedforward-Proportional-Integral control system allows to achieve high efficiencies. However, the derivative control module usually increases the noise in the output voltage.

The output value of the Feedforward-Proportional-Integral-Derivative control system is equal to the sum of the individual output values of the four control sub-modules. It is also possible to utilize only some of these sub-modules to control a DC/DC converter. Popular control methods are FP Control, which does not utilize the integral and derivative control submodule, or FPI Control, which does not utilize the derivative submodule. For these implementations, only the required submodules need to be implemented.

$$V_{Out,FPID}(t) = V_{Out,F}(t) + V_{Out,P}(t) + V_{Out,I}(t) + V_{Out,D}(t)$$

Three methods are available in order to determine the values of the PID gain factors. Employing the Ziegler-Nichols Frequency Domain (ZNFD) method, the gain factors are determined in three steps. In the first step, the integral and derivative control modules need to be disabled. Following this, the value of the proportional gain factor needs to be determined at which the oscillation in the converter output voltage transitions into a critical state. The critical state is defined by a continuous oscillation that is neither decaying nor building up. The value of the proportional gain factor in critical converter state is referred to as K_{ZNFD} in the following. In addition to this, also the period time of the oscillation in critical state needs to be determined, which is referred to as T_{ZNFD} in the following. Employing these two factors, Table 4.4 is available to define the proportional, integral and derivative gain factors in the third step. Other similar methods for the determination of gain factors are the Damped Oscillation Method (DOM) and the Good Gain Method (GGM).

Table 4.4 Ziegler-Nichols Frequency Domain method for gain factor determination [253] [260]

Controller type	K_F	K_P	K_I	K_D
P	1	$K_P = \frac{1}{2} \cdot K_{ZNFD}$	-	-
PI	1	$K_P = \frac{45}{100} \cdot K_{ZNFD}$	$T_I = \frac{8}{10} \cdot T_{ZNFD}$ $K_I = \frac{K_P}{T_I}$	-
PID	1	$K_P = \frac{6}{10} \cdot K_{ZNFD}$	$T_I = \frac{5}{10} \cdot T_{ZNFD}$ $K_I = \frac{K_P}{T_I}$	$T_D = \frac{125}{1000} \cdot T_{ZNFD}$ $K_I = K_P \cdot T_D$

Sequential Switching Shunt Regulators (S3R) require different control systems than switching DC/DC converters because these rely on a different operational principle. Solar generator sections are connected and disconnected sequentially to the main bus by Sequential Switching Shunt Regulators in order to balance the power supplied to the main bus with the load power demand. The electrical power transferred from a solar generator to the main bus is regulated by advancing the total number of solar cell strings that are connected to the main bus. Sequential Switching Shunt Regulators are usually controlled with reference to the bus voltage. A set of error amplifiers is employed in the control of an S3R. A simplified implementation of a control system for a 3-Channel S3R is visualized in Figure 4.51.

The Main Error Amplifier (MEA) is employed as a central module for the calculation of the voltage error between the measured main bus voltage and the main bus reference voltage. The output of the MEA is equal to the difference between both voltages.

$$V_{Error} = V_{MEAS.,Bus} - V_{Ref.,Bus}$$

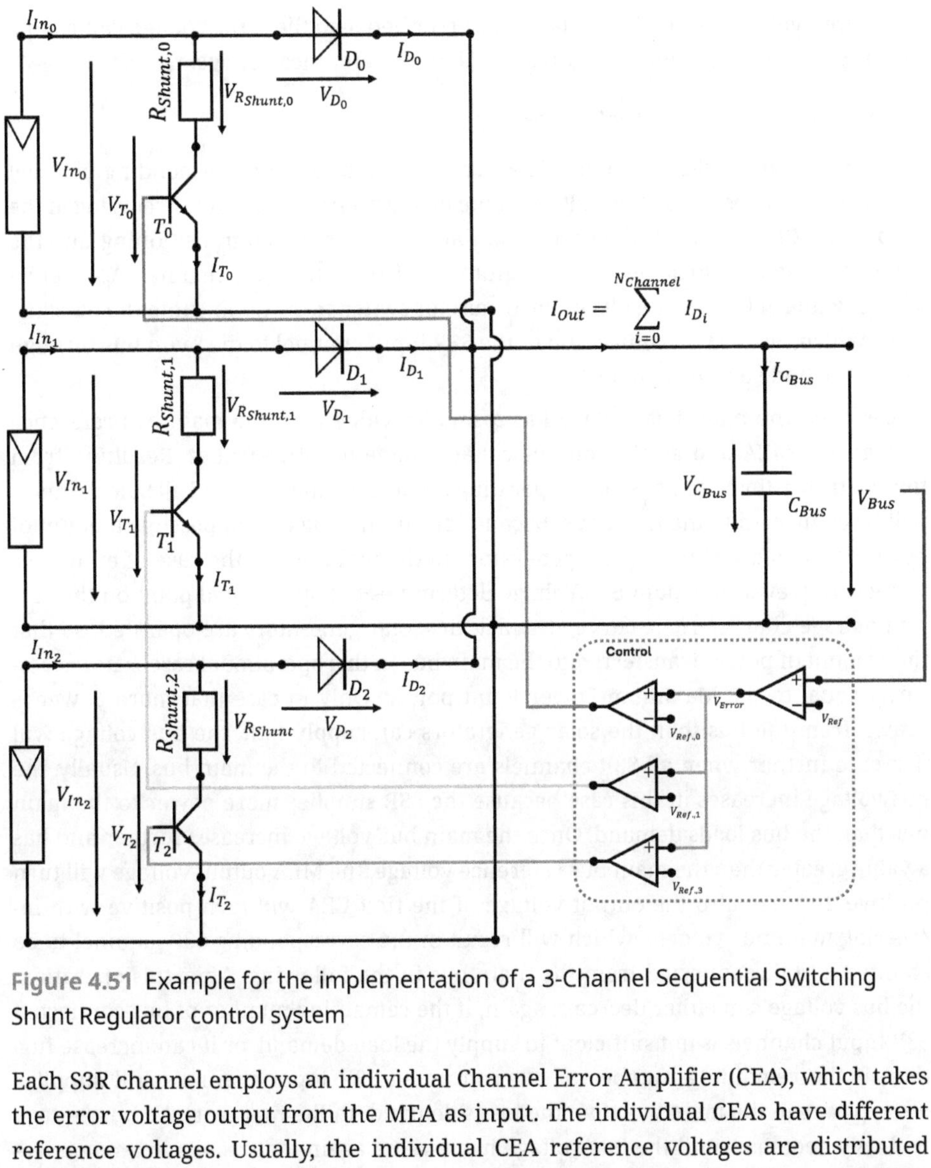

Figure 4.51 Example for the implementation of a 3-Channel Sequential Switching Shunt Regulator control system

Each S3R channel employs an individual Channel Error Amplifier (CEA), which takes the error voltage output from the MEA as input. The individual CEAs have different reference voltages. Usually, the individual CEA reference voltages are distributed homogenically between a threshold error voltage below which no actions take place and the maximum main bus error voltage at which all S3R channels are disconnected from the main bus.

$$V_{Ref.,Channel} = [V_{Ref.,1}, V_{Ref.,2}, \ldots, V_{Ref.,N}]$$

$$V_{Ref.,Channel} = \left[\Delta V_{Min}, 1 \cdot \frac{\Delta V_{Max} - \Delta V_{Min}}{N_{channel} - 1} + \Delta V_{Min}, 2 \cdot \frac{\Delta V_{Max} - \Delta V_{Min}}{N_{channel} - 1} \right.$$

$$\left. + \Delta V_{Min}, \ldots, \Delta V_{Max} \right]$$

The output voltage of a CEA is equal to the error between the error voltage value from the output of the MEA and the individual channel reference voltage.

$$V_{Error,Channel} = V_{Error} - V_{Ref.,Channel}$$

The output pins of the individual CEA are connected to the corresponding channel transistor switches. A positive voltage value in a satisfying magnitude occurring at the output of a CEA results in the corresponding transistor switch transitioning into the conduction state, which results in a shunting of the S3R input channel. A negative voltage value at the CEA results in the transistor switch transitioning into the blocking state, which results in a connection of the S3R input channel to the main bus through the series diode of the channel.

In case that the main bus voltage has a smaller value than the main bus reference voltage, the MEA and all CEA output voltages have negative values. Resulting from this, all transistor switches will be operating in the blocking state and all S3R channels will be connected to the main bus. Because of this, the maximum possible amount of power is transferred from solar generators to the main bus in the case of an undervoltage. However, the Main Bus Voltage determines the operational point on the current-voltage characteristic curve at which the solar generators are operated, so that the amount of power transferred to the main bus in this operational case is not necessarily equal to the Maximum Power Point power. Only in case that more power is drawn from the bus than the solar generators can supply to it, the bus voltage will decrease further when all S3R channels are connected to the main bus. Usually, the bus voltage increases in this case because the S3R supplies more power to the main bus than the bus loads demand. Once the main bus voltage increases further and has a value greater than the main bus reference voltage, the MEA output voltage will turn positive. In turn, also the output voltage of the first CEA will turn positive with increasing main bus voltage, which will result in the corresponding S3R channel to be shunted and disconnected from the main bus in the following. After this operation, the bus voltage can either decrease again, if the remaining number of bus-connected S3R input channels is not sufficient to supply the load demand, or it can increase further, if the remaining number of bus-connected S3R input channels still supplies more power to the main bus than the loads demand. In the first case, the bus voltage will only decrease until the bus voltage has a value again that results in a reconnection of the previously disconnected S3R input channel, and in the second case, the remaining S3R channels will be shunted and disconnected from the main bus sequentially until the bus voltage starts to decrease again.

As a result of this control method, all S3R input channels will be connected to the main bus in the case of undervoltages, and no S3R input channel will be connected to the main bus in the case of overvoltages above the maximum bus voltage. In all other cases in between these two extreme conditions, a static number of S3R input channels will be connected to the main bus constantly, a static number of S3R input channels will be shunted and disconnected from the main bus constantly, and only one S3R input channel will be sequentially switched between the bus-connected states and shunted states in static load conditions.

There are also more advanced control methods for S3R. It is possible, for example, to operate one of the S3R input channels in a Pulse-Width Modulation (PWM) way like a DC/DC converter in order to achieve a better voltage regulation.

Maximum Power Point Tracking (MPPT) can be defined as the method of operating power electronics in a way that, at the input, a voltage in the magnitude of the Maximum Power Point voltage of the connected solar arrays occurs. Compared to other operational methods, this method allows to increase the power yield from solar arrays. The location of the Maximum Power Point on the characteristic curve of a solar array varies with the operational conditions defined by parameters like irradiance, temperature, or degradation. It needs to be monitored and tracked dynamically during the operation of a solar array. A number of different algorithms and methods is available for the tracking of the Maximum Power Point. When an MPPT algorithm is implemented in a control system, the converter input voltage should oscillate around the Maximum Power Point voltage of the solar array.

MPPT algorithms work by conducting measurements and advancing the regulator operation with reference to this. They either advance the value of the duty cycle directly or of the control parameter. Depending on the load demand, MPPT algorithms advance the operational point between the Open Circuit Point (OCP) and the Maximum Power Point (MPP). It can be differentiated between MPPT algorithms requiring external measurements and those only requiring internal measurements from the converter.

The MPPT algorithm “Perturb and Observe” (P&O) is the most widely applied MPPT algorithm. This algorithm requires measurements of the converter or regulator output voltage and current as input. It controls the converter output reference voltage with reference to the converter output power. In the first step, the measurements are employed to calculate the converter output power as the product of output voltage and current. In case that the present value of the output power is smaller than the output power during the previous simulation step, the converter output reference voltage is decreased by one control increment, and in case that the present value is greater than or equal to the value of the output power during the previous simulation step, the converter output reference voltage is increased by one control increment. The P&O MPPT algorithm is visualized in Figure 4.52. Other popular MPPT algorithms are “Incremental Conductance” and “Zero Crossing“.

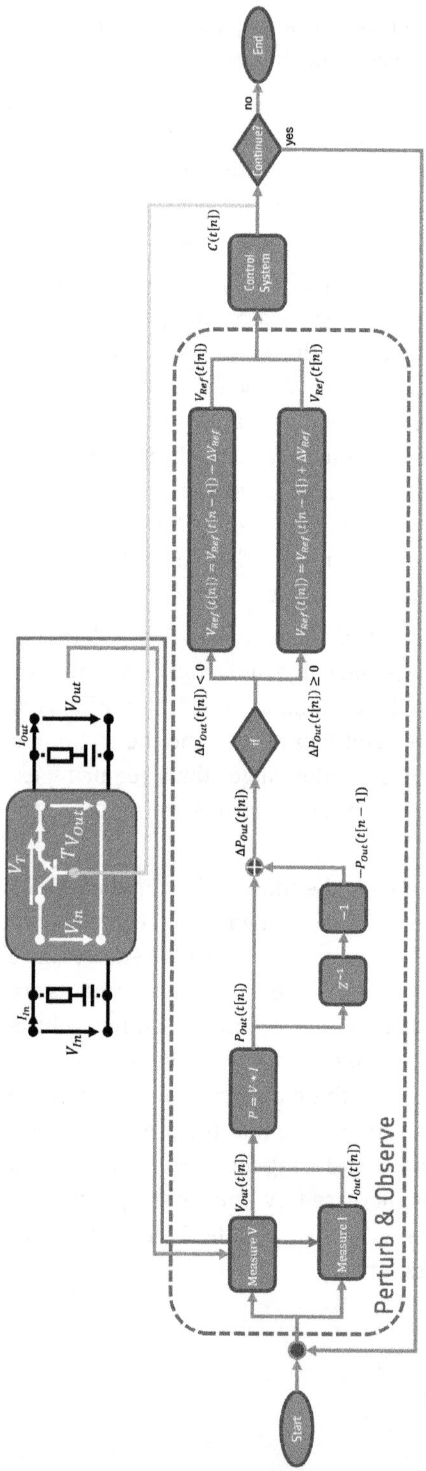

Figure 4.52
Functional overview of the MPPT algorithm "Perturb and Observe" (P&O)

MPPT algorithms have problems in the operation with shadowed solar arrays, which result in non-optimal tracking results. As the power-voltage characteristic curve of a shadowed solar array can have more than one peak, it is possible that the MPPT algorithm tracks a local Maximum Power Point instead of the global one. A simple way to enable the Maximum Power Point Tracking on shadowed solar arrays is to measure the complete characteristic curve from time to time and analyze it to identify the global Maximum Power Point.

Domain control is a widely applied method for the control of DC buses in spacecraft Electrical Power Systems. Employing this method, the operational state of different EPS sub-system is controlled with reference to the Main Bus Voltage (MBV). For a system that is controlled employing domain control, a domain needs to be defined in which it is in operation and outside of which it is not in operation. Space EPSs usually differentiate between the three domains battery discharge, battery charge, and solar generator shunting. An example for a 3-domain controlled spacecraft EPS is visualized in Figure 4.53. An example for the design of the control domains is visualized in Figure 4.54.

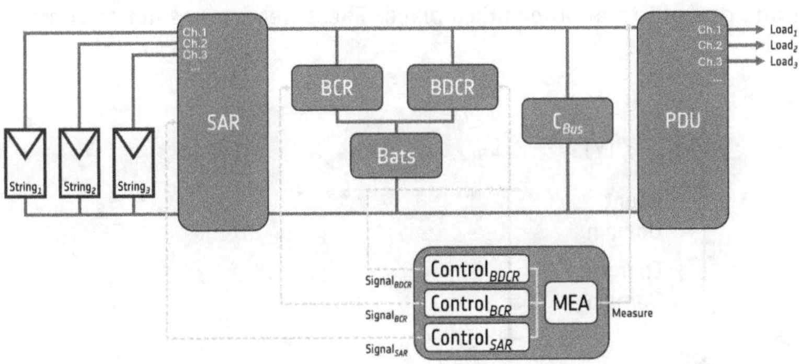

Figure 4.53 Example for a spacecraft EPS controlled by a 3-domain control

An EPS is operated in the discharge domain when the bus voltage has a smaller value than the reference voltage (Figure 4.54). Such low bus voltages result from a condition, in which the load demand is greater than the power generation capacity of the solar generators. The discharge domain is defined by negative MEA output voltages. When an EPS is operated in the discharge domain, the Solar Array Regulator (SAR) transfers as much power as possible to the main bus, the Battery Discharge Regulator (BDCR) is in operation, and the Battery Charge Regulator (BCR) is out of operation. Primary power is transferred to the bus from the solar generators, if solar energy is available, and additional secondary power is transferred from the energy storages to the bus.

The EPS is operated in the charging domain when the bus voltage has a value that is greater than the main bus reference voltage and smaller than the threshold voltage of the solar generator shunting domain (Figure 4.54). A bus voltage in this domain results from a condition in which the power generation capacity of solar generators is greater than the load demand, and in which excess power can be stored in energy storages. When an EPS is operated in the charge domain, the Solar Array Regulator (SAR) transfers the maximum possible power to the main bus, the Battery Discharge Regulator (BDCR) is out of operation, and the Battery Charge Regulator (BCR) is in operation. Primary power is transferred from the solar generators to the main bus, and power is drawn from the bus by the loads and energy storages.

The EPS is operated in the solar generator shunting domain when the bus voltage has a value greater than the threshold voltage of the shunting domain (Figure 4.54). A bus voltage in this domain results from a condition in which the power transferred to the bus from solar generators is greater than the sum of the load demand and the energy storage demand. The power transferred to the bus from solar generators is reduced when the EPS is operated in the shunting domain. The Battery Discharge Regulator (BDCR) is operated in disabled state, and the Battery Charge Regulator (BCR) is only operated in enabled state when the energy storage is not in fully charged state. In between the different domains, dead zones are usually implemented in which no transition from one domain to another takes place. These dead zones act as control hysteresis.

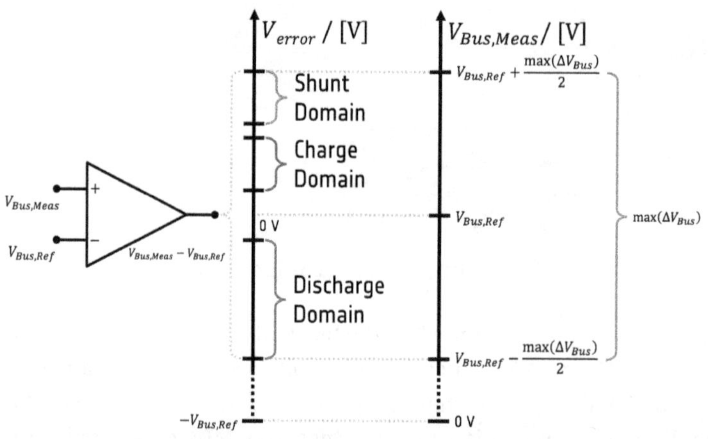

Figure 4.54 Visualization for the domain design in a 3-domain control

It is advantageous to implement a fourth domain for greater negative voltage errors below the battery discharge domain. Once the EPS transitions into this operational domain, different loads should be disconnected from the EPS in order to decrease the load demand and to prevent the bus voltage from collapsing. This operation is typically referred to as load shedding. It is also recommended to subdivide the load shedding domain into different subdomains, so that low-priority loads are disconnected before high-priority loads.

4.11 Battery Storages

Batteries are the most common energy storages in space Electrical Power Systems (EPSs) [48] [97] [98] [109]. Battery storages are non-linear electrochemical systems. There are two types: primary batteries and secondary batteries. The chemical reaction employed by a primary battery is not reversible, and following this, primary batteries cannot be recharged. Contrary to that, secondary batteries employ a reversible reaction, so that this type of batteries can be recharged. A photo of the old ISS nickel-hydrogen batteries is shown in Figure 4.55. A photo of the new ISS lithium-ion batteries is shown in Figure 4.56.

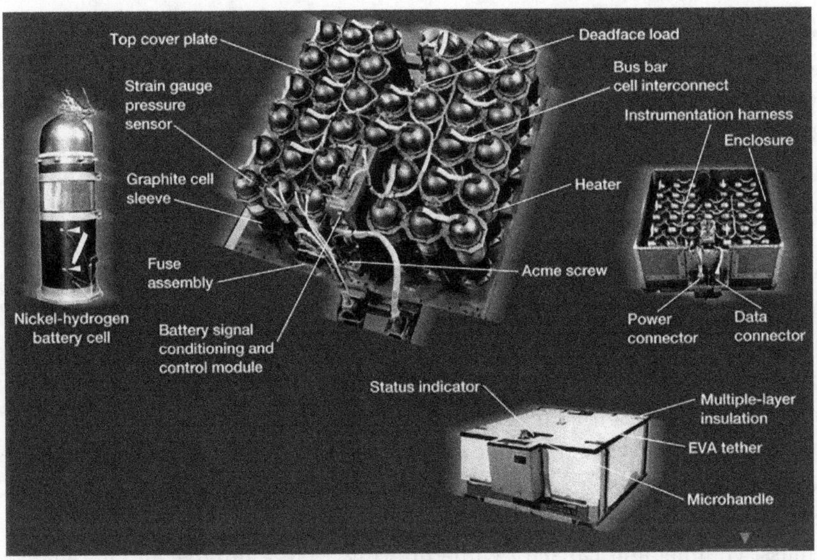

Figure 4.55 Internal view of an ISS nickel-hydrogen battery ORU (image: © NASA [267])

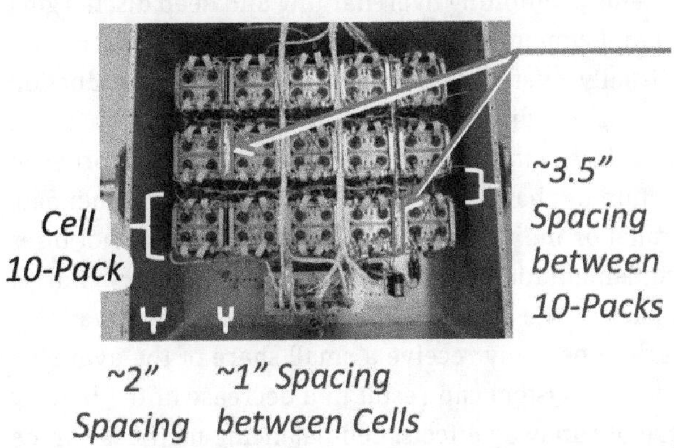

Figure 4.56 Internal view of an ISS lithium-ion Battery ORU (image: © NASA, Dalton/Balcer [270])

Batteries can be classified as short and medium duration storages and only have low power storage capabilities. Different types of batteries are available satisfying high power, high energy or high cycle life requirements. Current research activities are carried out in the fields of low temperature batteries, high temperature batteries, non-flammable batteries, solid-state batteries, and lithium-air batteries.

Single battery cells can be designed in a number of different shapes. Usually, a distinction is made between prismatic, spiral and elliptical cells, and some batteries can also be integrated in so-called pouches. For the application in space, battery cells are usually integrated in metal cases.

Battery cells consist of two electrodes, a chemical or polymer electrolyte, and an ion conducting separator membrane that prohibits the direct movement of electrons between the two electrodes.

The number of charge/discharge cycles that a secondary battery can carry out through its operational life, is depending on a number of variables. Examples for these variables are the employed secondary battery technology, the battery temperature, the depth of discharge during operation, or the utilized charging method. A secondary battery can usually complete between several hundreds and several thousands of charge/discharge cycles, while primary batteries can only be discharged once. Common use cases for primary batteries are Launch Vehicles and backup power sources because these are usually only in operation for a short period of time. Secondary batteries are usually implemented as energy storages in solar power-based EPSs. In this case, they are charged by solar generators during sunlit periods and discharged during eclipse periods. Subsequently, batteries must be designed in such a way that they can supply the spacecraft loads for the duration of one eclipse period. For the implementation in a space system, secondary batteries need to meet certain requirements. They need to be reliable, robust, safe and withstand extreme temperatures and high energy radiation. In addition to this, they also need to have a good balance of system mass, volume, and capacity while offering a large number of charge/discharge cycles.

The integration of additional systems prohibiting overcharging and deep discharging offers advantages because the employment of these minimizes the degradation resulting from these conditions. Usually, even battery cells from the same production line and of the same type have slightly different characteristics, which can cause charge imbalances when battery cells are implemented in a battery pack. In order to prohibit imbalances, the application of charge balancing methods is recommended. Batteries can be charged in parallel or individually. Parallel charging does not offer flexibility, but offers a simple implementation. It can lead to unequal battery charge, where one battery or a small part of the batteries receives most of the available charge and the rest of the batteries does only receive a small share of the available charge. Unequal charging of a battery system can result in a decrease of the battery life and it can also result in thermal runaway effects. Cell-balancing methods can be implemented into battery systems in order to prohibit unequal charging in battery systems.

Batteries work with an electrochemical reaction. As a by-product of the chemical reaction, electrical currents can be observed at the battery terminals. Energy is stored as chemical energy inside the chemicals of a battery. Secondary batteries have two electrochemical main reactions, while primary batteries only have one main reaction. When electrical energy is supplied to a secondary battery, one electrochemical reaction is occurring and the electrical energy is stored in the product chemicals of the reaction. Once the electrical energy is drawn from the battery during discharging, another electrochemical reaction is happening inside the battery. This reaction employs the product chemicals of the charging reaction as the reactant chemicals, and delivers the reactant chemicals of the charging reaction as product chemicals. The electrochemical reaction requires the emission or absorption of energy in the magnitude of the reaction enthalpy. During one reaction, energy is emitted, and during the reverse reaction, energy is absorbed.

Charging:

$$AB_{Chem.} + \Delta H_{Chem.} \longrightarrow A_{Chem.} + B_{Chem.}$$

Discharging:

$$A_{Chem.} + B_{Chem.} \longrightarrow AB_{Chem.} + \Delta H_{Chem.}$$

Energy that is released during a chemical reaction at constant pressure is defined as the reaction enthalpy. It can be determined from the enthalpy of the chemicals involved in the reaction.

$$\Delta H_{Chem.} = H_{Chem.,A} + H_{Chem.,B} - H_{Chem.,AB}$$

The total reaction enthalpy cannot be drawn from a battery because a share of the reaction energy converted into thermal energy during the electrochemical reaction. In order to calculate the amount of energy that is converted into thermal energy during the reaction, the entropy change needs to be calculated.

$$\Delta S_{Chem.} = S_{Chem.,A} + S_{Chem.,B} - S_{Chem.,AB}$$

The amount of thermal energy, that is absorbed or released during the electrochemical reaction, is usually referred to as thermal energy transfer and it can be calculated for thermodynamically reversible reactions employing the second law of thermodynamics. It defines the amount of absorbed or released thermal energy based on the temperature and the entropy change. Depending on the utilized battery cell chemistry, thermal energy is absorbed or released during charging and vice versa during discharging. For example, NiCd batteries feature a positive ΔQ_{Therm} value, which results in emissions of thermal energy during charging.

$$\Delta Q_{Therm.} = T \cdot \Delta S_{Chem.}$$

The amount of thermal energy released during a chemical reaction is released as a side product of the reaction. Subsequently, the available amount of energy after the reaction is reduced by the emission of thermal energy. The available amount of energy after the reaction can be calculated utilizing the Gibbs equation. The resulting value is usually referred to as free enthalpy or Gibbs potential.

$$\Delta G_{Chem.} = \Delta H_{Chem.} - \Delta Q_{Therm.} = \Delta H_{Chem.} - T \cdot \Delta S_{Chem.}$$

The electrons involved in the chemical reaction occurring inside a battery can be utilized in electrical circuits. Per molecule, usually one or two electrons are involved in the majority of chemical reactions. The amount of electrical energy that has been supplied to a battery over time can be calculated as the integral of the power supplied to it over time.

$$\Delta W_{Bat,El} = \int_{t_0}^{t_0+\Delta t} P_{Bat}(t)\, dt = \int_{t_0}^{t_0+\Delta t} V_{Bat}(t) \cdot I_{Bat}(t)\, dt = V_{Bat} \cdot I_{Bat} \cdot \Delta t$$
$$= V_{Bat} \cdot \Delta Q_{Bat} = V_{Bat} \cdot n_e \cdot q_e \cdot N_A$$

The voltage that can be measured at the battery terminals results from the entropy and enthalpy change. Hence, the amount of electrical energy and the free enthalpy both need to feature equal values. This circumstance can be utilized in order to determine the terminal voltage of a battery in thermal equilibrium state.

$$\Delta W_{El} = \Delta G_{Chem.}$$

$$V_{Bat,Therm.} \cdot n_e \cdot q_e \cdot N_A = H_{Chem.} - T \cdot \Delta S_{Chem.}$$

$$V_{Bat,Therm.} = \frac{\Delta H_{Chem.} - T \cdot \Delta S_{Chem.}}{n_e \cdot q_e \cdot N_A}$$

The battery voltage is a property of the employed battery technology. Variations in the battery cell design influence parameters like the battery capacity, but the value of the battery voltage cannot be influenced by the cell design.

The capacity of a battery can be defined as its ability to store a certain amount of electrical energy. Usually, not the complete battery capacity is utilized during operation. Instead, a certain operational domain is defined by a maximum and a minimum State of Charge (SoC), in which the battery is usually operated. This is done in order to enhance the cycle life of a battery cell. A special type of power electronics, so-called charge regulators and discharge regulators, can be employed in order to operate a battery. These are utilized to terminate the discharge operation, once the minimum SoC has been reached, and to terminate a charging operation, once the maximum SoC has been reached, as well as to control the charge and discharge operations.

Electrochemical effects have to be considered in the design of battery systems. A common effect is that with the discharge or charge current, the reaction speed inside a battery varies. The operation with small currents leads to slow reactions and the operation with high currents leads to high reaction speeds. Small currents allow to utilize more of the active electrode material, and following this, enable to extract more charge from a battery or to store more charge inside the battery. Contrary to this, high currents feature the potential to degrade batteries.

The current drawn from a battery is usually referred to as the discharge current and it is counted positively in the direction from the battery cell to the battery terminals. The battery charge current is counted positively from battery terminals to battery cells. Negative values for both types of currents indicate the reverse direction of the current flow.

$$I_{Bat,Discharge}(t) = -I_{Bat,Charge}(t)$$

The amount of charge that has been extracted from a battery during a certain time interval, can be calculated utilizing the integral of the discharge current over time. In the following, the amount of charge that is remaining inside a battery at a certain point in time can be calculated from the charge stored in the battery at the beginning of the discharge operation and the amount of the charge extracted during the operation. Positive discharge currents result in a decrease of the battery capacity, and negative discharge currents result in an increase of the battery capacity.

$$Q_{Bat}(t) = Q_{Bat}(t_0) - \int_{t_0}^{t} I_{Bat,Discharge}(\tau)\, d\tau$$

The battery voltage does only allow drawing conclusions regarding the battery state to a limited extent. Because of this, the amount of charge supplied to and drawn from a battery needs to be monitored. A parameter defining the amount of charge that is stored inside a battery at a certain point in time with reference to the total battery capacity is the parameter State of Charge (SoC). The SoC is calculated as the division of the charge, stored inside the battery at a certain point in time, by the total battery capacity. The parameter is usually indicated unitless as a ratio between zero and one. A value of zero indicates a completely discharged battery and a value of one indicates a completely charged battery. The definition in the unit percent is also common.

$$SoC(t) = \frac{Q_{Bat}(t)}{Q_{Bat,0}} = \frac{Q_{Bat}(t_0) - \int_{t_0}^{t_0+\Delta t} I_{Bat,Discharge}(\tau) d\tau}{Q_{Bat,0}}$$

A complementary measure is the parameter Depth of Discharge (DoD). It defines the amount of charge that has been extracted from a battery, with reference to the battery capacity. The DoD is usually indicated unitless as a ratio between zero and one.

A value of zero indicates that the battery is in a completely charged state, and a value of one indicates that the battery is in a completely discharged state. The definition in the unit percent is also common.

$$DoD(t) = 1 - SoC(t) = 1 - \frac{Q_{Bat}(t)}{Q_{Bat,0}} = 1 - \frac{Q_{Bat}(t_0) - \int_{t_0}^{t_0+\Delta t} I_{Bat,Discharge}(\tau)d\tau}{Q_{Bat,0}}$$

A maximum DoD or a minimum SoC value has to be defined during battery design, above which the discharge of a battery cannot be carried out further. In order to enable a long-life duration, a good operational strategy is to limit the amount of charge that is regularly discharged from the battery. With decreasing DoD values, the amount of possible charge/discharge cycles is increasing exponentially. A discharge above a DoD of 0.8 is usually defined as deep discharge [266]. Pulse discharging and pulse charge/discharge cycles usually do not account for the total number of charge/discharge cycles. As pulse discharging, discharging operations are defined, during which less than 0.1% of the battery charge is extracted [97].

The time-variable battery capacity is defined as the amount of electrical charge that can be discharged from a battery at a certain point of time. It is measured as the integral of the discharge current over time during one complete discharge operation. At the beginning of life, the time-variable battery capacity should be equal to the nameplate capacity, but its value decreases with time, resulting from degradation. An example for the decrease in the useable capacity of batteries on board of a spacecraft are shown in Figure 4.57. These plots correspond to the batteries of the four spacecraft of the CLUSTER mission.

$$Q_{Bat,var}(t) = \int_{t_0}^{t_0+T_{EoDC}} I_{Bat,Discharge}(\tau)d\tau$$

$$Q_{Bat,np}(t) = Q_{Bat,var}(t = BOL)$$

The nominal battery energy can be defined as the amount of electrical energy that can be drawn from a battery when it is discharged from full charge to full discharge state. It can be calculated by the integral of the electrical discharge power over time.

$$E_{Bat,Nom}(t) = \int_{t_0}^{t_0+T_{EoDC}} P_{Bat}(\tau)d\tau = \int_{t_0}^{t_0+T_{EoDC}} V_{Bat}(\tau) \cdot I_{Bat}(\tau)\, d\tau$$

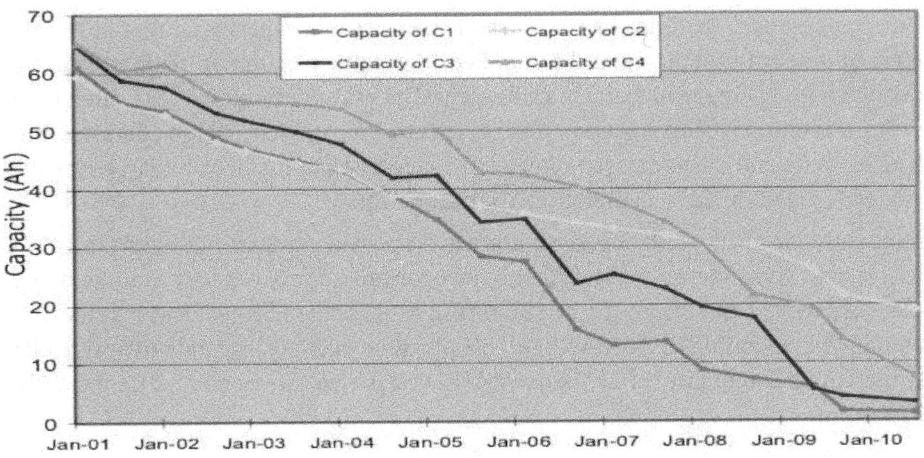

Figure 4.57 Degradation of battery capacity of the CLUSTER spacecrafts during their mission (image: © ESA/CLUSTER, B. Sousa [273])

The current that is supplied to or drawn from a battery, can be indicated as a current value in ampere. However, battery currents can also be indicated utilizing the parameter c-rate. The c-rate is a parameter defining the battery current with respect to the current value that would discharge a battery exactly after one hour. Discharging a battery with a c-rate of 1 will discharge the battery after a period of 1 hour.

$$C_{Rate} = \frac{I_{Bat,Discharge}}{I_{Bat,1h}}$$

$$I_{Bat,1h} = Q_{Bat,np} \cdot \frac{1}{h}$$

A number of different voltage parameters is utilized for batteries. The terminal voltage is the voltage that can be measured at the terminals of a battery. The nominal voltage is a parameter that is usually stated in manufacturer datasheets and it is defined as the voltage that a battery features at its terminals under nominal operating conditions. However, during operation, the battery voltage varies with the SoC. In full discharge state, a battery has the value of the full discharge voltage, and in full charge state, a battery has the value of the full charge voltage. For discharge operations, a minimum voltage is usually defined. In case that a battery voltage has the value of the minimum voltage, it should not be discharged further. This voltage value is sometimes also referred to as cut-off voltage. Utilizing a battery technology that features a comparatively flat characteristic curve, featuring a minimum of voltage variation between full charge state and full discharge state offers advantages, especially when the battery voltage is coupled with the power bus voltage.

During operation, batteries lose a share of their useable capacity, so that only a decreased amount of charge can be supplied to a battery or drawn from it after the degradation. A degraded battery cell reaches its end of charge voltage after a decreased amount of charge was supplied to it during a complete charge operation, and it reaches its end of discharge voltage after a decreased amount of charge was discharged from it during a complete discharge operation.

A measure for the degradation of a battery is the parameter State of Health (SoH). The SoH is defined as the ratio between the time variant battery capacity and the nameplate capacity at the Beginning of Life (BOL). At BOL, the SoH features a value of 1, and during operation, it decreases. A SoH of 0.8 or 80% corresponds to a degraded battery at its End of Life (EOL) [109] [143] [153]. It is also common to indicate the SoH in percent.

$$SoH(t) = \frac{Q_{Bat,var}(t)}{Q_{Bat,0}}$$

The SoH of a battery is influenced negatively by a number of variables such as the magnitude of the charge and discharge currents, the total number of charge/discharge cycles, the total operational life duration of the battery, the battery temperature, the Depth of Discharge and the degree of overcharge.

Two degradation effects can be observed in batteries. Batteries degrade independent from their operation with time and they also degrade with an increasing number of charge/discharge cycles. The parameter that relates the calendric degradation to a loss in battery capacity is the calendar life. It can be utilized to estimate the total number of days in operation before a battery reaches its EOL state. The parameter value can be determined from the product of the square root of the mission duration in days and a degradation parameter. The degradation parameter itself is battery technology-specific and can depend on the temperature in degree Celsius and the operational State of Charge in percent. Different battery-specific equations are available for the value determination of the degradation parameter.

$$\alpha_{CF,Calender} = k_{CF,a} \cdot \sqrt{n_{Bat,days}}$$

$$k_{CF,a} = \begin{cases} 0.0009 \cdot T_{Bat\,Cell}^2 - 0.0129 \cdot T_{Bat,Cell} + 0.1533 & \text{Saft VES, Saft VL, GS Yuasa} \\ (3.722 \cdot 10^{-6} \cdot SoC^2 + 7.504 \cdot 10^{-4} \cdot SoC) \cdot e^{0.0155 \cdot T_{Bat,Cell}} & \text{Saft VES16} \\ (8.883 \cdot 10^{-6} \cdot SoC^2 + 1.554 \cdot 10^{-3} \cdot SoC) \cdot e^{0.0173 \cdot T_{Bat,Cell}} + 0.0009 & \text{ABSL18650} \end{cases}$$

The parameter cycle life is a measure for the battery degradation resulting from the charge/discharge cycling during the battery operation. It relates the total number of charge/discharge cycles that have been accumulated during the battery life to a loss in battery capacity. It can be utilized to estimate the total number of charge/discharge cycles that a battery can carry out before it reaches the EOL state. The value of the cycle life parameter can be determined from the product of the square root of the total number of charge/discharge cycles and a degradation parameter. The degradation parameter is battery technology-specific and depends on the operational Depths of Discharge in percent. A number of different equations is available for different batteries.

$$\alpha_{CF,Cycle} = k_{CF,c} \cdot \sqrt{n_{Bat,cycles}}$$

$$k_{CF,c} = \begin{cases} 0.005 \cdot DoD + 0.021 & \text{Saft VES, Saft VL, GS Yuasa} \\ 0.0019 \cdot DoD + 0.0017 & \text{Saft VES16} \\ 0.004 \cdot DoD + 0.0009 & \text{ABSL18650} \end{cases}$$

The parameter capacity fading summarizes the degradation resulting from the calendric aging and from the charge/discharge cycling. It is calculated as the sum of the cycle life and the calendar life parameters. The parameter capacity fading can be used to scale the battery capacity at the beginning of life to its remaining value at a later point in time during the mission.

$$\alpha_{CF}(t) = \alpha_{CF,Calendar}(t) + \alpha_{CF,Cycle}(t)$$

$$Q_{Bat,var}(t) = Q_{Bat,np} \cdot \left(1 - \frac{\alpha_{CF}(t)}{100\%}\right)$$

The total number of charge/discharge cycles is limited because the battery voltage decreases with an increasing degradation. This decrease in the battery voltage is resulting from an increase in the internal resistance of the battery. After a certain number of charge/discharge cycles and the corresponding decrease in the battery voltage, a part of the voltage-capacity characteristic curve of the battery will lie below the useable voltage level. As the battery cannot be discharged further below this voltage level, a part of the battery capacity cannot be used in this case.

Different equations for the calculation of the maximum number of charge/discharge cycles with reference to the operational maximum depth of discharge are available for nickel-cadmium, nickel-hydrogen and lithium-ion batteries. Nickel-cadmium and nickel-hydrogen batteries can be operated at DoDs of up to 85% [97]. For lithium-ion batteries, maximum DoDs between 70% and 75% are recommended, but also DoDs up to 85% are possible in case that countermeasures for the charge/discharge cycling induced losses in battery capacity are taken into account [97]. The following equation uses the DoD indication in percent.

$$\max(n_{Bat,Cycles}) = \begin{cases} 207{,}829 \cdot e^{-0.0272 \cdot DoD} & \text{NiH}_2 \text{ and NiCd} \\ 890{,}000 \cdot e^{-0.0547 \cdot DoD} & \text{Li-Ion} \end{cases}$$

High continuous discharge c-rates above 1 C or high pulse discharge c-rates above 2 C can lead to cell damage with standard cells [97]. High-power cells can usually tolerate c-rates between 4 C and 11 C [97]. During charging, c-rate should be limited to 0.8 C [97].

Battery cell temperatures below the operational temperature limit result in an increase in the internal resistance of a battery, and temperatures above the operational limit result in a decreased cycle life. In addition to this, mechanical conditions can degrade batteries.

The efficiency of a battery can be expressed in terms of the supplied and drawn charge or energy. The parameter ampere-hour efficiency is a measure for the discharged battery charge, with reference to the supplied battery charge and the parameter watt-hour efficiency is a measure for the discharged energy with reference to the supplied energy. The ampere-hour efficiency is calculated as ratio between the integral of the battery current over a discharge operation and the integral of the battery current over a charge operation. The watt-hour efficiency is calculated as ratio between the integral of the battery power over a discharge operation and the integral of the battery power over a charge operation. The battery power itself is calculated as the product of battery terminal voltage and current. The factor −1 is employed in both equations because battery currents are only counted positively in one direction.

$$\eta_{Bat,Ah} = \frac{Q_{Bat,Discharge}}{Q_{Bat,Charge}} = \frac{\int_{t_0}^{t_0+T_{EoDC}} I_{Bat,Discharge}(t)dt}{-\int_{t_0+T_{EoDC}}^{t_0+T_{EoC}+T_{EoC}} I_{Bat,Discharge}(t)dt}$$

$$\eta_{Bat,Wh} = \frac{E_{Bat,Discharge}}{E_{Bat,Charge}} = \frac{\int_{t_0}^{t_0+T_{EoDC}} V_{Bat}(t) \cdot I_{Bat,Discharge}(t)dt}{-\int_{t_0+T_{EoDC}}^{t_0+T_{EoC}+T_{EoC}} V_{Bat}(t) \cdot I_{Bat,Discharge}(t)dt}$$

The parameters specific energy and specific power are measures for the amount of energy or power that can be stored inside a battery per unit mass. The parameter power density is a measure for the amount of power that can be stored inside a battery per unit volume. These parameters can be utilized in order to compare different batteries against each other or in order to determine the mass or volume of a possible battery system for a given power or energy.

$$E_{Bat,Specific} = \frac{E_{Bat,Nom}}{m_{Bat}}$$

$$P_{Bat,Specific} = \frac{P_{Bat,Nom}}{m_{Bat}}$$

$$P_{Bat,Density} = \frac{P_{Bat,Nominal}}{V_{Geo,Bat}}$$

The voltage of a battery varies with its State of Charge, with the magnitude of the discharge or charge current, as well as with the system temperature, but it does not have a linear progression with the State of Charge. It reaches its maximum value in full charge state and decreases with the amount of extracted charge until the battery reaches its minimum value in full discharge state. Both the full discharge state and the full charge state are defined by a certain voltage that can be measured at the battery terminals. The discharge capacity is usually employed to indicate the amount of charge stored inside a battery. It is defined as the amount of charge that has been extracted from a battery, since it reached the full charge state.

The battery terminal voltage has a different progression during charging and discharging with reference to the discharge capacity. During charging, the characteristic curve can be subdivided into 2 domains. In the first domain, a linear increase of the battery voltage can be observed with an increasing State of Charge. At the end of the charging operation, the inclination of the characteristic curve increases rapidly in the second domain, until the battery reaches its end of charge state. Theoretically, a battery can be charged further, although it has already reached its end of charge voltage. This operation is defined as an overcharging. In full charge state, all reactive material in the battery has already reacted and there are no reactants left for reactions in order to utilize the supplied electrical energy. Because of this, the supplied energy can only be dissipated and contribute to the thermal energy of the battery, or it can be utilized for secondary reactions. Overcharging a battery can also lead to battery cell damage and must therefore be prevented. The ideal voltage-capacity characteristic curve of a battery during discharge operation features a much flatter progression and it can be subdivided into 3 domains. At the beginning of the discharge operation, the voltage drops exponentially until the curve transitions into flat and approximately linear progression. Because of this, the domain between the full charge state and the transition from the exponential voltage drop to a flat linear progression is usually referred to as the exponential domain. The domain in the middle of the characteristic curve, featuring a flat and approximately linear progression, is usually referred to as the nominal domain. At the end of the nominal domain, the voltage starts to transition to a rapid decrease with ongoing discharge. This domain is usually referred to as the end of discharge domain. The progression of the battery terminal voltage of an ideal battery is visualized for one charge/discharge cycle in Figure 4.58.

The main reaction of the battery is dominantly happening in the nominal domain. The occurrence of certain side reactions increases in the exponential domain and end of discharge domain.

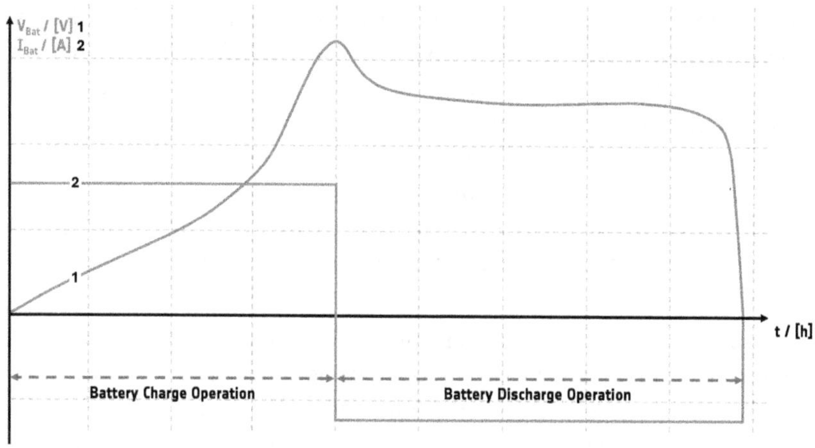

Figure 4.58 Example for a typical charge and discharge characteristic curve of the battery voltage

A number of characteristic points can be defined on the discharge voltage-capacity characteristic curve by a pair of a voltage value and a discharge capacity value. Full charge is the first characteristic point. In this characteristic point, the battery features a discharge capacity of zero ampere hours and the full charge voltage. The second characteristic point is defined at the transition between exponential domain and nominal domain and it is usually referred to as the end of the exponential domain. The characteristic point at the transition from the nominal domain to the end of discharge domain is usually referred to as the end of the nominal domain. A battery has its minimum voltage and a discharge capacity in the magnitude of the battery capacity in the full discharge characteristic point. This characteristic point is sometimes also referred to as the end of discharge characteristic point.

Full charge:

$$Q_{Bat,Discharge} = 0\ \text{Ah},\ V_{Bat} = \text{V}_{\text{Bat,FC}}$$

End of the exponential domain:

$$Q_{Bat,Discharge} = Q_{Bat,Exp},\ V_{Bat} = V_{Bat,Exp}$$

End of the nominal domain:

$$Q_{Bat,Discharge} = Q_{Bat,Nom},\ V_{Bat} = V_{Bat.Nom}$$

Full discharge:

$$Q_{Bat,Discharge} = Q_{Bat,0},\ V_{Bat} = V_{Bat.FDC}$$

The full charge voltage, the nominal voltage, and the full discharge voltage are usually available from manufacturer datasheets. The voltage and the discharge capacity in the characteristic points "end of the exponential domain" and "end of the nominal domain" are usually not indicated in manufacturer datasheets. The value of the discharge capacity in the characteristic point "full discharge" can be defined employing the nameplate capacity. Usually, the values of the remaining parameters have to be extracted from discharge characteristic curves. However, not all manufacturers offer them in their datasheets. The progression of the battery voltage during one charge operation and during one discharge operation is shown in Figure 4.59.

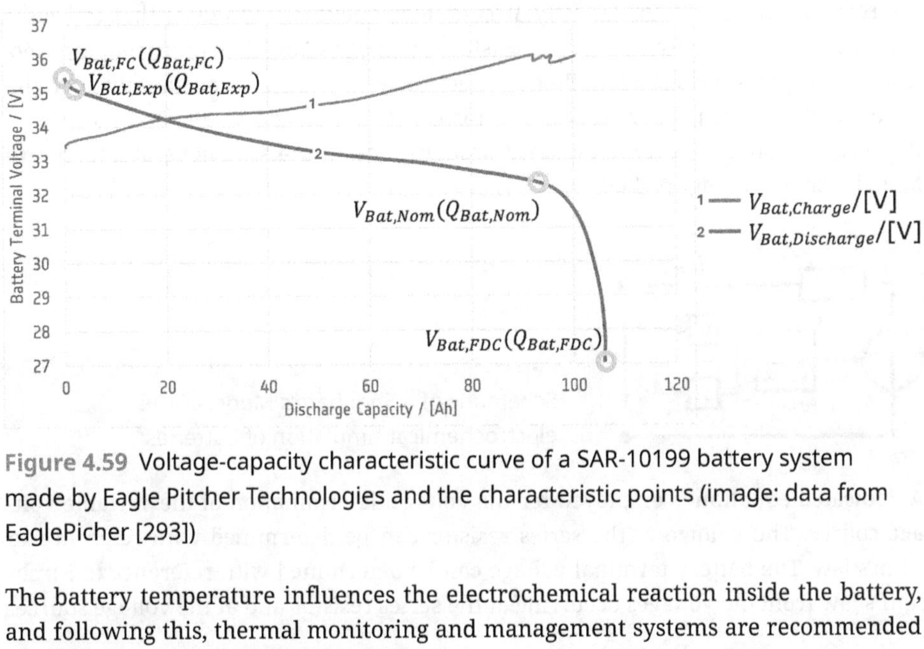

Figure 4.59 Voltage-capacity characteristic curve of a SAR-10199 battery system made by Eagle Pitcher Technologies and the characteristic points (image: data from EaglePicher [293])

The battery temperature influences the electrochemical reaction inside the battery, and following this, thermal monitoring and management systems are recommended for implementation in battery systems.

Battery models are available in different levels of complexity. With the model complexity level, the required efforts in the battery characterization and in the model fitting usually increase, but also the accuracy of a battery model usually increases. There are simple battery models that can be implemented employing parameters available from datasheets or the characteristic points. The implementation of more complex models usually requires a great extent of laboratory measurements. Expensive equipment such as caloric chambers or spectrum analyzers is required for this purpose.

Electrochemical models employ equivalent reference circuits for the simulation of the terminal characteristics and are well suited for the simulation of the behavior of a battery connected to an electrical grid. The accuracy with which the equivalent circuit is behaving like a battery depends on the level of complexity of the utilized equivalent circuit. A number of different equivalent circuits are available for the simulation of batteries. They consist of basic electrical parts such as capacitors, resistors, diodes, as well as voltage and current sources. Voltage sources are usually implemented as variable voltage sources that change the voltage level with reference to the State of Charge. In order to simulate the battery capacity, capacitors can be integrated into the equivalent circuits and self-discharge can be modelled employing a parallel resistor. Temperature- and state-of-health-related effects can also be modeled employing resistors.

Shepherd's Model is a battery model that allows to simulate the battery voltage with reference to the discharge capacity. It is an equivalent circuit model that utilizes a variable voltage source and a series resistor. Shepherd's Model assumes that the electrodes feature porous material, that the electrolyte resistance is constant through the complete discharge process, that the cell is charged with constant current, and that the polarization is a linear function. A visualization of the schematic of Shepherd's Model is shown in Figure 4.60.

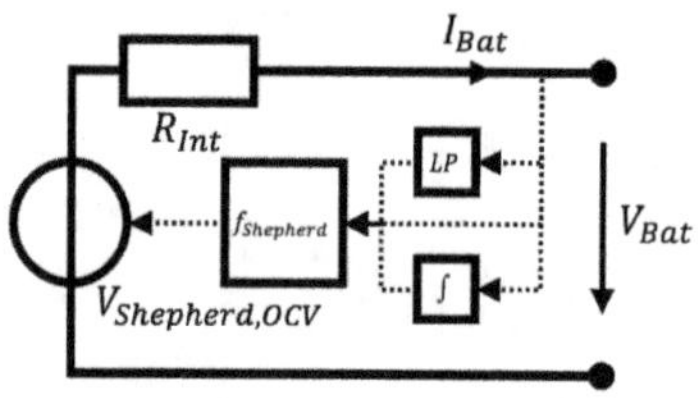

Figure 4.60
Schematic of a Shepherd's Model of the electrochemical simulation of batteries

A dedicated equation is employed for the voltage determination of the variable voltage source. The voltage at the series resistor can be determined with reference to Ohm's law. The battery terminal voltage can be determined with reference to Kirchhoff's law from the voltages occurring at the series resistor and at the voltage source.

$$V_{Bat}(Q_{Bat,Discharge}, I_{Bat}) = V_{Shepherd,OCV}(Q_{Bat,Discharge}, I_{Bat}) - R_{Int} \cdot I_{Bat}$$

The voltage equation of the variable voltage source of the original Shepherd's Model utilized a constant battery voltage, the battery nameplate capacity, a coefficient for the polarization resistance, the discharge capacity, as well as the battery current for the determination of the voltage at the variable voltage source. This equation allowed to simulate the nominal zone and the voltage breakdown at the end of the nominal zone of the voltage-capacity characteristic curve of batteries, but it did not cover the exponential zone.

$$V_{Shepherd1,OCV}(Q_{Bat,Discharge}, I_{Bat}) = V_{Shepherd,0} - K_{Shepherd} \cdot \frac{Q_{Bat,np}}{Q_{Bat,np} - Q_{Bat,Discharge}} \cdot I_{Bat}$$

A later version of the Shepherd's Model featured an additional summand in the voltage equation of the variable voltage source, which has been utilized in order to simulate the voltage drop occurring in the exponential zone. This term utilizes a factor for the amplitude of the exponential domain, an inverse time constant for the exponential domain, as well as the battery nameplate capacity and the discharge capacity. Advanced versions of this battery model have, for example, been utilized in the SimPowerSystems toolbox of the Matlab/Simulink software.

$$V_{Shepherd2,OCV}\left(Q_{Bat,Discharge}, I_{Bat}\right) = V_{Shepherd,0} - K_{Shepherd} \cdot \frac{Q_{Bat,np}}{Q_{Bat,np} - Q_{Bat,Discharge}} + A_{Shepherd} \cdot e^{-B_{Shepherd} \cdot Q_{Discharge}}$$

The parameters of the different versions of the Shepherd's Model have originally been determined by Shepherd with reference to two discharge characteristic curves. However, recent implementations of battery models based on the later version of the Shepherd's Model only utilize one discharge characteristic curve to fit the model.

The characteristic points full charge, end of the exponential domain, and end of the nominal domain, were employed in order to fit the model parameters to a given discharge characteristic curve. The amplitude of the exponential zone was defined as the voltage difference between the characteristic points full charge and end of the exponential domain. The inverse time constant of the exponential zone was defined by the division of a curve fitting coefficient by the value of a discharge capacity in the characteristic point end of the exponential domain. A value of 3 is recommend for this curve fitting coefficient.

$$A_{Shepherd} = V_{Bat,FC} - V_{Bat,Exp}$$

$$B_{Shepherd} = \frac{\alpha_{Shepherd}}{Q_{Bat,Exp}} = \frac{3}{Q_{Bat,Exp}}$$

The value of the constant battery voltage summand can be determined from the battery voltage equation in the characteristic point referred to as full charge. This characteristic point has the advantage that the discharge capacity has a value of zero ampere hours, which enables simplifications in the voltage equation.

$$V_{Bat,FC} = V_{Shepherd,0} - K_{Shepherd} \cdot \frac{Q_{Bat,np}}{Q_{Bat,np} - 0\,Ah} + A_{Shepherd} \cdot e^{-B_{Shepherd} \cdot 0\,Ah} - R_{Int} \cdot I_{Bat}$$

$$V_{Shepherd,0} = V_{Bat,FC} + K_{Shepherd} - A_{Shepherd} + R_{Int} \cdot I_{Bat}$$

In the following, the value of the polarization resistance coefficient can be determined employing battery voltage equation in the characteristic point end of the nominal domain.

$$V_{Bat,Nom} = V_{Shepherd,0} - K_{Shepherd} \cdot \frac{Q_{Bat,np}}{Q_{Bat,np} - Q_{Bat,Nom}} + A_{Shepherd} \cdot e^{-B_{Shepherd} \cdot Q_{Bat,Nom}} - R_{Int} \cdot I_{Bat}$$

$$K_{Shepherd} = \left(1 - \frac{Q_{Bat,np}}{Q_{Bat,Nom}}\right) \cdot \left(V_{Bat,Nom} - V_{Bat,Exp} - A_{Shepherd} \cdot e^{-B_{Shepherd} \cdot Q_{Bat,Nom}}\right)$$

Battery voltage and current requirements can be met by utilizing parallel and series implementations of single battery cells in a battery system. The battery terminal voltage can be defined as the sum of the individual voltages of the battery cell connected in series to strings. The maximum output current can be defined as the sum of the maximum string currents. The schematic of the battery system is visualized in Figure 4.61.

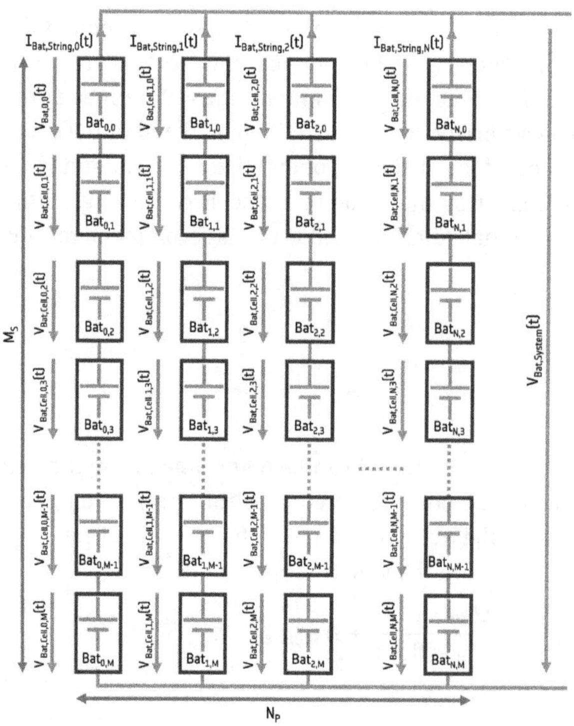

Figure 4.61
Schematic of a battery system

The series and parallel schematic is usually defined by the following nomenclature: $M_S\, sN_P\, p$. The parameter M_S is defined as an integer number to indicate the number of battery cells connected in series and the parameter N_P is defined as an integer number to indicate the number of battery cells connected in parallel.

$$V_{Bat,System}(t[n]) = \sum_{i=1}^{M_S} V_{Bat,Cell}(t[n])$$

$$\max(I_{Bat,System}(t)) = \sum_{i=1}^{N_P} \max(I_{Bat,Cell}(t))$$

Overcharging can lead to an increase in battery temperature and an increased battery temperature can lead to a pressure build-up, which in turn can lead to explosions. Nickel-hydrogen batteries are tolerant to overcharging in the magnitude of up to 20% and nickel-cadmium batteries are tolerant to overcharging in the magnitude of 5% to 10% [97]. Contrary to that, an overcharging of lithium-ion batteries by between 0.1 V and 0.3 V can lead to decreases in the cycle life of up to 80% [97]. Overcharging lithium-ion batteries can also result in thermal runaway effects that can heat battery cells up to 150 C [151]. At this cell temperature, the separator membrane starts to melt, leading to short circuit conditions between the electrodes, which can result in explosions in the following. Deep discharging with lithium-ion batteries can lead to permanent cell damage. In addition to this, the operation at low SoCs can alter electrode structures that lead to decreases in the useable capacity. Subsequently, special safety measures have to be implemented with lithium-ion batteries in order to prevent overcharging and deep discharging.

A degraded battery cell reaches the end of charge voltage after a smaller amount of charge has been stored inside the battery than in the undegraded case. Batteries implemented in a battery system degrade non-uniformly and because of this, batteries in a certain series and parallel implementation reach their EoC voltage at different points in time and after different amounts of charge have been supplied to them. When the first battery cell reaches its EoC voltage, the charge stored in all cells is equal. However, because its useable capacity has a smaller value in the most degraded battery cell than the useable capacity of the remaining battery cells, the State of Charge only has a value of 100% for the most degraded battery cell. Because of this, the State of Charge of the less degraded cells will have a smaller value than the State of Charge of the most degraded cell and the other cells will have lower voltages than the most degraded battery cell. The battery string voltage is determined from the sum of the individual battery cell voltages in a battery string. However, all battery strings have to have the same voltage in a parallel implementation. Because of this, non-equally charging in battery systems can result in high internal currents. In order to prevent these, the employment of cell balancing methods is recommended. Passive cell balancing interrupts a battery charge operation, once a cell reaches its EoC voltage. After that, a resistor is connected in parallel to the cell. This resistor discharges the cell again, until it features the same voltage as the other cells. Once the voltages are equalized, the resistor is disconnected again and the battery charge operation is continued. This approach usually requires a number of iterations until all batteries are fully charged. The excess charge is dissipated and converted into thermal energy. Contrary to that, when utilizing active cell balancing methods, the excess energy is stored in electrical components like parallel capacitors and, after that, it is transferred to other battery cells. It is recommended to apply cell balancing methods when lithium-ion batteries are utilized. The employment of these methods enables discharging to a greater depth of discharge. The duration of the operational period of a battery can also be extended by employing these methods.

Lithium-thionyl-chloride, silver-zinc, lithium-mono-fluoride and lithium-sulphur-dioxide are common primary battery technologies for space applications. As secondary batteries, nickel-cadmium, nickel-hydrogen and lithium-ion batteries are especially popular in space. Nickel-cadmium was the dominant secondary battery technology until nickel-hydrogen batteries were introduced in space in the 1980s. Since 2016, no spacecraft has been manufactured relying on nickel-cadmium batteries [48] [97]. Currently, lithium-ion batteries are the dominant battery technology [48] [97] [268]. Possible future technologies for the employment on board of spacecraft are lithium-air, lithium-polymer, lithium-sulphur, or zinc-air batteries.

Lithium-ion batteries offer significant reduction of the battery volume in the magnitude of 65% and of the battery mass in the magnitude of 50%, compared to nickel-cadmium batteries, while providing the same capacity to the spacecraft EPS [48] [98] [109]. In addition to this, also the operational temperature domain of lithium-ion batteries is wider than the operational temperature domains of nickel-cadmium and nickel-hydrogen batteries. Only a minimum of secondary reactions takes place in lithium-based batteries and following this, the self-discharge rate of lithium batteries is also very low. The casing of lithium-ion batteries can be designed as cylindrical, prismatic or pouch. In addition to this, lithium-ion batteries have a high cycle life, no memory effect and are tolerant to fast charging methods.

A number of different lithium metal oxides can be utilized as the positive electrode in lithium-ion batteries. Examples for possible positive electrode materials are lithium cobalt oxide, lithium iron phosphate and lithium nickel mangan cobalt oxide. As material for the negative electrode, usually graphite is utilized, but also other materials such as lithium-titanate can be utilized. The membranes are usually manufactured from polymer foils, such as polyethylene or polypropylene, featuring a thickness of around 30 μm [151]. Salts and polymers can be utilized as electrolyte materials. Lithium batteries are based on the effect of lithium intercalation into the molecular structure of the electrodes. Lithium is released from one electrode during charging as ions. These lithium ions are transported through the membrane to the other electrode and intercalated there. In the following, the charging and discharging reaction of a lithium mangan based battery is shown as an example and the reactions for other types of lithium-based battery technology can be found in the works of Böcker [153], McKissock et al. [269], and Mertens [151].

Charging

- Positive electrode: $LiMn_2O_4 \rightarrow Li_{1-x}Mn_2O_4 + x \cdot Li^+ + x \cdot e^-$
- Negative electrode: $n \cdot C + x \cdot Li^+ + x \cdot e^- \rightarrow Li_xC_n$

Discharging

- Positive electrode: $Li_{1-x}Mn_2O_4 + x \cdot Li^+ + x \cdot e^- \rightarrow LiMn_2O_4$
- Negative electrode: $Li_xC_n \rightarrow n \cdot C + x \cdot Li^+ + x \cdot e^-$

Netto reaction

- Charging: $LiMn_2O_4 + n \cdot C \rightarrow Li_{1-x}Mn_2O_4 + Li_xC_n$
- Discharging: $Li_{1-x}Mn_2O_4 + Li_xC_n \rightarrow LiMn_2O_4 + n \cdot C$

The terminal voltage of a lithium-ion battery is determined from the difference of the chemical potential of the electrodes. Following this, battery cells employing different electrode material combinations also feature different characteristic curves. Usually, the chemical voltage potential of the two electrodes is measured for each electrode with reference to a reference electrode made from metallic lithium.

$$V_{Battery}(\text{SoC} = 100\%) = V_{Cathode}(\text{SoC} = 100\%) - V_{Anode}(\text{SoC} = 100\%)$$

The progression of the terminal voltage with reference to the DoD of different types of lithium-ion batteries is shown in Figure 4.62.

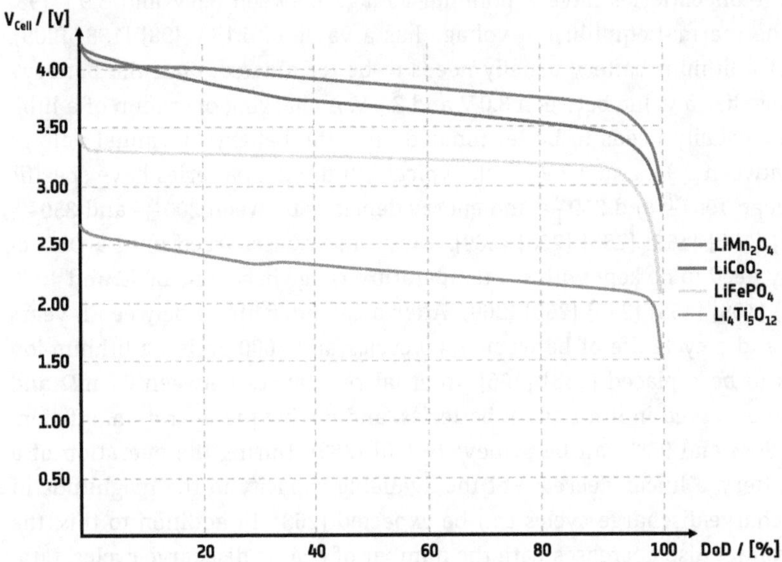

Figure 4.62 Discharge characteristic curves of different lithium-ion battery technologies (image: data from K. Mertens [151], M. Wohlfahrt-Mehrens, M. Sterner und I. Stadler)

Nickel-cadmium NiCd batteries have a cell voltage of 1.2 V under nominal operating conditions, which increases to 1.5 V at the end of the charge state and a thermal equilibrium voltage of 1.299 V [153]. During charge and discharge operations, high variations in the cell voltage can be observed. This is also the reason, why charge monitoring with NiCd batteries is challenging. NiCd batteries feature a charge efficiency of 83% [97]. Usually, a temperature domain between 5 °C and 20 °C is indicated for NiCd battery operation, but this battery technology can even be operated between −30 °C and 25 °C [98] [153] [268]. With NiCd batteries, specific energies between 44.4$\frac{\text{Wh}}{\text{kg}}$ and 50$\frac{\text{Wh}}{\text{kg}}$ can be achieved [153]. Power densities can feature values between 200$\frac{\text{W}}{\text{kg}}$ and 1000$\frac{\text{W}}{\text{kg}}$ and the energy density can feature a value of 100 $\frac{\text{Wh}}{\text{l}}$ [98] [153] [268]. Nickel-cadmium batteries can be operated for more than 30,000 cycles [268].

Nickel-hydrogen batteries achieve energy densities between 35$\frac{\text{Wh}}{\text{kg}}$ and 57$\frac{\text{Wh}}{\text{kg}}$ [97] [98] [153] [268]. Different batteries can be differentiated into Individual Pressure Vessels (IPV) that achieve lower efficiencies than Common Pressure Vessels (CPV), and the energy densities achievable with CPV are lower than the ones that can be achieved with Single Pressure Vessels (SPV). Batteries made of this technology can be operated within a temperature range from −10 °C to 30 °C [97] [98] [152] [268]. Depending on the pressure vessel approach that is being employed, a NiH_2 battery has a nominal voltage between 1.25 V and 2.5 V [97]. At the end of charge state, the cell voltage increases to between 1.55 V and 3.1 V [97]. Charge efficiencies between 83% and 83.5% of can be achieved with NiH_2 batteries [97].

Typical lithium-ion batteries have a nominal voltage between 3.6 V and 3.9 V [98] [153] [269]. The thermal equilibrium voltage has a value of 4.15 V [98] [153] [269]. Discharging of a lithium battery usually needs to be terminated, once the battery's terminal voltage has a value between 3.0 V and 2.5 V. A charging operation of a lithium-ion battery usually needs to be terminated once the battery's terminal voltage has a value between 3.9 V and 4.2 V [269]. Typical lithium-ion batteries have specific energies between 100$\frac{\text{Wh}}{\text{kg}}$ and 250$\frac{\text{Wh}}{\text{kg}}$ and energy densities between 200$\frac{\text{Wh}}{\text{l}}$ and 350$\frac{\text{Wh}}{\text{l}}$ [97] [98] [101] [109] [153] [256] [268] [269]. They can be discharged until a DoD of 100% and they need to be kept within a temperature range between −30 °C and 40 °C [97] [98] [101] [109] [153] [256] [268] [269]. After a calendar life of between 5 years and 20 years and a cycle life of between 1000 cycles and 5000 cycles, a lithium-ion battery needs to be replaced [153] [256]. Internal resistances between 0.8 mΩ and 120 mΩ can be observed in lithium-ion batteries, and during the operation, efficiencies between 80% and 95% can be achieved [256] [269]. During the operation of a lithium-ion battery, a linear decrease of the available capacity in the magnitude of 15% per 500 charge/discharge cycles can be expected [268]. In addition to this, the open circuit voltage also decreases with the number of charge/discharge cycles. Lithium-ion battery cells are tolerant to gamma radiation. In response to the exposure to 18 Mrad gamma radiation, a decrease in the available capacity in the magnitude of 10% can be expected [268]. A comparison regarding the technical parameters of different battery technologies that was summarized from a number of sources, is shown in Table 4.5.

Table 4.5 Technical parameters for different battery technologies [97] [98] [101] [109] [153] [256] [268] [269]

	Unit	Ag-Zn	NiCd	NiH_2	Li-Ion	Li-Polymer
end of charge voltage	V		1.5	1.55 to 3.1	3.9 to 4.2	4.1 to 4.2
nominal voltage	V		1.2	1.25 to 2.5	3.6 to 3.9	3.6 to 3.8
end of discharge voltage	V				2.5 to 3.0	2.5 to 3.0
charge efficiency	%		83.0	83.0 to 83.5	80.0 to 96.0	
equilibrium voltage	V		1.299		4.15	
specific power	$\frac{W}{kg}$		200 to 1000		200 to 400	
specific energy	$\frac{Wh}{kg}$	60 to 130	44.4 to 50	35 to 57	100 to 250	130 to 190
energy density	$\frac{Wh}{l}$	200	100	80	200 to 350	
temperature range	°C	−10 to +25	−30 to +25	−10 to +30	−30 to +40	
calendar life	Year	< 1	> 5	5 to 10	4 to 20	
cycle life		< 100	> 30,000	> 40,000	1000 to 5000	
parasitic resistance	mΩ				0.8 to 120	20 to 60

During the operation of batteries, efforts need to be taken into the charge control and into the charging procedure. This is beneficial and offers to increase the operational life of a battery. For some battery technologies, like lithium-ion batteries, it is even required in order to prevent damaging. The application of a voltage-temperature method is sufficient to monitor the State of Charge of nickel-cadmium and nickel-hydrogen batteries, while for lithium-ion batteries, the battery charge needs to be accounted accurately.

A number of different charge protocols are defined in the norm DIN 41772 [256]. Utilizing the protocol decreasing current, a battery is charged without any charge regulation. The battery is directly connected to a source of electrical power instead [256]. Following this, the battery is charged uncontrolled [256]. In the beginning of a charge operation, relatively high currents will be drawn by the battery and the amount of charge current will decrease during the charge operation [256]. The power source is disconnected once the battery terminal voltage has the value equal to the end of charge voltage [256]. Utilizing the protocol constant current, the battery is connected to a power source and a constant current is supplied to the battery [256]. The charge current is actively controlled [256]. In response to the supplied charge, the battery terminal voltage increases and the charge operation is terminated once the terminal voltage has a value equal to the end of charge voltage [256]. Contrary to that, utilizing the protocol constant voltage, a constant voltage is applied to the battery terminals [256]. The battery will draw a certain current in response [256]. In the beginning, of the charging operation, high currents will be drawn, but from the beginning to the end, the current drawn by the battery will decrease [256]. The charge operation will be terminated once the battery current reaches a value of zero ampere [256]. The protocol constant current—constant voltage can be seen as a mixture of the previously discussed protocols [151] [256]. Utilizing this protocol, the battery is charged with constant current first, until it reaches a State of Charge between 50% and 80% [256]. From this point on, the battery is charged with a constant voltage until the charge current reaches a value of zero ampere [151] [256].

4.12 Regenerative Fuel Cells

Regenerative Fuel Cells (RFCs), a combination of Fuel Cells (FCs) and electrolyzers, are a promising technology for the utilization as energy storages on a lunar base. Electrolyzers utilize the chemical process electrolysis in order to split up a reactant chemical into gaseous product materials with the use of electrical energy. Two supplied reactant gases react inside a Fuel Cell to a product chemical and electrical energy is released as by-product. In case that the product chemical of a certain Fuel Cell type is identical with the supply reactant of a certain electrolyzer type and the product gases of the Electrolyzer type are identical with the supply reactants of the Fuel Cell type, both technologies types can be connected together to a Regenerative Fuel Cell. Regenerative Fuel Cells (RFCs) can be seen as rechargeable chemical energy storages. During the operation of Fuel Cells or electrolyzers, reactant materials need to be supplied to it and reaction products need to be drawn from it continuously. In order to utilize Regenerative Fuel Cells as energy storages, reactant chemicals and product chemicals need to be stored in separate tanks that are also part of the energy storage system. The volume and the internal pressure of these tanks determine the capacity of the energy storage system. Additional pumps and valves need to be implemented for the operation of each system. Thermal management systems also need to be implemented together with Fuel Cells and electrolyzers because thermal energy is released as a by-product of the chemical reaction happening inside these systems. Regenerative

Fuel Cells utilizing a reaction employing hydrogen and oxygen are advantageous for the implementation on a lunar base because these reactant gases can also be utilized as vehicle propellant or crew supplies. An overview over the different system is shown in Figure 4.63.

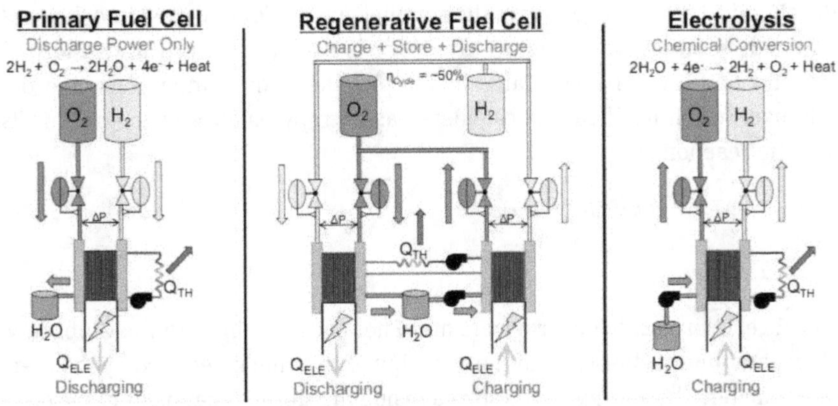

Figure 4.63 Overview and comparison of the primary fuel cell, the regenerative fuel cell and the electrolyzer technologies (image: © NASA, Jakupca [38])

Regenerative Fuel Cells utilize the electrolyzer subsystem in order to store energy. The Fuel Cell subsystem is utilized in order to generate electrical energy from the product chemicals of the electrolyzer. Because charge and discharge operations do not occur at the same time, always either the Fuel Cell subsystem or the electrolyzer subsystem will be in operation at a time during the operation of a Regenerative Fuel Cell, but both subsystems will not be in operation simultaneously. In nominal operation conditions in solar-powered EPS, the electrolyzer subsystem will be in operation during sunlit periods and the Fuel Cell subsystem will be in operation during eclipse periods.

Chemical energy storages feature the advantage, that they can store energy for long periods of time. High capacities and long discharge periods can be achieved by the implementation of a suitable volume of the storage tanks.

Inside a Fuel Cell, an oxidation reaction is taking place, and inside an electrolyzer, a redox reaction is taking place. These reactions can be subdivided into individual anode and cathode reactions. As a result of supplied electrical power, the reactions inside an electrolyzer are taking place, and as a result of supplied reactant chemicals, the reactions inside a fuel call are taking place. A by-product of fuel cell reactions is an electrical potential gradient between anode and cathode that can be measured at the fuel cell terminals as a voltage, and that can be utilized to draw electrical power from the fuel cell.

The total amount of energy that is supplied to the chemical reaction inside electrolyzers and Fuel Cells in the form of the reactant chemicals is referred to as enthalpy of reaction. It results from the enthalpy balance of the chemicals involved in reactions.

$$\Delta H_{Chem.} = H_{Chem.,A} + H_{Chem.,B} - H_{Chem.,AB}$$

During the chemical reactions, a part of the supplied energy is converted into thermal energy as a side-product. This amount of energy is usually referred to as thermal energy transfer and it can be calculated by the product of the temperature and the entropy change, which itself can be calculated as entropy balance of the chemicals involved in the reaction.

$$\Delta S_{Chem.} = S_{Chem.,A} + S_{Chem.,B} - S_{Chem.,AB}$$

$$\Delta Q_{Therm.} = T \cdot \Delta S_{Chem.}$$

The chemical reaction inside electrolyzers and Fuel Cells is subject to the Gibbs law. The available amount of energy from the enthalpy of reaction is reduced by the thermal energy transfer. The energy that is stored inside the storage chemicals of the electrolyzer, and the amount of energy that can be drawn from a Fuel Cell is defined by the free reaction enthalpy, which is also referred to as Gibbs potential. It can be calculated as the difference between the enthalpy of reaction and the thermal energy transfer.

$$\Delta G_{Chem.} = \Delta H_{Chem.} - \Delta Q_{Therm.} = \Delta H_{Chem.} - T \cdot \Delta S_{Chem.}$$

Water or hydrocarbons can be utilized as a basis for the chemical reaction inside a Regenerative Fuel Cell. The process of electrolysis utilizes supplied electrical energy in order to split up water into product chemicals. Electrolysis processes can achieve efficiencies between 65% and 85% [109] [323]. The reaction speed of the electrolysis process varies with the amount of supplied current.

While electrolysis is a widely applied process on Earth, in space, with the absence of gravity, problems with the detachment of gaseous bubbles from the electrodes occur during the electrolysis process. Reasoned by this, the thermal losses of electrolyzers increase and the RFC efficiency decreases with decreasing gravitational force. A possible countermeasure are surface structure advancements on the electrodes. On the International Space Station, an electrolysis process was implemented in a water recovery system. For the application in space, PEM electrolyzers have a satisfying Technology Readiness Level (TRL) and have already been demonstrated at a relevant scale.

In Regenerative Fuel Cells (RFCs), the energy is stored inside the product chemicals of the electrolysis process. These have to be stored in tanks, which can be utilized as long-term energy storages. Pressurized storages, liquid cryogenic storages, chemical hybrid storages, and graphite nanostructure storages are available chemical storage technologies. Energy densities of $3\frac{\text{Wh}}{\text{l}}$ can be achieved utilizing RFC energy storages on the Earth with reactant tanks under normal pressure conditions [256]. A storage pressure of 250 bar increases the possible energy density to values of up to $750\frac{\text{Wh}}{\text{l}}$ [143] [256]. In cryogenic state, energy densities of even up to $2400\frac{\text{Wh}}{\text{l}}$ can be achieved [143] [256]. Hydrogen needs to be cooled down to −253 °C in order to get into cryogenic state, which requires energy in the magnitude of $10.5\frac{\text{kWh}}{\text{kg}}$ [109]. The equipment demand for liquification of the reactant gases of the Fuel Cell can be reduced by placing the cryogenic tanks in the cold-biased, permanently shadowed areas of lunar craters. In case that cryogenic storages shall be utilized, technologies for the pressurization of the stored chemicals also need to be implemented. Mechanical or electrochemical processes can be employed for the pressurization process, but electrochemical processes feature higher efficiencies. Hydrogen storages are also subject to self-discharging at daily rates between $0.003\frac{\%}{\text{day}}$ and $0.1\frac{\%}{\text{day}}$ [109] [256]. Regenerative Fuel Cells cannot be discharged completely. During the operation, the Depth of Discharge should be situated in the domain between 40% and 60% [256]. Another possible systematic approach to store the product hydrogen is methanization. This process employs the chemical reaction of the product hydrogen with carbon dioxide to methane. With methane storages, up to 3 times higher energy densities can be achieved than with hydrogen storages [256].

Fuel Cells employ an oxidation reaction and as a side-product, electrical energy is released. The efficiency of a Fuel Cell can be defined as the ratio between the electrical output power and the product of the caloric value of the supplied reactants and their mass flow rates. This parameter can also be defined as the product of the thermodynamic efficiency, the voltage efficiency, the current efficiency, the efficiency of the reaction from the supply reactant to the electrical energy, and the system efficiency.

$$\eta_{FC} = \frac{P_{El}}{H_U \cdot \frac{dm}{dt}} = \eta_{FC,rev} \cdot \eta_{FC,V} \cdot \eta_{FC,I} \cdot \eta_{FC,FG} \cdot \eta_{FC,Sys}$$

The thermodynamic efficiency of a Fuel Cell is defined as the division of the maximum amount of energy that can be drawn from the system, which is equal to the Gibbs potential, by the supplied energy in the form of reactant chemicals, which is equal to the enthalpy of reaction.

$$\eta_{FC,rev} = \frac{\Delta G_{Chem.}}{\Delta H_{Chem.}} = \frac{\Delta H_{Chem.} - T \cdot \Delta S_{Chem.}}{\Delta H_{Chem.}} = 1 - T \cdot \frac{\Delta S_{Chem.}}{\Delta H_{Chem.}}$$

The reversible cell voltage is the maximum cell voltage that a Fuel Cell can achieve. It is defined by the Gibbs potential, the faraday constant, and the Z number. The terminal voltage of a real Fuel Cell is lower than the reversible cell voltage. The difference is resulting from internal power dissipation. The voltage efficiency is defined as the division of the actual terminal voltage by the reversible cell voltage.

$$V_{FC,Rev} = -\frac{\Delta G_{Chem.}}{n_Z \cdot F}$$

$$\eta_{FC,V} = \frac{V_{FC,Terminal}}{V_{FC,Rev}}$$

The theoretical maximum current of a Fuel Cell is defined by the mass flow rate, the Z number of the reactant, the faraday constant, and the molar mass of the reactant. The output current of a real Fuel Cell has a smaller value than the maximum possible Fuel Cell current. The reason for this circumstance is internal power dissipation. The current efficiency is defined as the ratio between the real fuel cell terminal current and the theoretically maximum possible Fuel Cell current.

$$I_{FC,theo} = \frac{\frac{dm}{dt} \cdot n_Z \cdot F}{M}$$

$$\eta_{FC,I} = \frac{I_{FC,Terminal}}{I_{FC,theo}}$$

Fuel Cells can only utilize a part of the supplied reactants. The fuel gas efficiency is a measure for the share of the supplied reactants that is utilized by the chemical process.

Additional subsystems, like thermal control systems or pumps, are required for the operation of Fuel Cells. The power required for their operation accounts for the system losses. The system efficiency is a measure for the amount of power that is demanded by Fuel Cell subsystems during operation, with reference to the total Fuel Cell output power. State of the art Fuel Cell implementations achieve efficiencies between 56% and 60% [109] [256]. Power losses occur mainly during the conversion processes.

The voltage that occurs at a Fuel Cell when it is not supplying electrical energy is defined as the open circuit voltage. Different types of voltage losses occur inside a Fuel Cell that reduce the usable voltage. There are penetration losses, losses at the parasitic internal resistance, and diffusion losses.

$$V_{FC,Terminal}(I_{FC,Terminal}) = V_{FC,OC} - \Delta V_{FC,P}(I_{FC,Terminal}) - \Delta V_{FC,R}(I_{FC,Terminal}) - \Delta V_{FC,Diff}(I_{FC,Terminal})$$

Penetration losses occur at low currents and are resulting from the transport of electrons through the interfacing area between electrolyte and electrodes. Voltage losses increase linearly with increasing terminal current and are resulting from the parasitic inner resistance of a Fuel Cell. During high-demand operation, when the current has increased to a magnitude close to the maximum output current, the output voltage starts to drop. This voltage drop is resulting from the properties of a Fuel Cell. The reaction speed inside the Fuel Cell increases with increasing power demand. Subsequently, during high-demand operation, the reaction speed inside the Fuel Cell will also be high in order to supply the demanded power. As a result of this, the demand for supplied reactants increases. The voltage drop occurs when the reaction speed reaches an order of magnitude where it is not possible anymore to supply enough reactants to the Fuel Cell. Resulting from a lack of supplied reactants, the fuel cell voltage breaks down in this case.

Alkaline Fuel Cells (AFCs) utilize hydrogen as the reactant and oxygen as the oxidant. Potash lye is employed as electrolyte. AFCs require operational temperatures in the domain from 20°C to 90°C [109]. This type of fuel cells achieves high efficiencies at low temperatures [109]. However, AFCs require a supply with pure oxygen because the carbon dioxide content of breathing air would alter the performance of the AFC.

- cathode reaction: $\frac{1}{2}O_2 + H_2O + 2e^- \longrightarrow 2OH^-$
- anode reaction: $H_2 + 2OH^- \rightarrow 2\,H_2O + 2e^-$

Polymer Electrolyte Fuel Cells (PEFCs) employ hydrogen as the reactant and oxygen as the oxidant. The oxidant can be supplied in the form of pure oxygen and in the form of breathing air. PEFC employ a proton conducting membrane as the electrolyte. This membrane is usually manufactured from perfluorinated or sulfonated polymers. Electrodes are manufactured from coal or metals coated with platinum. At the anode, hydrogen molecules react with hydrogen ions, and electrons are released as a by-product. The hydrogen ions will then transit through the membrane and react at the cathode with the supplied oxygen to water. The chemical reaction inside PEFCs requires a temperature of 80°C [109].

- cathode reaction: $2H^+ + \frac{1}{2}O_2 + 2\,e^- \longrightarrow H_2O$
- anode reaction: $H_2 \rightarrow 2H^+ + 2e^-$

Proton-Exchange Membrane Fuel Cells (PEMFCs) employ hydrogen as the reactant and oxygen or breathing air as the oxidizer. The electrolyte is implemented by a proton-conducting membrane. Perfluorinated or sulfonated polymers are possible membrane materials. Electrodes are usually manufactured from platinum-coated metals or coal. PEMFCs require operational temperatures in the domain from 60°C to 120°C [109]. Regenerative Fuel Cells based on PEM technology have been taken into account as energy storages for a lunar base within the Constellation program.

- cathode reaction: $2H^+ + \frac{1}{2}O_2 + 2e^- \longrightarrow H_2O$
- anode reaction: $H_2 \rightarrow 2H^+ + 2e^-$

Direct Methanol Fuel Cells (DMFCs) can utilize methanol as a reactant and breathing air or pure oxygen as an oxidizer. They rely on a proton-conducting membrane as electrolyte. Operational temperatures of DMFCs need to feature values between 60°C and 130°C [109].

- cathode reaction: $\frac{3}{2}O_2 + 6H^+ + 6e^- \longrightarrow 3H_2O$
- anode reaction: $CH_3OH + H_2O \rightarrow CO_2 + 6H^+ + 6e^-$

Phosphoric Acid Fuel Cells (PAFCs) utilize phosphoric acid as an electrolyte. Hydrogen is employed as the reactant and breathing air or pure oxygen can be employed as the oxidant. Polytetrafluoroethylene is utilized as an electrode material. Temperatures between 160 °C and 220 °C are required during the operation of PAFCs [109].

- cathode reaction: $2H^+ + \frac{1}{2}O_2 + 2e^- \longrightarrow H_2O$
- anode reaction: $H_2 \rightarrow 2H^+ + 2e^-$

Molten Carbonate Fuel Cells (MCFCs) utilize a ceramic matrix filled with molten alkali carbonate as an electrolyte. This type of Fuel Cell can be supplied with hydrogen or hydrocarbonates as reactants and with oxygen or breathing air as oxidizer. The electrodes of an MCFC have to be manufactured from nickel. The operation of an MCFC requires temperatures between 600 °C and 650 °C [109]. Because of this, MCFCs can supply both electrical energy and thermal energy.

- cathode reaction: $CO_2 + \frac{1}{2}O_2 + 2e^- \longrightarrow CO_3^{-2}$
- anode reaction: $H_2 + CO_3^{-2} \longrightarrow H_2O + CO_2 + 2e^-$

Solid Oxide Fuel Cells (SOFCs) employ zirconium oxide that has been stabilized with yttrium as ion-conducting ceramic membrane. Hydrocarbons or hydrogen can be employed as reactants and breathing air or pure oxygen can be utilized as oxidizer. A Solid Oxide Fuel Cell requires temperatures in the domain between 800 °C and 1000 °C during the operation [109].

- cathode reaction: $\frac{1}{2}O_2 + 2e^- \longrightarrow O^{2-}$
- anode reaction: $H_2 + O^{2-} \longrightarrow H_2O + 2e^-$

Fuel cell-based energy storages were utilized in crewed missions such as Skylab, Apollo, Gemini, and the Space Transportation System (STS). Popular are those Fuel Cell technologies that employ hydrogen and oxygen as the supply reactants and a reaction with water as the product material. The spacecraft of the Apollo program utilized Fuel Cell Units (FCUs) for employing a reaction between hydrogen and oxygen. On board of the STS, an advanced version of the FCUs from the Apollo Program was utilized. In total, a number of 6 Fuel Cell Units have been implemented on board of an STS. Each of these Fuel Cell Units consisted of a series connection of 32 single fuel cells [97]. Three of these Fuel Cell Units were connected in series to a Fuel Cell String (FCS) and two of these FC Strings were connected in parallel to a Fuel Cell System (FCS) [97]. Each Fuel Cell System was capable of supplying 12 kW of electrical power continuously to the 28 V power bus of the STS [97] [98]. During peak demand periods, it was possible to increase the power supply to up to 16 kW [97] [98]. As a by-product of the chemical reaction inside the STS Fuel Cells, $0.36\frac{kg}{kWh}$ of water were generated [98]. During STS missions, 104 kg of water were generated per day [98]. The STS Fuel Cell System required a start-up time of 15 min, but allowed instantaneous shutdown [98]. It was developed for an operational period of 2400 h [98]. The efficiency of the

Fuel Cell System ranged from 80% during low demand periods to between 50% and 60% during increased demand periods [98].

Regenerative Fuel Cells require separate electrolysis and Fuel Cell Units in order to optimize each unit. Because of this, both functionalities cannot be integrated into a single unit. RFCs offer high energy storage capacities, high power and long discharge periods. Because of this, RFCs can be classified as an enabling technology for lunar surface exploration.

Possible systematic options are PEM Fuel Cells, Alkaline Fuel Cells and Solid Oxide Fuel Cells. However, only limited efforts have been undertaken in the research of Fuel Cells for the utilization in space since the end of the STS program. In addition to this, the availability of space-ready Fuel Cells was also terminated with the end of the STS program. Unfortunately, Fuel Cells developed for Earth-based applications cannot be utilized for the exploration of space because the internal flow characteristics are different in these two applications. In addition to this, space-certified electrolyzers that can work under high pressure are currently not available. Currently, only designs for Regenerative Fuel Cells in low Technology Readiness Levels (TRLs) exist. Research in the field of Regenerative Fuel Cells is carried out at different institutes. For example, at NASA's Glenn Research Center, an RFC demonstrator was developed and tested. This demonstrator achieved a power of 100 W and an efficiency of 47.4% [152]. An image and an overview schematic of this demonstrator is shown in Figure 4.64.

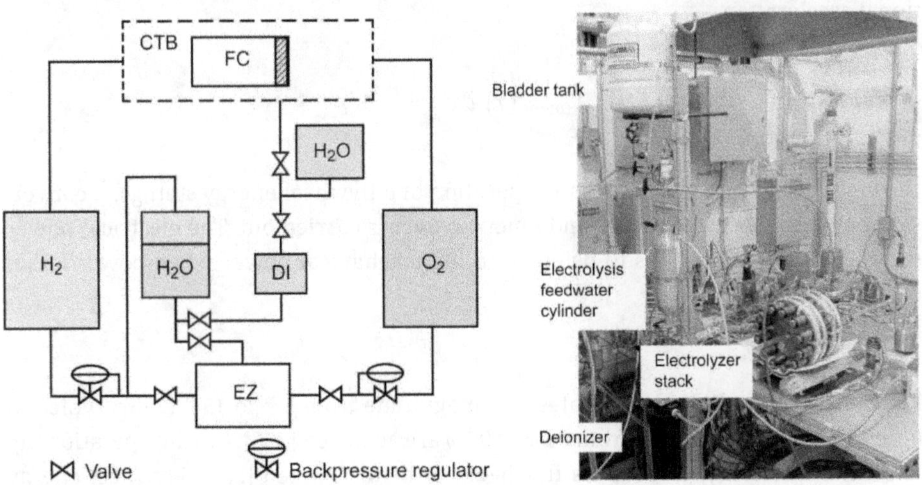

Figure 4.64 Regenerative Fuel Cell demonstrator at NASA's Glenn Research Center 2020 (image: © NASA, Bennett et al. [152])

4.13 Flywheels

Mechanical energy storages can store energy as kinetic energy or potential energy. Potential energy storages usually employ weights or fluids that are transported against a gravitational field in order to store energy. Kinetic energy can be stored in accelerated masses. The employed movement can be translational, for example in pendulums, or rotational, for example in flywheels. Flywheel-based energy storages convert electrical energy into kinetic energy and store it in the movement of a rotating mass. The electrical power supplied to a flywheel can be defined as the product of voltage and current and the energy stored in a flywheel energy storage can be calculated employing the time integral over the electrical power.

$$P_{El} = V \cdot I$$

$$E_{El} = \int_{t_0}^{t_0+\Delta t} P_{El}(\tau)\, d\tau = \int_{t_0}^{t_0+\Delta t} V(\tau) \cdot I(\tau)\, d\tau$$

The rotational kinetic energy of a flywheel is defined by its moment of inertia and the rotational velocity. It is also equal to the time integral of the kinetic power at a flywheel. Rotational kinetic power is defined as the product of torque and rotational velocity. The torque is defined as the product of the moment of inertia and the rotational acceleration.

$$E_{Kin,Rot} = \frac{1}{2} \cdot J \cdot \omega^2 = \int_{t_0}^{t_0+\Delta t} P_{Kinetic}(\tau)\, d\tau$$

During the charging or discharging operation of a flywheel energy storage, a conversion between electrical power and kinetic power is carried out. The electrical power supplied to a flywheel has to be equal to the mechanical power when power losses can be neglected.

$$P_{El} = P_{Kin}$$

Flywheels have self-discharge rates in a magnitude from $5\frac{\%}{\mathrm{h}}$ to $15\frac{\%}{\mathrm{h}}$ [256]. Typically, efficiencies in a magnitude from 80% to 95% are achieved [256]. During operation, up to 75% of the energy stored in a flywheel can be discharged [256]. Flywheel energy storages have a calendar life of up to 15 years and a cycle life in the magnitude of multiple millions of charge/discharge cycles [256]. State-of-the-art flywheel energy storages achieve energy densities from $80\frac{\mathrm{Wh}}{\mathrm{l}}$ to $200\frac{\mathrm{Wh}}{\mathrm{l}}$ and power densities of $10\frac{\mathrm{W}}{\mathrm{l}}$ [256]. The specific energy of a flywheel energy storage typically has a value in the domain from $5.56\frac{\mathrm{Wh}}{\mathrm{kg}}$ to $20.83\frac{\mathrm{Wh}}{\mathrm{kg}}$ [153].

There have also been plans to replace the batteries of the International Space Station with a Flywheel Energy Storage System (FESS), but these plans have been shifted towards the employment of lithium-ion battery technology. The proposed FESS had two flywheels with integrated charge and discharge units.

4.14 Power Bus Loads

The electrical power required to supply payloads and subsystems through all mission phases is defined as the operating power budget. The analysis of the operating power budget requires the differentiation of different operational phases with different combinations of systems that are in operation during these different phases. For example, during spacecraft operation, the orbital determination system and the communication system usually need to be in operation constantly, but certain payloads, such as scientific instruments, are only in operation for a certain period of time in order to minimize the power consumption. The power budget has to be calculated and analyzed for each of the different phases separately. The transient behavior of the power bus also needs to be taken into account during systems design because otherwise other systems connected to the bus could be harmed. Loads are usually supplied by Direct Current (DC) voltages in the magnitude of 28 V (DC) in space EPS [98]. However, DC supply voltages have been utilized in space implementations in the domain between 5 V (DC) and 270 V (DC) [98]. In addition to this, for some loads also single phase Alternating Current (AC) systems with a voltage of 115 V_{RMS} at a frequency of 60 Hz or 3-phase AC systems with a voltage of 140 V_{RMS} or 440 V_{RMS} at a frequency of 400 Hz have been utilized [98]. Loads can be connected to a power bus directly, if the bus satisfies the load input requirements. In case that a load needs to be protected against voltage or current variations and in case that the load input requirements, such as the voltage level or the voltage variability, does not match the power bus specifications, additional power electronics also need to be employed. Pulse currents are demanded by certain loads, such as radar systems, and need to be taken into account during EPS design.

4.15 Bus Protection

High currents can flow and low voltages can occur in electrical error situations. High currents have a great potential to harm all types of electrical equipment in Electrical Power Systems (EPSs). Electrical error situations can also cause drops in the bus voltage. Subsequently, safety measures against fault currents and voltage drops have to be implemented.

Electrical fault protection strategies have to be applied in a way that the power bus can recover the bus voltage after a voltage breakdown without any interventions from ground operators, as long as enough electrical energy is provided by the generative systems. The focus of electrical fault protection lies on the detection, isolation and correction of electrical faults. Different methods are available: The power bus can be protected from failed systems by isolating them with traditional electrical fuses, active current limiting circuits or voltage clamps. These systems are triggered by a certain condition and disconnect the electrical connection between the power bus and a failed load in response in order to keep the bus voltage from collapsing.

Electrical fuses isolate failed loads and disconnect them from the power bus permanently by melting the series connection between the power bus and the failed load. The melting conditions need to be precisely defined, so that the right fuse for a certain system and mission can be identified in order to prevent accidently triggering a fuse during operation. In case that fuses are implemented in a spacecraft EPS, other active disconnecting systems need to be implemented in a way that they are acting with a delay of 50 ms after a fault condition has been detected because fuses need a short amount of time to melt [97] [256]. In addition to this, the high currents required for triggering electrical fuses result in a huge amount of power required to melt a fuse, which has to be supplied by a power bus and can also cause bus voltage drops. Because of this, a requirement of ESA is that fuses shall be avoided within ESA systems [97].

Active Current Limiting (ACL) circuits limit the current through them once it reaches a value above a certain threshold. Different Active Current Limiting circuits are available for the application in space. Latching Current Limiting (LCL) circuits allow currents of up to 160% of the nominal currents [97]. In case that this limit is exceeded, LCL circuits disconnect the electrical connection between load and power bus. This type of circuit can be actively controlled from ground operators and can be reset into the conductive state again, after it has been triggered. Foldback Current Limiters (FCL) tolerate overcurrents between 10% and 25% of the nominal current [97]. In case that this limit is exceeded, the FCL circuit will actively reduce the voltage and the current. FCL circuits return to the nominal conducting state without any intervention from ground operators after the disappearance of electrical fault conditions.

Voltage clamps protect against undervoltage conditions by disconnecting the electrical connection to loads. In nominal conditions, voltage clamps are conducting.

Electrical loads have to be disconnected from fully regulated power buses in case that a drop in the bus voltage in the magnitude of 10% of its nominal value occurs [97]. The threshold voltage for triggering the load disconnection needs to be defined individually for other bus topologies. It needs to be defined in a way that enough electrical energy is still available in the EPS to set the space system into a safe state when the bus voltage is equal to the threshold voltage.

4.16 Grounding Concept

An electrical ground potential needs to be implemented into spacecraft EPSs. A common concept is the Distributed Single Point Grounding (DSPG) concept. Employing this concept, all back cables have to be connected to a single common ground point. This common ground point has to be connected to the space system structure. In addition to the supply of electrical systems with wires, it is also possible to supply them with current rails. In this case, the space system structure is employed instead of a back cable. An increased conductive coupling occurs between the current rails and the system structure. An increased voltage noise can also be observed with this approach.

4.17 Thermal Energy Storage

Thermal energy could be stored in special thermal energy storages on the lunar surface. During the lunar night, the thermal energy stored in the thermal storage can be utilized in two ways: Thermal energy can be supplied directly to a lunar base from a thermal storage, and electrical energy can be supplied to a lunar base EPS when additional generator systems have been implemented.

The lunar surface material regolith can be utilized as a thermal storage medium. Thermal energy can be supplied to regolith with the employment of probes or fluidic loops, together with a working medium. Hot fluids can be utilized as working mediums that transfer thermal energy to regolith while passing through the fluidic loop.

Thermal energy can be stored as sensible heat and latent heat. Sensible heat storages store energy in the temperature of the storage medium and latent heat storages store energy in the phase change of the storage medium. Latent heat storages offer advantages, but with regolith, only the phase change between solid and liquid can be employed, because the melting temperature of regolith features a value between 1373 K and 1653 K [138].

In order to utilize the energy stored inside a thermal storage for the generation of electrical energy, dynamic power generators, like Stirling engines, or static power generators, like thermoelectric generators, need to be utilized for the conversion between the two types of energy.

Research in the field of regolith-based thermal energy storages is also carried out by ESA. An image of their technology demonstrator in a vacuum chamber is shown in Figure 4.65.

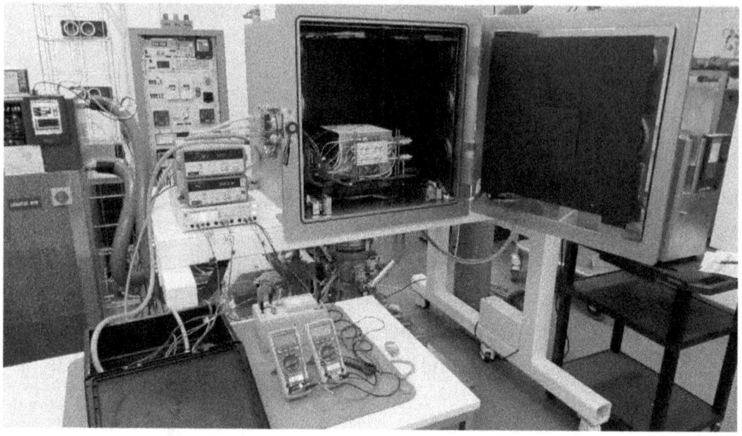

Figure 4.65 Thermal energy storage based on processed regolith in a vacuum chamber (image: © ESA/Azimut Space [329])

4.18 Thermal Control Systems

Spacecraft subsystems feature different requirements for the system temperature. Usually, a certain temperature domain consisting of the range between a certain minimum and maximum system temperature requirement is defined. For some subsystems, survivable and operational temperature domains can be differentiated. Subsystems can only be operated at system temperatures within the operational temperature domain. As long as a subsystem is out of operation when its system temperature has a value outside the operational temperature domain but inside the survivable temperature domain, no additional measures need to be taken into account. Temporary or permanent system damage has to be expected in the case that the system temperature exceeds the survivable temperature domain.

The implementation of Thermal Control Systems (TCSs) is highly recommended when it is expected that the system temperature exceeds the survivable temperature domain. Some systems, like power electronics, are subject to power dissipation during their operation. In order to prevent an overheating of such a system, the dissipated energy needs to be constantly drawn from the system during the operation by a TCS. In addition to this, the TCS implementation also allows operating a system during periods when it would have to be out of operation when no TCS would have been implemented.

TCSs control the temperature of spacecraft subsystems. By the utilization of TCSs, it can be assured that a system temperature has a predefined reference value or that it has a value within a predefined temperature domain.

TCSs can be implemented as passive or as active systems. Passive TCSs can be implemented by the utilization of surface coatings that advance the emissive and absorptive surface properties for thermal radiation. Active TCSs can employ a number of active subsystems that are implemented to cool subsystems by transporting thermal energy away from a spacecraft to space, to transport thermal energy through a spacecraft between different subsystems, or to supply a spacecraft subsystem with thermal energy in order to heat it up. Electrical heater units are usually implemented in order to supply thermal energy to subsystems. High-capacity heat conductors, active loops employing a working medium or heat pipes can be utilized in order to transport thermal energy through a spacecraft. In case that thermal energy is distributed inside a spacecraft, the waste heat from one system can be utilized in order to heat another system up. When the available waste heat cannot be utilized by other subsystems, it has to be radiated into space employing radiators. Passive radiators consist of materials featuring high emissive properties. Active radiators additionally employ active systems to control the amount of thermal energy that is transported from the inside of a spacecraft to the radiator systems. This functionality can be implemented utilizing the employment of an active cooling loop that is connecting hot spacecraft subsystems with external radiators. Another possibility is to integrate a passive radiator surface under a shutter unit inside a spacecraft. Utilizing the shutters, the radiator can be completely exposed, partly exposed, or not exposed to space in order to control the amount of thermal energy that is radiated into space by the radiator.

4.19 Space-Based Solar Power

Space-Based Solar Power (SBSP) can be defined as the systematic approach of harvesting solar energy in orbit and transmitting it to a receiver on the surface of a celestial body. Systematic options can be subdivided into those concepts utilizing concentrating mirrors to focus solar energy on a receiver on the surface of a celestial body and into those concepts utilizing traditional or concentrating photovoltaic generators on board of a satellite in orbit in order to convert solar energy into electrical energy, together with a Wireless Power Transmission (WPT) link to transmit energy from the satellite to a receiver on the surface.

Different SBSP system concepts have already been developed. Research in the field of SBSP has been focused on the development of system concepts, on the development of subsystems and on technology demonstration. Recent research focuses on the development and optimization of WPT links. Figure 4.66 shows the SBSP concept, and a visualization of a SBSP satellite in Earth orbit is shown in Figure 4.67.

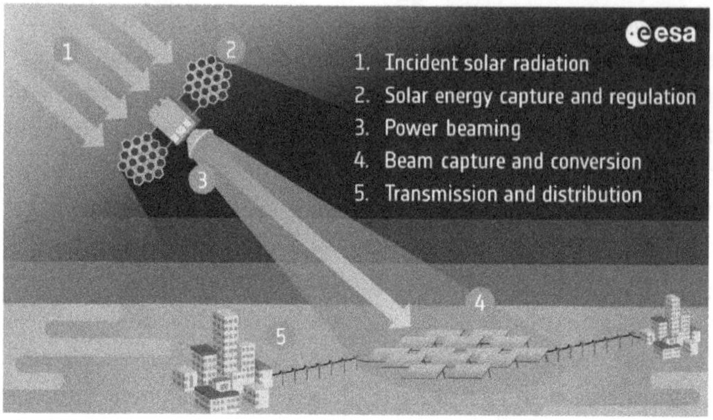

Figure 4.66 Topology of a Space-Based Solar Power system (image: © ESA [338])

Figure 4.67 ESA concept for an in-orbit Space-Based Solar Power demonstrator in Earth orbit (image: © ESA [339])

Resulting from the absence of atmospheric effects, photovoltaic generators achieve higher conversion efficiencies in space, while the utilization of WPT links depicts transmission losses. The application of SBSP in Geosynchronous Earth Orbit (GEO) features the advantage that satellites in GEO appear at a constant location in the sky over a receiver location on the Earth and that satellites in GEO are not subject to eclipse conditions. Subsequently, a receiver on the surface of Earth can be supplied without interruptions by a solar satellite in GEO.

In lunar applications, SBSP offers advantages to supply lunar surface systems and a lunar base during the lunar night. Because stationary lunar orbits do not exist, a solar satellite in lunar orbit will not appear at a constant position in the lunar sky and it will rise above the lunar sky and sink below it in reoccurring periods. Compared to the surface of celestial bodies, the solar power generation period duration increases with increasing orbit altitude. Resulting from the long duration of the lunar night in most lunar locations, the utilization of traditional solar energy and energy storage-based EPSs often results in EPS dimensions and masses, which lead to unfeasible EPS concepts. The utilization of solar satellites has the advantage that solar energy can be supplied to the lunar surface, especially during local night periods. Subsequently, the application of SBSP systems offers the possibility to reduce EPS mass, volume, and costs. Because of this, SBSP can be seen as an enabling technology for lunar exploration. The supply of a base at the lunar poles also offers the advantage that a satellite in a polar orbit does not necessarily need to be subject to eclipse conditions when the orbit is chosen adequately.

The supply of lunar surface systems with SBSP, such as robotic systems operated in lunar craters, has the advantage, that they can be supplied wirelessly with energy during operation, so that they do not need to return to a recharge station and so that no power transmission infrastructure would be required to power these systems. In Earth-based applications, the WPT link of SBSP systems would be subject to atmospheric effects, which would limit the WPT efficiency. Due to the absence of a significant atmosphere on the Moon, these effects do not need to be taken into account in lunar applications.

Economic reasons are the main reason for the unfeasibility of SBSP systems because SBSP systems usually feature huge dimensions, which result in high launch costs. Besides these, the environmental impact, international acceptance, and societal desirability are issues in the application of SBSP on Earth.

The losses occurring in the WPT link generally increase with an increasing transmission distance. Low altitudes have the advantage of smaller transmission losses and smaller receiver sizes, while high altitudes have the advantages of longer transmission durations and longer power generation periods. Solar satellites for lunar applications can be operated in lunar orbit and at the Lagrange points of the Earth-Moon system. For the application of SBSP, equatorial and polar orbits are of potential interest. The application of solar satellites in altitudes above 800 km is recommended because lower altitudes have shorter contact and power transmission periods to a ground receiver. Lower altitudes result in a decrease of the contact and power transmission period duration. Continuous coverage from non-stationary orbits can only be guaranteed employing a constellation of multiple solar satellites in an orbit. Solar satellites can also be operated at the Lagrange points. The Lagrange points L1, L2, L3 and L4 are suited for solar satellites with a focus on lunar applications. Because Lagrange points feature a stable location in the Earth-Moon system and because of the bonded rotation of the Moon, solar satellites operated at the Lagrange points appear at a stable position in the lunar sky. Subsequently, also the area on the lunar surface, that has contact conditions to a Lagrange point satellite, does not change with time. L1 and L2 can be classified as unstable Lagrange points and need to be orbited in halo orbits, but have the advantage of smaller transmission distances to the lunar surface compared to the other Lagrange points. L3 and L4 have a much longer transmission distance to the lunar surface, which decreases the WPT efficiency, but they are highly stable Lagrange points. However, all Lagrange points have comparatively high distances to the lunar surface.

During contact periods between a solar satellite and a ground receiver, the energy required to operate the lunar surface systems during one contact and one Loss of Contact (LOC) period needs to be received by a WPT ground receiver. For the supply of electrical systems during LOC periods, EPS need to feature energy storages.

Space-Based Solar Power systems consist of a number of different subsystems. The first major subsystem is the solar power collector, which is part of the solar satellite. It can be implemented utilizing space photovoltaics. The utilization of additional concentrators offers the advantage of higher conversion efficiencies between solar power and electrical power, as well as smaller generator sizes. The utilization of Maximum Power Point Tracking (MPPT) is recommended for the power regulation of photovoltaic-based generators in order to maximize the power yield. In case that an orbit has been chosen that does not feature continuous contact conditions, energy can only be transmitted to a ground receiver during contact periods. When no energy storages have been implemented on board of the solar satellite, the power, that can be transmitted to the surface receiver, is limited to the power generation during contact periods. The implementation of energy storages on board of a solar satellite offers the advantage that energy can be generated constantly, as long as solar energy is available in the solar satellite location, and transmitted to the surface receiver during contact periods. The second major subsystem is the Wireless Power Transmission (WPT) link, which consists of a transmission system and a receiving system. Microwave- and laser-based power transmission are popular WPT technologies. The majority of system designs focuses on microwave-based WPT. Popular microwave transmission frequencies are 2.45 GHz, 5.8 GHz and the x-band [330] [332] [333] [337]. Antenna sizes decrease and transmission efficiencies increase with increasing transmission frequencies. With increasing transmission distance, antenna sizes increase and transmission efficiencies decrease. Microwave-based power transmission has the advantages of high-power applications, long-distance power transmission and high conversion efficiencies between wirelessly transmitted power and electrical power. Integrated sandwich modules exist that consist of photovoltaic generators on the front side, microwave transmitters at the backside and DC to RF conversion units in between. The utilization of laser-based WPT is also popular among SBSP concepts and has the advantages of higher transmission efficiencies and smaller transmitter and receiver sizes. Solar cells can be utilized as laser-based WPT receivers. Standard solar cells have the advantage that they can convert both direct solar power and laser-based transmitted power into electrical power. Solar cells specifically optimized as receivers for laser-based WPT links have the advantage of higher conversion efficiencies for lased-based transmitted power. The combination of receiving antenna and subsystems for the conversion from wirelessly received power to electrical power is usually referred to as rectenna. Rectenna dimensions can reach magnitudes in the kilometer scale. Because of this, these are usually implemented employing an array of smaller rectennas.

5 Electrical Power Systems for Human Outposts in Space

The power requirements of a future lunar base will be very similar to the requirements of the International Space Station (ISS) today. As the Electrical Power System (EPS) implemented on board of the ISS is currently the most powerful EPS in human-made spacecrafts, a future EPS for lunar exploration within the Artemis program will most likely be based on the ISS EPS. Because of this, the ISS EPS will be introduced in this chapter and possible systematic options for lunar EPSs will be discussed.

5.1 Electrical Power System of the International Space Station

The EPS of the ISS can be subdivided into the EPS of the international segment and the EPS of the Russian segment. Both EPSs are interconnected by dedicated DC/DC Converter Units (DDCUs) that enable a power transfer between both EPSs. The American to Russian Converter Unit (ARCU) is utilized to transfer electrical power from the international to the Russian segment and the Russian to American Converter Unit (ACU) is utilized to transfer electrical power from the Russian segment to the international segment. Solar energy is utilized as primary energy source in both EPSs. In the beginning of the design phase of the ISS EPS, it was considered to utilize an Alternating Current- (AC-)based EPS, but for the final system, it was decided to utilize a Direct Current- (DC-)based EPS. The user voltage level of the international segment has a value of 124.5 V (DC) and the user voltage level of the Russian segment has values of 28 V (DC) and 120 V (DC). The power generation capacity of the international segment has a value of 76 kW under normal operation conditions and more during optimal operational conditions. The Russian segment has a power generation capacity of 29 kW that sum up to a total generation capacity of 105 kW under normal conditions.

The EPS of the international ISS segment consists of Solar Array Wings (SAWs), battery storages, DC/DC converters, switching units, different power channels, and a cable harness. The total mass of the ISS EPS has a value of 80 T. Between Solar Array Wings and Main Bus Switching Units (MBSUs), cables featuring a length between 50 m and 60 m are implemented. The schematic of the different elements of the ISS EPS are shown in Figure 5.1 as well as in Figure 5.4.

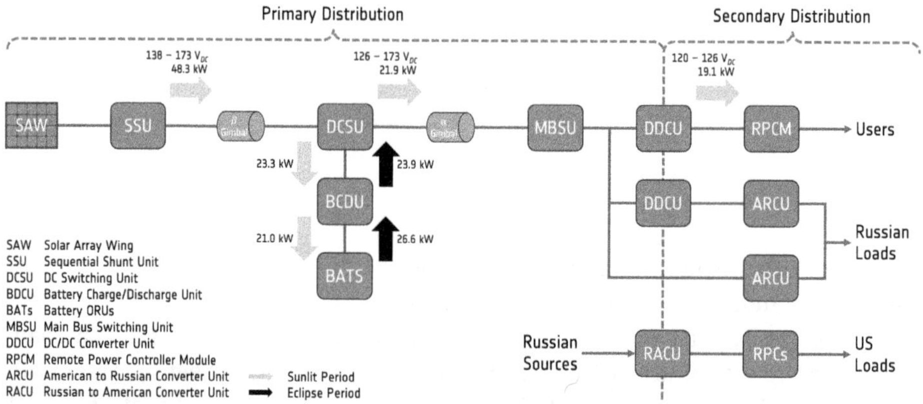

Figure 5.1 Simplified schematic of the Electrical Power System (EPS) of the International Space Station (ISS) (image: data from Gietl et al. [328])

The international ISS EPS can be subdivided into a primary power bus and a secondary power bus. Sometimes the primary bus is also referred to as transmission bus because it is utilized to transfer electrical power from the Solar Array Wings (SAWs) and batteries mounted on both ends of the Integrated Truss Structure (ITS) to the Main Bus Switching Unit (MBSU). In this case, the secondary power bus is referred to as the main bus, since it is available to user loads. The purpose of the primary bus is to generate, store, and distribute electrical energy. The bus voltage of the primary bus has a nominal value of 160 V (DC), whereby it can vary between 115 V (DC) and 173 V (DC). The Solar Array Wings are connected to Sequential Shunt Units (SSUs) and the batteries are connected to the Battery Charge Discharge Units (BCDUs). The SSU and the BCDUs are connected to the Direct Current Switching Units (DCSUs). Electrical power is transferred from the DCSUs through the alpha gimbals to the Main Bus Switching Unit (MBSU) in the Integrated Truss Structure (ITS). The MBSU consists of a set of relays, each of which is capable of interrupting up to 350 A. Even larger fault protection relays are implemented in the primary bus. The MBSU features 8 input channels. One SAW, one battery system, as well as SSU, DCSU, and BDCU are connected to each of the input channels of the MBSU. DC/DC Converter Units (DDCUs) and Remote Power Controllers (RPCs) are connected to the output channels of the MBSU. Because of this, the MBSU is integrated as a primary switching point between the primary bus and the secondary bus. In emergency situations, it is also possible to combine multiple channels of the MBSU. The primary bus is connected to the secondary bus via DDCUs that themselves are supplied by the MBSU. The DDCUs are utilized to step the voltage down to the user voltage in the magnitude of 124.5 V ± 1.5 V (DC) and to isolate loads connected to the secondary bus from the primary bus. The secondary bus has been implemented as a fully regulated bus. Each DDCU is capable of handling up to 6.25 kW and transferring it to user loads via Remote Power Controller Units (RPCs). Within the DDCUs, current limiting in the magnitude of 150% has been integrated. On board of the ISS, DDCUs with a total power of 400 kW are integrated. The RPCs are integrated in order to control the power flow from the DDCUs to the user loads. In overcurrent or undervoltage conditions, the RPCs disconnect loads from the DDCU to which a failed load is connected to provide fault isolation close to a failed load. 6 different types of RPCs are implemented in the ISS EPS for currents from 3.5 A to 65 A and some of these also have integrated current limiting systems. RPCs with a total power of 2 MW are integrated on board of the ISS. The ISS EPS is controlled by a network of different computers that are linked by the standard MIL-STD-1553. These computers provide functionalities like solar array tracking, thermal control, and energy storage control as well as power balance control.

The ISS employs solar energy as primary source of energy. During one orbit, the ISS experiences sunlit conditions for 90 minutes, followed by an eclipse period for the duration of 35 minutes. Solar cells are utilized in the ISS EPS as generative subsystems. These are integrated in 8 Solar Arrays Wings (SAWs). Each SAW consists of 2 solar array blankets. Each solar array blanket consists of 32,800 matched crystalline Silicon c-Si solar cells featuring an efficiency of 14.5% at BOL and an expected life duration of 15 years. Each solar cell has a front area size of 8 cm^2 and is covered with cover glass. The solar cells are interconnected by flexible circuits that are integrated in a Kapton/Copper/Kapton laminate. Each Solar Array Wing consists of 82 solar cell strings, and each solar cell string consists of 400 solar cells connected in series. Bypass diodes are integrated in the SAWs. For this integration, the solar cells of the solar cell strings are subdivided into sets of 8 cells connected in series, and to each of these sets, one bypass diode is connected anti-parallel. Each solar array string has a voltage of 160 V (DC) at the Maximum Power Point at BOL and each Solar Array Wing (SAW) has a peak power of 31 kW. After 15 years in orbit, the peak power reduced to 26 kW as it has initially been expected. At BOL, a total power generation capacity of 248 kW has been installed. The SAWs have been implemented as flexible deployable structures and have been rolled in for their transport to the ISS. When rolled out, the length of one SAW is 32 m. Figure 5.2 shows a part of the EPS in the ITS with deployed SAWs.

Figure 5.2 Astronauts during an Extravehicular Activity on the ISS ITS truss: Different EPS subsystems like the Alpha and Beta Rotational Joints, the Solar Array Wings, the radiators or the battery ORUs of the IEA systems can be seen (image: © NASA [342]).

The replacement of the original ISS SAWs with new SAWs started in mid-2021. The new integrated SAW system is referred to as iROSA (ISS Roll Out Solar Array). Each of these units will be installed on the mast canisters of the original SAWs and the new SAWs will be tilted to the original ones by 10°. A rolled out iROSA system is 18.2 m long and 6 m wide. An iROSA system has an output power of 20 kW [344]. An image of a rolled-out and installed iROSA system is shown in Figure 5.3.

Figure 5.3 Installed and rolled-out iROSA system on the ISS (image: © NASA TV [344])

On each end of the Integrated Truss Structure (ITS), 4 Solar Array Wings (SAWs) are integrated. The SAWs are connected to the Integrated Equipment Assembly (IEA) system via the beta gimbal rotational joints. On each end of the ITS, 2 IEA systems are integrated. The IEA system is an integrated system with a total mass of 17 T per system. Besides others, each of these systems consists of 2 solar array regulators, 12 battery ORU systems, 6 battery charge discharge regulators, 1 Electronic Control Unit (ECU) as well as a thermal control system for the thermal management of batteries and power electronics. The IEA systems are connected to the ISS EPS via roll-rings that are providing a continuous connection. These are integrated into the alpha gimbal motorized joints. Figure 5.4 shows a systematic view of one IEA System.

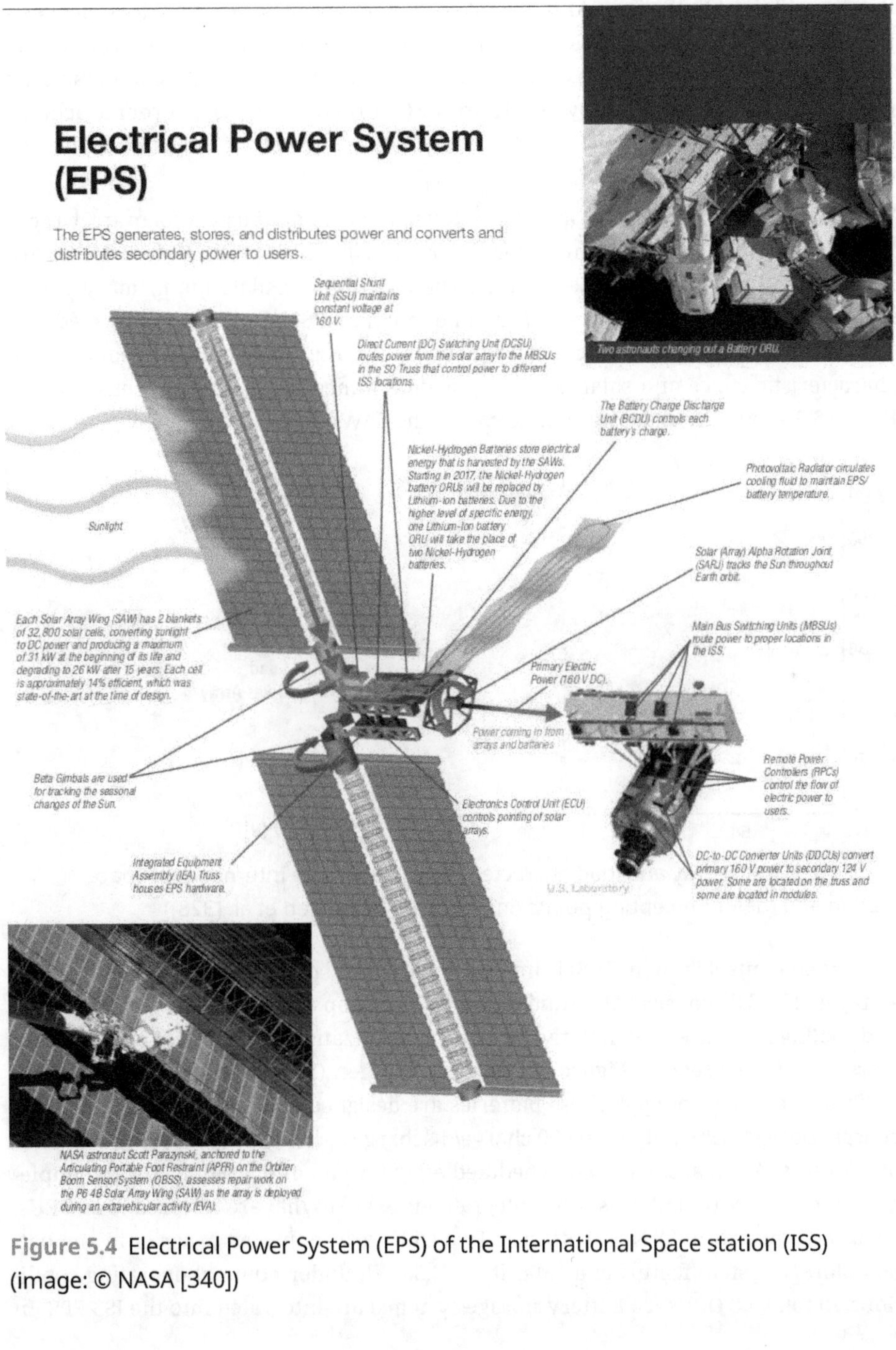

Figure 5.4 Electrical Power System (EPS) of the International Space station (ISS) (image: © NASA [340])

A dual axis tracking mechanism is implemented for all Solar Array Wings, utilizing two rotational motorized joins, the so-called alpha- and beta-gimbals. These gimbals are sometimes also referred to as Solar Alpha Rotational Joint (SARJ) and Solar Beta Rotational Joint (SBRJ). The SARJs are integrated on both ends of the ITS, in between the ends of the ITS and the IEAs, in order to rotate the SAWs around the axis of the ITS. The SBRJs can rotate the SAWs around their axis. The movement of the motors in the rotational joints is controlled by the Electronic Control Units (ECUs) in order to achieve optimal solar tracking results. Both types of rotational joints are indicated in the systematic overview in Figure 5.4.

The Solar Array Wings (SAWs) are connected to the primary bus via solar array regulators. Sequential Switching Shunt Regulators (S3Rs) utilizing solid state switches are implemented as Solar Array Regulators. The S3R units regulate the primary power from the SAWs. For each of the 8 SAWs, one individual S3R unit is implemented and each unit is operated at a switching frequency of 20 kHz. The operational point on the characteristic curve of a solar cell string is determined by the primary bus voltage. Figure 5.5 shows the characteristic curves of the SAWs and of the ISS load.

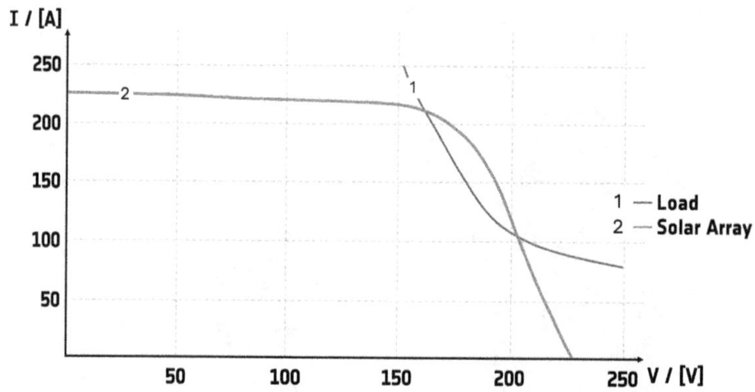

Figure 5.5 Solar array and load characteristic curves of the International Space Station and their intercepting points (image: data from Gietl et al. [328])

A Thermal Control System (TCS) is implemented for the power electronics and energy storages. The TCS consists of a fluid-based cooling loop and a radiator. The radiators and cooling loops are shown in the overview visualization in Figure 5.4 and the radiators can also be seen in Figure 5.2. Nickel-hydrogen (NiH_2) batteries were initially utilized as energy storages. These batteries are designed for a nominal Depth of Discharge (DoD) of 35% and for 40,000 charge/discharge cycles during an operational life of 6.5 years. A replacement was scheduled after 5 years of operation. For the implementation, the NiH_2 batteries were integrated into boxes that are referred to as ORUs. Each ORU consists of 38 NiH_2 batteries. Two ORUs are connected in series to one battery storage system featuring a capacity of 2.5 kWh under nominal operating conditions. In total, 48 ORUs (24 battery storage systems) are integrated into the ISS EPS. In order to determine the State of Charge (SoC), individual pressure and temperature monitoring systems are implemented. Initially, it was intended to replace the NiH_2 batteries after they reached their End of Life (EOL) with a Flywheel Energy Storage System (FESS), but these plans were shifted towards the lithium-ion technology. One lithium-ion ORU allows to replace two NiH_2 ORUs connected in series. The battery replacement started in 2017. A photo taken during the battery replacement is shown in Figure 5.6. The technical parameters of the old NiH_2 batteries and the new lithium-ion batteries are summarized in Table 5.1.

Table 5.1 Technical parameters of ISS NiH_2 and lithium-ion batteries [148]

	ISS NiH_2	Li-Ion
rated capacity	81 Ah	134 Ah–144 Ah
energy density	65 Wh/kg	150 Wh/kg
nominal cell discharge voltage	1.25 V	3.6 V
self-discharge rate	7%/day at 20 °C	0.05%/day
cycle life in LEO	60,000–75,000 cycles at 20–30% DoD	58,000 cycles at 20–25% DoD
specific cycle life	6.5 years at 35% DoD	10 years at ISS power level
storage life	4 years	6 years
overcharge	tolerant	controlled by 2 FT design
total energy storage	8 kWh (2 ORUs combined)	15 kWh
battery weight	337.5 kg	188.2 kg

Figure 5.6 Replacement of old nickel-hydrogen ORUs with new lithium-ion ORUs (image: © NASA [345])

In order to connect the battery storage systems to the ISS EPS, 24 combined Battery Charge Discharge Units (BCDUs) are integrated into the ISS EPS. These BCDUs utilize a Weinberg buck-boost topology and offer electrical isolation between the batteries and the primary bus. During discharge, the BCDUs have a discharge power of 4.4 kW under nominal operating conditions. The Weinberg BCDUs are operated at a switching frequency of 80 kHz. On the primary bus side, the voltage varies between 115 V (DC) and 173 V (DC) during operation, and on the battery side, the voltage varies between 121.8 V (DC) and 124.2 V (DC).

The Columbus laboratory is supplied by two independent DDCUs. Each DDCU supplies 1 of 2 redundant Power Distribution Units (PDUs). These PDUs supply Columbus with a total power of up to 18 kW, from which 13.5 kW can be accounted for payload power. Columbus systems are supplied with 120 V (DC) and 28 V (DC). Payloads are supplied with 120 V (DC). The external facilities outside the Columbus laboratory can be supplied over two power lines featuring a power transmission capacity of up to 1.25 kW per power line. A schematic of the internal power distribution system inside the Columbus laboratory is visualized in Figure 5.7.

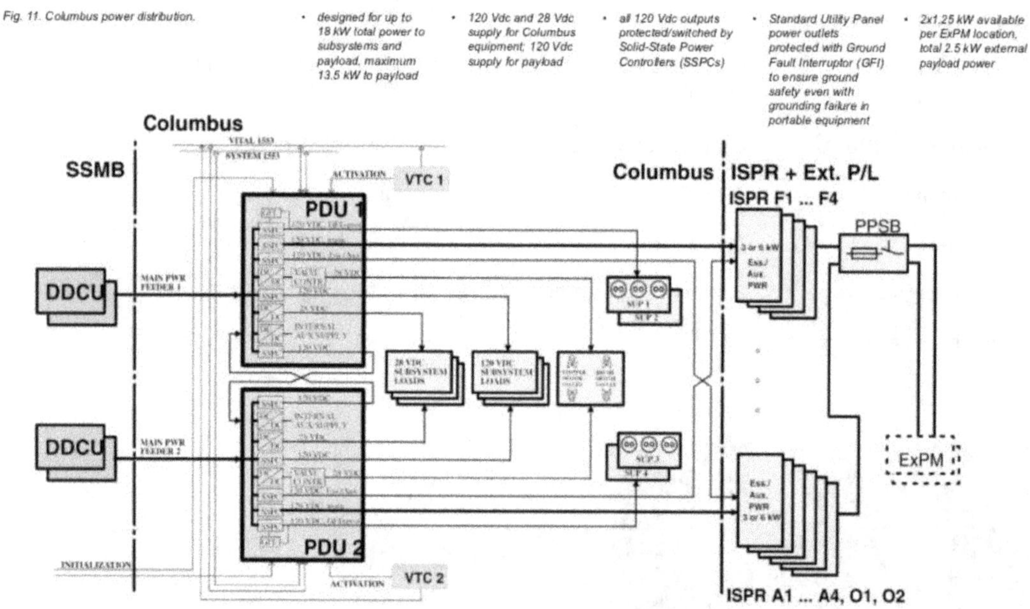

Figure 5.7 Electrical Power System of the Columbus laboratory on board of the International Space Station (image: © ESA [341])

The SAWs are in contact with the ionized space plasma. Due to the high Maximum Power Point voltage of the ISS SAWs, arching between the SAWs and the space environment is possible if no countermeasures are implemented. As a countermeasure, a plasma contactor is integrated on the ISS ITS that is constantly releasing a plume of ionized xenon plasma. The plasma is acting as a low impedance conducting bridge between the space environment and the ISS.

5.2 Electrical Power System Options for a Possible Lunar Base

Between 1989 and 2017, numerous studies were carried out with a focus on the development of an optimal design for the Electrical Power System (EPS) of a possible lunar base. Some of the most interesting studies are the Exploration Systems Architecture Study (ESAS) carried out by NASA in 2005 and the study Power Systems Concepts for a Lunar Outpost carried out by Khan et al. in 2006 [75]. Available studies discuss EPS concepts that are similar to the EPS of the International Space Station (ISS), but that are adapted to the lunar environment. The EPS concepts proposed by Mason and Bloomfield as well as Khan et al. employ nuclear fission reactors together with photovoltaic power generators and energy storages [75] [94]. The different concepts proposed by the ESAS study are based on nuclear power generation, a combination of solar generators with Regenerative Fuels Cell storages as well as hybrid EPS employing different energy sources together. Different subsystems are already under development for the application on the lunar poles, such as nuclear fission reactor systems or vertically deployable solar arrays. Visualizations of different subsystem concepts are shown in Figure 5.11, Figure 5.10, and Figure 5.12. Visualizations of different EPS concepts are shown in Figure 5.8 and Figure 5.9.

Figure 5.8 Artwork of a lunar surface power system based on solar energy and regenerative fuel cell system (image: © NASA, Hickman et al. [74])

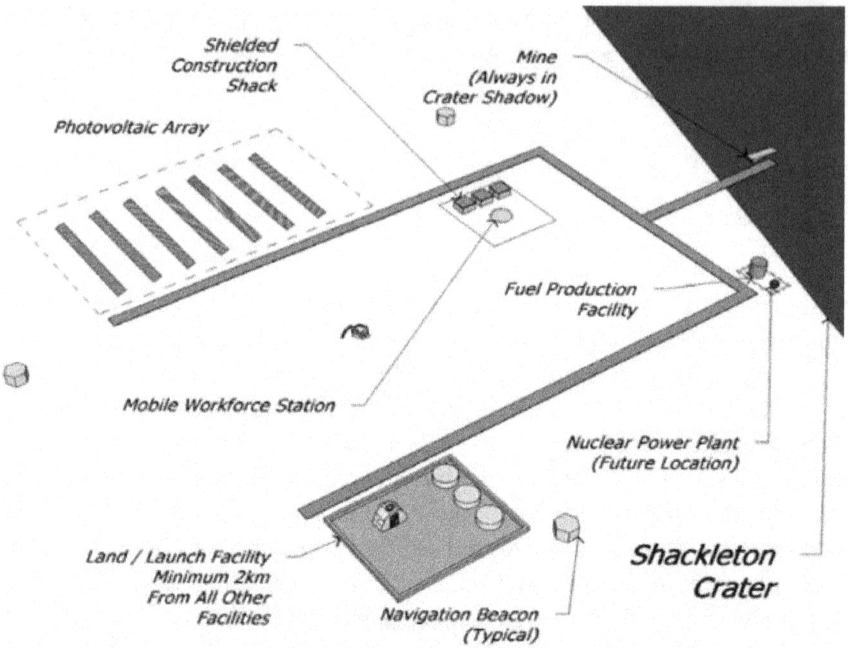

Figure 5.9 Visualization of a lunar Electrical Power System based solar and nuclear power (image: © NASA, Khan et al. [75])

Khan et al. also did a detailed assessment of the selected available technologies and developed a technology scoring model in order to estimate which of these technologies would be most appropriate for the application in a lunar EPS [75]. They estimated scores between 1 and 3 for each technology in different scoring categories like constructability, maintainability, reliability, adaptability, system safety or technology risks, and calculated final technology scores as a sum of the individual category scores to find the most appropriate technologies for lunar application. The results of their assessment are summarized in Table 5.4 for energy storage technologies, in Table 5.5 for AC- and DC-based transmission, and in Table 5.6 for the transmission voltage level.

A lunar Electrical Power System (EPS) will feature generative systems, energy storage systems, power management, and distribution systems, as well as subsystems that are supplied by the EPS and therefore account as EPS loads. In addition to this, backup and emergency systems have to be integrated into the EPS.

A lunar base EPS will be implemented incrementally over longer periods of time and EPS subsystems will be delivered to a lunar base construction site with regular cargo and crew missions. Following this, the topology of the lunar Power bus also needs to allow to add systems incrementally after they have been delivered to the lunar surface. The power bus topology needs to be able to adapt to different and dynamic load conditions, since the type and number of systems connected to the power bus will change over time. Due to the environmental conditions on the lunar surface, the subsystems of a lunar base EPS need to be able to survive lunar day and night phases.

A future lunar base will be in operation for a minimum period of 10 years. Between 10 kW and 150 kW of electrical power needs to be accounted for the operation of a lunar base. When resource mining and ISRU activities shall be included in lunar base activities, additional 30 kW to 50 kW of electrical power needs to be accounted. Additional power needs to be accounted for heating systems in order to keep the temperature of a lunar base during the lunar night at an acceptable level. Thermal energy could be supplied by electrical heaters or by nuclear heaters, but also the by-product thermal energy from nuclear generator systems could be utilized. Initially, the EPS of a lunar base will feature a power capacity of between 25 kW and 83 kW. The EPS capacity will increase to 100 kW or more during later construction phases. The application of demand-side management strategies is recommended in solar-powered lunar EPS. High power demand periods should be scheduled during sunlit periods, and low power demand periods should be scheduled during the lunar night. In this way, activities like ISRU activities would only be scheduled during the lunar day, but systems like the environmental control and life support system would be in operation continuously. The application of demand-side management strategies allows to reduce the load requirement to $3\frac{\text{kW}}{\text{Astronaut}}$ during the lunar night (Table 5.2).

Table 5.2 Lunar base power system power demand assumptions [15] [74] [144]

	Demand assumptions for a lunar base power system
power demand during lunar day	base power between 10 kW and 150 kW additional $> 3\frac{\text{kW}}{\text{Astronaut}}$
power demand during lunar night	$> 3\frac{\text{kW}}{\text{Astronaut}}$
ISRU activities	30–50 kW

A lunar base EPS design should offer the possibility of incremental growth. In this way, first EPS subsystems could be implemented during the initial phase for the supply of the lunar base in its initial state. As the power demand increases, the EPS could be expanded with additional units, or the initial units could be utilized as backup systems after additional systems have been implemented.

A lunar base can be supplied by a primary energy source independent from solar irradiance that can be utilized permanently. An example for such a primary source of energy are nuclear isotopes. Solar irradiance can also be utilized as a primary energy source when additional energy storages systems are utilized for the supply of the lunar base during periods where solar irradiance is not available. In this case, the irradiance conditions need to be defined accurately in order to define the required system parameters. Hickman et al. and Seedhouse recommend to implement a lunar EPS based on nuclear generators during the initial phase and expand the initial EPS with additional solar energy and energy storage systems during later phases [14] [74]. This offers advantages when high power demands are expected during initial phases of a lunar base. When high power demands are only expected in later implementation phases, then the initial implementation of a solar energy- and energy storage-based EPS that is expanded with nuclear-based systems or additional solar energy and energy storage systems during later phases, is also reasonable in the initial phase of a lunar base.

Photovoltaic and solar thermal generators are possible technological options for harvesting solar energy on the Moon. Due to their legacy in space application, photovoltaic generators are much more likely to be implemented on the lunar surface than solar thermal generators. Photovoltaic generators are lightweight, modular, and widely applied in space. Therefore, photovoltaic generators can be classified as a reliable space technology. This technology features relatively low development costs, since space-certified generators based on photovoltaics are already available, but degradation of the generation systems over time has to be expected. Resulting from the large collector size, solar generators cannot be delivered to the lunar surface in their final integration. Instead, generators have to be packed in order to fit in the payload bay of a transfer vehicle before they can be deployed on the lunar surface. A possible way to achieve this is the utilization of flexible solar arrays.

Solar generators are highly location-sensitive and only mass competitive with nuclear-based generators in the case of nearly perfect irradiance conditions. It is therefore important to optimize the incidence conditions of solar irradiance on solar generators in order to maximize the power yield. This can be achieved by the utilization of mounting and tracking systems. However, the construction of extensive mounting and tracking system on the lunar surface can be an engineering challenge. While dual axis tracking systems offer up to perfect solar tracking, extensive efforts are also required for the construction of those systems in most locations on the lunar surface. In polar locations, highly similar tracking results can be achieved with vertical solar generators that track the Sun by a rotation around their vertical axis. Because of this, the companies Lockheed Martin, Astrobotics and Honeybee Robotics have already been selected by NASA to develop such vertical solar generator systems. These vertical solar generators shall have a height of 9.75 m and they should be able to autonomously deploy and retract the solar generators repeatedly. Figure 5.10 shows such a vertical solar generator.

Figure 5.10 Rendering of the conceptual design of a vertical solar array for the lunar surface (image: © NASA [77])

Another approach that could the integrated by astronauts would be tilting the solar generators to the lunar equator by a static angle without any active motor system. This would significantly increase the incidence conditions compared to an uninclined generator, but the variations in the solar incidence angle during the lunar day would still be high. An interesting approach is the deployment of two solar generators together as a tent-shaped generator from which one generator would be facing in a western direction and the other one would be facing in an eastern direction. Such a system would be stable and could be deployed in a simple way. The most constant power yield with such a system could be achieved when the generators are each tilted by 60 degree to the lunar surface. An artwork of a tent shaped generator is shown in Figure 5.11.

Figure 5.11 Artwork of a self-deployable tent solar array (image: © NASA, Hickman et al. [74])

The development of the combination of a solar generator and storage system units with an output power of 25 kW during the lunar day and an output power of 12.5 kW during the lunar night was recommended by Hickman et al. [74]. They estimated a demand for 3 units of those systems.

Mintz et al. recommend to design solar generators in a way that they feature a voltage of 150 V (DC) and a current of 40 A [87]. When alternating technology shall be utilized for the lunar energy grid, Mintz et al. recommend converting the solar generator output voltage to 120 V (AC) with a frequency of 60 Hz [87].

Secondary batteries and Regenerative Fuel Cells (RFCs), supercapacitors, flywheels, as well as gravity or thermal energy-based energy storage systems are conceivable energy storage systems for a lunar EPS. Systematic options are reduced to secondary batteries and Regenerative Fuel Cells, resulting from the requirements of a lunar base. Table 5.3 shows a review of the technical parameters of different storage technologies and Table 5.4 shows the assessment of the different technologies.

Table 5.3 Design parameters of different energy storage technologies [75]

Technical Parameters	Cryogenic Fuel Cell	Batteries	Flywheels
DoD	100%	80%	100%
energy storage capacity requirement	$6449\ \text{kWh} \cdot \frac{100\%}{\text{DoD}} = 6449\ \text{kWh}$	$6449\ \text{kWh} \cdot \frac{100\%}{\text{DoD}} = 8062\ \text{kWh}$	$6449\ \text{kWh} \cdot \frac{100\%}{\text{DoD}} = 6449\ \text{kWh}$
storage mass	4300 kg–6449 kg	32,248 kg–80,620 kg	214,978 kg
energy density	1.0 kWh/kg–1.5 kWh/kg	0.1 kWh/kg–0.25 kWh/kg	0.03 kWh/kg

RFCs offer advantages as long duration energy storages in order to supply a lunar base with electrical energy for the duration of the lunar night in the magnitude of up to 13.5 days in most lunar locations. Because of this, RFCs are often considered as energy storage systems for the application on the lunar surface. Regenerative Fuel Cells employing a reaction between hydrogen and oxygen are of potential interest. These RFCs have an energy density of around $500\frac{\text{Wh}}{\text{kg}}$ when reactant chemicals are stored in gaseous state and of between $1\frac{\text{kWh}}{\text{kg}}$ and $1.5\frac{\text{kWh}}{\text{kg}}$ when reactant chemicals are stored in cryogenic state [74]. The equipment demand for the liquification of cryogenic reactants can be reduced by placing the cryogenic tanks in the cold-biased, permanently shadowed locations of craters. However, in this case the demand for piping and equipment would also be increased in order to transport the reactants from the bottom of a crater to the RFC system outside the crater.

Table 5.4 Scientific assessment of different energy storage technologies [75]

Evaluation criteria	Cryogenic Fuel Cell	Batteries	Flywheels
constructability	3	3	2
maintainability	2	2	2
reliability	3	3	2
adaptability	3	2	2
system safety	3	2	2
technology risks	2	2	1
technology score	16	14	11

Different battery technologies are available for the application in space and each of these technologies is featuring different boundary conditions and technology-specific parameters. Secondary batteries are well-suited for the application in space systems and are therefore often utilized in spacecraft EPSs. Because nickel-hydrogen batteries are not mass- and volume-compatible with lithium-ion batteries and because lithium-ion batteries offer a wider operational temperature range, nickel-hydrogen batteries do not need to be taken into account for future implementations and the implementation in a lunar EPS. Energy storage systems will be the mass- and volume-driving factor of a possible solar energy- and battery storage-based lunar EPS in which they would account for between 70% and 90% of the EPS mass [74].

Nuclear based generators offer advantages since they can continuously supply a lunar base with electrical power independently from the irradiance conditions at the lunar base location, while also balancing low mass and volume with high output power. For smaller power supply requirements in the range up to 2 kW, Radioisotope Thermoelectric Generators (RTGs) offer many advantages. Because RTGs are relatively small, they can also be integrated into lunar surface mobile robotic systems. In order to meet power supply requirements of up to multiple megawatts, nuclear reactors utilizing a nuclear chain reaction need to be employed. The nuclear fission reactor type has been the preferred generator type in the ESAS study. Nuclear fission reactors are robust generator systems with low sensitivities to locations on the lunar surface. Following this, NASA already requested the companies Lockheed Martin, Westinghouse, and IX to develop a nuclear fission reactor system for the lunar surface. This nuclear reactor system should have an electrical output power of 40 kW and it should be utilized on the lunar surface in locations where the electrical power generation with solar-based systems is challenging [76]. A rendering of such a nuclear fission reactor concept on the lunar surface is shown in Figure 5.12.

Figure 5.12 Rendering of the conceptual design of a nuclear fission reactor for the application on lunar surface (image: © NASA [76])

Alternating Current (AC) transmission systems and Direct Current (DC) transmission systems are possible systematic options for power transmission on the lunar surface. DC transmission systems features less power losses than AC transmission systems. However, power transformation and power conditioning can be implemented in a simpler way by employing AC systems because switches can switch at zero current and transformers can be utilized for the transformation between different voltage levels. The mass of AC systems usually has smaller values than the mass of DC systems. DC transmission systems are usually more complex than AC transmission systems, have more switching losses and require more efforts in fault protection. Power dissipation in DC transmission systems is mainly resulting from the internal parasitic resistance of wires. In AC transmission systems, the skin effect is the dominating power dissipation effect. Following this, the cables for different transmission methods need to be optimized with special regards of minimizing different power dissipation effects. In electrical fault conditions, high currents occur in DC transmissions systems, but during nominal operation, DC systems have less power dissipation.

It is recommended to lay power transmission lines directly on the lunar surface since this approach is relatively simple and can be carried out with a minimum number of additional systems. Burying the cables would be an engineering and construction challenge, especially because the subsurface material can be rocky. However, placing DC power transmission lines on the lunar surface will result in a conductive coupling between the lunar soil and the power lines that needs to be taken into account during systems design. AC transmission systems utilize a certain transmission frequency. The power converter mass decreases with increasing frequency, but the power line mass increases with increasing transmission frequency as a countermeasure for skin effect losses. Because of this, Khan et al. recommend to employ relatively low transmission frequencies [75]. An assessment of DC and AC transmission systems is summarized in Table 5.5.

Table 5.5 Scientific assessment of different energy transmission technologies [75]

Evaluation criteria	AC transmission	DC transmission
constructability	3	3
maintainability	2	3
reliability	2	3
adaptability	3	3
system safety	3	2
technology risks	2	3
technology score	15	17

Transmission voltages greater than 300 V are classified as high voltages in space applications. An increase in the transmission voltage results in a decrease in the transmission currents, while meeting the same power transmission requirement, which allows to design power lines in a lighter way. Minimum transmission system masses can be observed at voltages in the magnitude of 5 kV. For later phases of a lunar EPS, Khan et al. recommend the implementation of a high voltage transmission grid [75]. An assessment of high and low voltage levels is summarized in Table 5.6.

Table 5.6 Scientific assessment of different voltage levels [75]

Evaluation criteria	High voltage > 300 V	Low voltage < 300 V
constructability	3	2
maintainability	3	2
reliability	3	2
adaptability	3	2
system safety	2	3
technology risks	2	3
technology score	16	14

Power systems require a grounding concept. The lunar soil is not classified as a good conductor, and because of this, grounding utilizing the lunar soil cannot be taken into account for a lunar EPS.

The Power Management and Distribution System (PMAD) consists of the power transmission cables, power conditioning units, power conversion units, power distribution units, and switchgear. In order to take incremental construction and development of a lunar base into account in the PMAD system design, Khan et al. recommend to utilize a similar PMAD system as it has also been utilized on board of the ISS and upgrade it during later phases with additional power generation and energy storage systems [75].

Solar and nuclear generators have been classified as the most potential generator technologies for the lunar surface. Battery storages and Regenerative Fuel Cells are the most potential energy storage technologies for lunar exploration. Lunar EPSs will also feature a number of power converters and regulators as well as a Power Management and Distribution System and a transmission grid. A standard topology for a future lunar EPS is shown in Figure 5.13.

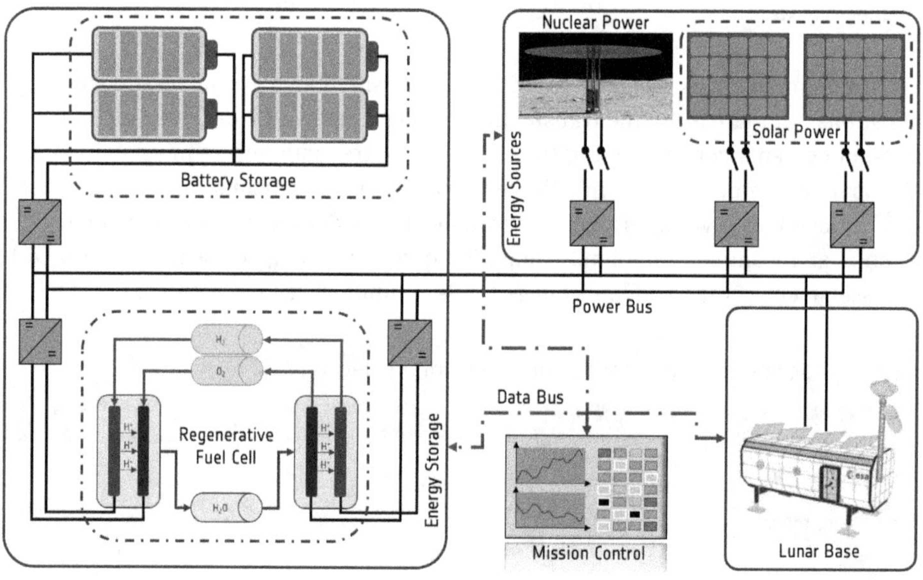

Figure 5.13 Topology of a possible lunar base power system (image of nuclear reactor: © NASA, image of habitat: © ESA [348] [349])

6 Design of an Electrical Power System for Lunar Application

Different design methods are available for different spacecraft and spacecraft subsystems. These design methods optimize system designs by a minimization of parameters, such as system mass, volume, or cost, and a maximization of parameters, such as the mission return, the modularity, or the system availability. Different system designs are usually developed and optimized during a number of iterations. In the end of the system's design process, an evaluation of the different system designs is carried out to compare the designs against each other and the design with the highest score value is chosen.

6.1 Standards and Requirements for Space Systems

A number of documents are available that define technological standards that have to be considered in the design of space systems [20] [350] [351] [352] [353] [354]. The Human-Systems Integration Requirements (HSIR) document CxP 70024 of the Constellation program released by NASA defines the requirements for human-rated space systems [354]. The documents "Power Standard, PSS-02-10, Volume 1, Issue 1" and "Rationale for the Power Standard, PSS-02-10, Volume 2, Issue 1," released by the Power and Energy Conversion Division at the European Space Agency, are of potential interest for the development of Electrical Power Systems for space systems [20] [351] [352]. The "Power Standard" includes standard definitions and requirements for space EPS and the "Rationale for the Power Standard" contains additional explanations for the definitions and requirements in the Power Standard document [20] [351] [352]. Definitions and requirements for power system interfaces can be found in the documents ECSS-E-ST-20-20C and HB-20-20-C [353] [355].

The bus voltage levels 28 V (DC), 50 V (DC) or 120 V (DC) can be utilized in space EPS [20]. For a future lunar base, 120 V (DC) is the most suitable voltage level because it is defined as the harmonized NASA/ESA power bus voltage for human-rated space systems [20]. A power margin of 10% shall be considered for average and peak power demand in the EPS design [20]. The power bus voltage shall be regulated in a way that possible variations are limited to ±1% of the bus voltage [20]. Voltage spikes resulting from regulators connected to the EPS shall be limited to ±1% of the bus voltage and these should decay with a time constant of 2 ms [20]. The power distribution inside a spacecraft shall be designed utilizing one or more power buses utilizing DC systems [352]. Power buses need to be designed single-point-of-failure-free [351].

6.2 Physical Lunar Base Construction and Design

A possible method for the construction of a lunar base is the combination of space station-like modules with inflatable modules in order to decrease the volume during the transport phase to the lunar surface (Figure 6.1). It is also recommended to bury lunar base modules under a 1.5 m thick layer of lunar regolith, which can act as protective layer against micrometeoroid impacts and radiation [94]. It is recommended to utilize robotic systems for the construction of a lunar base in order to limit the employment of astronauts in these activities.

Figure 6.1 Example of a lunar base concept consisting of ISS-like modules and inflatable modules buried under a layer of processed regolith (image: © ESA/ Foster + Partners [356])

The subsystems of a lunar EPS will be distributed over the lunar surface. Subsequently, the implementation of the power transmission grid over intermediate and longer distances will also be required [14]. Seedhouse estimates a total transmission length of 5 km [14].

The application of dust mitigation methods will be important on the lunar surface as a countermeasure for regolith accumulation on solar generators and radiators, as well as to minimize the contact of an astronaut crew with the lunar surface material. Dust mitigation methods could include technical dust mitigation and removal methods, such as the grounding of surfaces, and it could also include restricted areas. Close to solar generators and radiators, only a minimum of operational activities should be carried out. High dust facilities, such as landing infrastructure or ISRU facilities, need to be located in a sufficient distance from optical surfaces, so that these cannot be reached from the dust released close to high dust facilities. It is highly recommended to implement optical surfaces, such as solar generators and thermal radiators, at a distance of multiple kilometers from any location of operation in order to limit the accumulation of dust on these surfaces. Possible mining, soil processing, and ISRU facilities should be located 5 km away from a future lunar base. Propellant processing plants should be located in the same location as the other soil processing and ISRU facilities. The distance between a launch pad facility and the propellant processing plant shall be smaller than 1 km in order to minimize the propellant transport infrastructure [94]. Different distance requirements are summarized in Table 6.1.

Table 6.1 Distance requirements of lunar base facility design [14] [15] [74] [94]

	Distance
distance between base and ISRU facilities	5 km
distance between launch pad and ISRU facilities	1 km
distance between nuclear reactor and lunar base	2 km radius keep-out zone, needs to be buried
distance between RTG and lunar base	1 km radius keep out zone
total length of the EPS transmissions cable	5 km
distance between solar generators and areas of surface operation	multiple kilometers

In case that nuclear systems are employed in a lunar EPS, additional efforts need to be undertaken in shielding methods in order to protect the lunar surface crew from the potential hazardous risks of radiation. A relatively simple approach is to bury a nuclear generator under the lunar surface. Additionally, there would be a restriction zone around a nuclear generator in which the presence of the surface crew would be restricted in order to reduce radiation risks. Such a restriction zone would feature a minimum radius of 1 km for RTGs and of 2 km for nuclear reactors [15] [74]. Nuclear generators require radiator systems that are employed to dump thermal energy that has not been converted into electrical energy during low demand periods. The same dust mitigation requirements that apply for solar generators can also be utilized for radiators, since accumulated dust on the radiator would alter its ability to radiate thermal energy.

6.3 Spacecraft Design

A spacecraft can be subdivided into the parts "bus" and "payload." The "payload" part consists of all systems that contribute to the mission goal. Examples for a payload are crew cabins of crewed spacecraft or scientific instruments. The "bus" part includes all spacecraft subsystems that are implemented in the spacecraft in order to enable the operation of the payload during a mission. Examples for bus subsystems are thermal management systems, command and telemetry systems, or electrical power systems.

Spacecraft are designed utilizing a dedicated design process. In the first step, mission and systematic constraints as well as design requirements have to be analyzed. Requirements are determined by the mission architecture and the mission concept as well as the payload requirements. Based on the results of the first step, a design approach for the spacecraft system configuration is developed in the second step. In the third step, budgets for power, propellant, and mass are factored into the design. Based on this, preliminary subsystems are designed and the configuration of the spacecraft is developed in the following steps. The outcome of a development cycle is evaluated and discussed in the sixth step [98]. The design is enhanced throughout multiple iterations of this process.

6.4 Power Budget

Power budgeting is a method to estimate the total power demand by spacecraft subsystems during different mission phases. It needs to be determined for all mission phases [98]. For each mission phase, different system states and environmental conditions can be differentiated [98]. It is also recommended to add a safety margin to the power budget.

Khan et al. differentiate several construction phases of a possible lunar base [75]. For the initial operational phase of a lunar base, they estimated 43 kW for ISRU processes, 1 kW for general science operation, 5 kW for lunar surface rover charging and 25 kW for energy storage charging [75]. Including a safety margin of 10%, Khan et al. estimated a power budget of 81.4 kW for an initial phase lunar base [75]. The power budget estimation table used by Khan et al. can be found in Table 6.2 [75]. However, this power budgeting method is very broad and does not take into account different operational conditions.

Table 6.2 Power budget estimation table for different operational phases of a lunar base (Khan et al. [75])

Load	Nominal power / W			
	Phase 0	Phase 1	Phase 2	Phase 3
ISRU processes	43 kW	...	...	...
general science	1 kW	...	...	...
rover recharge	5 kW	...	...	...
energy storage recharge	25 kW	...	...	...
sum of required loads	**74 kW**	...	...	...
safety margin 10%	$74\,\text{kW} \cdot \frac{10\%}{100\%} = 7.4\,\text{kW}$	...	...	...
total load requirement	**81.4 kW**	...	...	...

It is expected that the power demand of a lunar EPS will increase over time and it is also expected that the power generation capacity of solar generators will decrease over time. Subsequently, safety margins need to be applied to the power demand of different units. A good approach is to apply different margins to different loads with respect to their development status. A recommendation is to apply a margin of 20% to the load demand of newly developed units, of 10% to the load demand of newly developed units utilizing a flight-proven design, and of 5% to the load demand of flight-proven units [155]. An additional margin of between 10% and 20% has to be applied to the total energy demand [75] [155]. Power dissipation in power converters and regulators can be considered with an additional safety margin of 5%, and a power availability at the End of Life can be ensured by adding an additional power margin of 5% [155]. An overview of the individual power margins is shown in Table 6.3.

Table 6.3 Table of safety margin parameters [75] [155]

	Safety Margin
newly developed units	20%
newly developed units where the design is passed on to a flight-proven unit	10%
flight-proven units	5%
margin for the sum of all loads	10–20%
margin accounting for power dissipation in power electronics	5%
margin ensuring power availability at the End of Life	5%

A dedicated process is available for the determination of the power budget. In the first step of the process, data regarding the nominal load demand of all loads connected to the spacecraft EPS needs to be collected. The time duration during which a certain load will be in operation needs to be estimated based on the operational schedule in the second step. For lunar EPS, it is recommended to differentiate between the operation during the lunar day and during the lunar night. The energy demand of a load can be calculated as the product of the nominal load power and the operational period duration. This has to be calculated for each load, each mission phase, and each operational period in the third step of the analysis. An individual safety margin needs to be added to the individual energy demands. In the fourth step, the total electrical energy demand of all bus loads has to be calculated for each mission phase as the sum of the individual system energy demands. In addition to this, safety margins need to be added in order to account for increases in the demand, the degradation during operation, as well as the power dissipation in converters and regulators. In the last step, the total power budget is determined by dividing the total load energy demand, including the safety margins, by the duration of the lunar day or night period. Table 6.4 can be utilized for the power budgeting process.

The power budget can be seen as the mean power demand during nominal operation periods. In addition to this, also the peak power demand needs to be determined. A simple approach for the estimation it is to multiply the power budget by a factor of 3 [98]. However, this method has a high degree of uncertainty. Instead, it is recommended to determine the peak power demand from an analysis of the individual peak power demands. A possible method for an analysis is to define the load power as a time-dependent load profile for each load, to calculate the total load profile as the sum of the individual load profiles and to utilize the method presented here together with the time-dependent load profile. In this way, also periods of high and low demand can be determined. For a more detailed analysis of the power budget, it is recommended to utilize specialized software tools.

Table 6.4 Table for the detailed estimation of the power budget for a lunar base in a certain operational phase (based on the work of Mertens [151] for applications on the Earth, advanced for lunar EPS)

Load	Nominal power [W]	Operational period duration [h]		Load energy demand [Wh]	
		Lunar day	Lunar night	Lunar day	Lunar night
load 1	$P_{L1} = x_1$	$T_{D1} = y_{D1}$	$T_{N1} = y_{N1}$	$E_{L1,D} = \left(1 + \frac{\alpha_{S,L1}}{100\%}\right) \cdot P_{L1} \cdot T_{D1}$	$E_{L1,N} = \left(1 + \frac{\alpha_{S,L1}}{100\%}\right) \cdot P_{L1} \cdot T_{N1}$
load 2	$P_{L2} = x_2$	$T_{D2} = y_{D2}$	$T_{N1} = y_{N2}$	$E_{L2,D} = \left(1 + \frac{\alpha_{S,L2}}{100\%}\right) \cdot P_{L2} \cdot T_{D2}$	$E_{L2,N} = \left(1 + \frac{\alpha_{S,L2}}{100\%}\right) \cdot P_{L2} \cdot T_{N2}$
load 3	$P_{L3} = x_3$	$T_{D3} = y_{D3}$	$T_{N3} = y_{N3}$	$E_{L3,D} = \left(1 + \frac{\alpha_{S,L3}}{100\%}\right) \cdot P_{L3} \cdot T_{D3}$	$E_{L3,N} = \left(1 + \frac{\alpha_{S,L3}}{100\%}\right) \cdot P_{L3}P_{L3} \cdot T_{N3}$
...	...	...	...	...	...
load N	$P_{LN} = x_N$	$T_{DN} = y_{DN}$	$T_{NN} = y_{NN}$	$E_{LN,D} = \left(1 + \frac{\alpha_{S,LN}}{100\%}\right) \cdot P_{LN} \cdot T_{DN}$	$E_{LN,N} = \left(1 + \frac{\alpha_{S,LN}}{100\%}\right) \cdot P_{LN} \cdot T_{NN}$
total energy budget [Wh]				$E_{L,D} = \sum_{i=1}^{N} E_{Li,D}$	$E_{L,N} = \sum_{i=1}^{N} E_{Li,N}$
general safety margin α_{SG}				$E_{SG,D} = E_{L,D} \cdot \frac{\alpha_{SG}}{100\%}$	$E_{SG,N} = E_{L,N} \cdot \frac{\alpha_{SG}}{100\%}$
power dissipation safety margin $\alpha_{S,Loss}$				$E_{S,Loss,D} = E_{L,D} \cdot \frac{\alpha_{S,Loss}}{100\%}$	$E_{S,Loss,N} = E_{L,N} \cdot \frac{\alpha_{S,Loss}}{100\%}$
EOL availability safety margin $\alpha_{S,EOL}$				$E_{S,EOL,D} = E_{L,D} \cdot \frac{\alpha_{S,EOL}}{100\%}$	$E_{S,EOL,N} = E_{L,N} \cdot \frac{\alpha_{S,EOL}}{100\%}$
total energy budget incl. safety margin [Wh]				$E'_{L,D} = E_{L,D} + E_{SG,D} + E_{S,Loss,D} + E_{S,EOL,D}$	$E'_{L,N} = E_{L,N} + E_{SG,N} + E_{S,Loss,N} + E_{S,EOL,N}$
total power budget incl. safety margin [W]				$P_{L,D} = \frac{E'_{L,D}}{T_D}$	$P_{L,N} = \frac{E'_{L,N}}{T_N}$

6.5 Solar Generator Design

Solar generators offer the advantage to generate electrical power in situ. This type of generators is usually integrated into spacecraft EPSs together with additional energy storages in order to enable the supply of electrical power to the spacecraft power bus during periods where primary energy generation is not possible. Usually, one sunlit/

eclipse cycle is taken as a reference for the solar generator design. The solar generator system design has to allow to generate the energy during one sunlit period that is required for the supply of the EPS during one sunlit/eclipse cycle. Solar generators are usually designed utilizing the End of Life (EOL) output parameters of the utilized solar cells. In this way, it can be assured that the solar generator can still provide the required power to the bus after the degradation during the scheduled duration of the mission. As a side effect, there will be more power available from solar generators than required by the loads before the EOL of the solar generator. Solar generators need to be designed with reference to the power budget. The electrical energy that a solar generator can generate during a sunlit/eclipse cycle can be calculated as the product of the electrical output power and the duration of the sunlit period. The electrical output power of a solar generator is not necessarily equal to the Maximum Power Point power. An operation at the Maximum Power Point can only be assumed when MPPT solar regulators are utilized. When Direct Energy Transfer solar regulators are utilized, the operational point of the solar cells needs to be defined with reference to the Main Bus Voltage for this analysis.

Two different efficiency parameters need to be defined in the design process. The first efficiency parameter accounts for the power dissipation of the power bus occurring during sunlit periods, when electrical power is transferred from solar generators to electrical loads and energy storages, and the second efficiency parameter accounts for the power dissipation occurring during eclipse periods, when electrical power is transferred from energy storages to electrical loads. A good approach for a rough estimation during early design phases is to utilize a power bus efficiency of 85% for sunlit periods and of 65% for eclipse periods for EPSs utilizing Direct Energy Transfer solar power regulation, as well as of 80% for sunlit periods and of 60% during eclipse periods for EPSs utilizing Maximum Power Point Tracking solar power regulation.

In the next step, these efficiencies are employed to determine the required energy generation requirement during one sunlit/eclipse cycle to satisfy the expected load demand. The energy demands during a sunlit period and during an eclipse period can be calculated as the product of the power budget for these periods and the period duration. The resulting values need to be scaled utilizing the associated efficiencies. The total energy generation requirement for a sunlit/eclipse cycle can be calculated as the sum of the energy generation requirements for the sunlit and for the eclipse period.

$$E_{SAW,S,E} = \left(\frac{T_S}{\eta_{Bus,S}} + \frac{T_E}{\eta_{Bus,E}} \right) \cdot P_{Load,Nom,S/E}$$

$$E_{SAW,S,E} = \frac{T_S}{\eta_{Bus,S}} \cdot P_{Load,Nom,S} + \frac{T_E}{\eta_{Bus,E}} \cdot P_{Load,Nom,E}$$

In a more detailed approach, the total efficiency of the power transfer from generator to loads or from generator through the energy storage to loads can be defined as the product of individual efficiency parameters of the implemented subsystems.

The amount of energy calculated in this way is the amount that needs to be generated by a solar generator during one sunlit period in order to meet the power budget requirement during one lunar day/night cycle. The solar generator EOL power required to generate this amount of energy can be calculated by the division of this amount of energy by the duration of the sunlit period [101] [359]. It needs to be ensured that the amount of power calculated in this way can be transferred from the generators during all operational periods. Since solar generators will degrade with their time in operation, the amount of power calculated here is the amount of power that needs to be available after the degradation during the operation at the End of Life (EOL) of the generative systems.

$$P_{SAW,EOL} = \frac{E_{SAW,S,E}}{T_S}$$

Dedicated methods are available in order to estimate the amount of solar generator power that needs to be installed at the Beginning of Life, so that the remaining solar generator power after the performance degradation during the mission is sufficient to supply the bus at the EOL. These methods include the calculation of a factor that accounts for the degradation of solar generators [101] [359]. A simple method for the degradation estimation utilizes an annual degradation factor to calculate the solar generator degradation with reference to the operational period duration for different solar cell types.

$$\alpha_{SAW,Deg} = \left[1 - \alpha_{SAW,\frac{Deg}{Year}}\right]^{T_{Mission,years}}$$

This parameter can be utilized in order to scale the required solar generator output power at the End of Life, and to determine the required solar generator output power that needs to be installed at Beginning of Life, so that the remaining solar generator power at End of Life is sufficient to meet the power budget requirement.

$$P_{SAW,\,BOL} = \frac{P_{SAW,EOL}}{\alpha_{SAW,Deg}}$$

Photovoltaic generators are manufactured from single photovoltaic cells that are integrated in a series and parallel integration. A number of photovoltaic cells are connected in series to strings in order to achieve a certain terminal voltage and a number of strings are connected in parallel to achieve the required generator current and power.

$$V_{SAW,BOL} = V_{SolarCell,BOL} \cdot M_S$$

$$I_{SAW,BOL} = I_{SolarCell,BOL} \cdot N_P$$

$$P_{SAW,BOL} = V_{SAW,BOL} \cdot I_{SAW,BOL} = V_{Cell,BOL} \cdot M_S \cdot I_{Cell,BOL} \cdot N_P$$

The number of photovoltaic cells that need to be connected in series in order to achieve the required string voltage can be calculated as the division of the required string voltage by the terminal voltage of a single photovoltaic cell.

$$M_S = \frac{V_{SAW,BOL}}{V_{SolarCell,BOL}}$$

The number of required strings in a parallel implementation can be calculated as deviation of the required generator current by the current of a single photovoltaic cell. It is also possible to calculate this number as the division of the required output power at BOL by the product of the required terminal voltage and current of one photovoltaic generator.

$$N_P = \frac{I_{SAW,BOL}}{I_{Cell,BOL}} = \frac{1}{I_{SolarCell,BOL}} \cdot \frac{P_{SAW,BOL}}{V_{SAW,BOL}}$$

The resulting values need to be rounded up to the next greater integer number in order to ensure that the required output parameters can be achieved. The solar generator power that can be achieved with the series and parallel integration can be calculated as the product of the photovoltaic cell power with the total numbers of parallel strings and series cells.

$$P_{SAW,BOL} = P_{SolarCell,BOL} \cdot M_S \cdot N_P$$

The required size of the solar array front side can be calculated as the product of the front surface area of one photovoltaic generator with the total number of integrated generator units. It is also possible to estimate the size of the solar array front side employing the load power demand, the power bus efficiency, and the irradiance parameters.

$$A_{SAW,Collector} = A_{SolarCell,Collector} \cdot n_{Series} \cdot n_{parallel} = \frac{P_{Load,Nom}}{\eta_{BUS} \cdot E_0}$$

Additional systems like frames and structural elements also need to be integrated into the design, and because of this, the total front side area of a solar array is usually greater than the total front side area of the integrated photovoltaic cells. A packaging factor can be employed in order to determine the size of the solar array front side with reference to the front side of the integrated photovoltaic cells. Similar estimations can be made for the generator mass and volume employing dedicated factors relating the mass or the volume of a solar generator to its output power.

$$A_{SAW} = \frac{A_{SAW,Collector}}{\alpha_{SAW,PF}}$$

$$M_{SAW} = \frac{P_{SAW,BOL}}{\alpha_{SAW,W2M}}$$

$$V_{SAW,Geo} = \frac{P_{SAW,BOL}}{\alpha_{SAW,W2V}}$$

6.6 Energy Storage Design

Energy storages are integrated into spacecraft EPSs in order to supply the power bus during periods where the primary source of energy is not available. Such periods can occur during nominal operational periods, when the source is not able to supply the bus demand or when it is not available at all. Another possible case would be a failure in the generative systems. Eclipse conditions are an example in solar power EPSs for conditions where the primary source of energy is not available. In addition to this, energy storages can also be employed in order to transfer secondary power to the bus during high power demand periods. In high demand periods, a condition can occur in which the complete amount of primary power supplied to the bus by the generative systems is smaller than the load demand. In this case, additional secondary power needs to be supplied to the bus from energy storages in order to satisfy the load demand.

Energy storages need to be designed in order to meet certain requirements from the bus and the connected loads, and they also need to be designed with regards to other elements of the Electrical Power System (EPS). The design needs to ensure that an energy storage can supply the required power during all operational periods without interruptions. Following this, the battery capacity needs to be determined with reference to the load profile. A battery design needs to be selected that allows the availability of the system through the complete scheduled operational period of the mission. The duration of the charge and discharge periods, the number of scheduled charge/discharge cycles, the required Depth of Discharge (DoD), and the charge rates are important parameters in the energy storage design.

The first step in the energy storage design is the determination of the required amount of energy that needs to be stored inside an energy storage during operation, so that it can be discharged for the operation during a discharge period. It can be defined as the time integral of the discharge power during the discharge period. In static operation conditions, it is equal to the product of the discharge power and the discharge duration.

$$E_{Bat,Discharge} = \int_{t_0}^{t_0+T_{Discharge}} P_{Bat,Discharge}(\tau)\, d\tau = P_{Bat,Discharge,Nom} \cdot T_{Discharge}$$

Power is dissipated in the cable harness and in optional discharge regulators. In order to meet the power demand requirement, the energy dissipated during the discharge period also needs to be supplied by the energy storage. An efficiency parameter for the discharge operation can be defined that accounts for the accumulated losses occurring in the EPS during the discharge operation where secondary power is transferred from the energy storage to electrical loads. In the energy storage design, this efficiency parameter is employed in order to scale the energy requirement of the energy storage.

$$E'_{Bat,Discharge} = E_{Bat,Discharge} \cdot \frac{1}{\eta_{Bat,Discharge}}$$

Energy storages usually cannot be discharged completely. Because of this, a part of the energy stored inside the energy storage always needs to remain inside the storage. The minimum DoD value during operation can be utilized as a scaling parameter in the determination of the storage energy.

Another important parameter in the energy storage design for Earth-based applications is the parameter Days of Autonomy (DoA). Energy storages designed for island-grid applications on the Earth are designed for the supply during one night and to be recharged on the following day. The DoA parameter accounts for the possibility that a complete recharge of the storage is not possible during the following day because of a limited availability of solar energy. In space engineering, this parameter is usually not utilized, since the irradiance conditions during a mission can be defined accurately. However, besides atmospheric conditions limiting the access to solar energy do not occur on the lunar surface, it is possible that generative systems can be out of operation temporarily, that their power yield is temporarily decreased, or that emergency situations occur. The DoA parameter can be utilized as a scaling parameter in the storage energy determination in order to take the occurrence of such situations into account in the energy storage design. An additional factor has been implemented in the following equation in order to allow the definition of the DoA parameter in Earth days. In order to account for additional energy demands during emergency situations, usually a certain amount of energy is added to the storage energy, which is referred to as the backup energy.

$$E_{Bat} = E'_{Bat,Discharge} \cdot \frac{1}{DoD} \cdot \left(DoA \cdot \frac{24 \frac{h}{Earth\ Day}}{655.72 \frac{h}{Lunar\ Day}} \cdot \frac{1}{Lunar\ Day} \right) + E_{BackUp}$$

The energy storage capacity can be determined with reference to the storage energy. It can be calculated as the division of storage energy by the nominal storage voltage.

$$Q_{Bat,EOL} = \frac{E_{Bat}}{V_{Bat.Nom}}$$

The capacity calculated in this way is the capacity of an energy storage system. Energy storage systems are usually manufactured from single energy storage units/cells. These are connected in series to strings to achieve the required voltage, and multiple strings are connected in parallel to achieve the required capacity. Depending on the utilized bus topology, only one energy storage system featuring the complete required storage capacity or multiple energy storage systems featuring smaller individual capacities can be connected in parallel to the power bus.

Batteries degrade during operation, and because of this, the battery degradation during the mission needs to be taken into account in systems design. Usually, battery systems featuring a greater capacity than required are implemented at BOL in order to account for the degradation during the mission. The required battery capacity is scaled up in a way that at EOL, the required battery capacity is still available. Modelling the battery degradation is a complex task because batteries are electrochemical systems, and, because of this, the estimation of the required BOL battery capacity is a complex task. A simple approach is available for initial design estimations. For a more detailed simulation, manufacturers often offer dedicated software. In order to estimate the battery degradation during the initial design phase, the parameter capacity fading is utilized. Capacity fading is a scaling parameter that allows to calculate the battery capacity after a certain amount of time in operation and a certain number of charge/discharge cycles. It can also be utilized to estimate the required battery capacity that needs to be installed at the BOL in order to satisfy the capacity requirement at the EOL.

$$Q_{BAT,BOL} = \frac{Q_{Bat,EOL}}{1 - \frac{\alpha_{CF}}{100\%}}$$

6.7 Power Regulator Design

Power regulators are an incremental part of the spacecraft Electrical Power System (EPS). The purpose of a power regulator is to control the power flow from the primary energy source to the power bus. The sum of the load power demand and the power transferred to the energy storages during recharge operation needs to be equal to the power transferred to the power bus from the primary energy source. If both amounts of power are not balanced, the bus voltage will increase or decrease in response. In case that less power is transferred from the source to the bus than power is demanded from the bus by loads and energy storages, the bus voltage will decrease in response, and vice versa. The purpose of a power regulator is to control the amount of power transferred from the primary energy source to the bus and balance it with the power demand of loads and energy storages in order to stabilize the bus voltage.

A dedicated process is available for the decision on a certain power regulator design. The first step in the process of determining the primary power regulation approach is to make a decision on the type of primary source that shall be utilized by the spacecraft EPS. Spacecraft usually employ solar power. In the next step, it needs to be decided which type of regulator should be utilized together with the source of primary energy. For the regulation of solar energy, the two regulator approaches, Direct Energy Transfer (DET) and Maximum Power Point Tracking (MPPT), are conceivable. A regulator needs to be designed with regards to the source characteristics and the load profile. In the third step, a decision needs to be made on the power bus topology in order to optimize the regulator for the bus topology in the fourth step. The required order of control over the bus voltage usually limits the number of possible bus topologies.

6.8 Fault Protection Design

Fault protection systems are integrated into spacecraft Electrical Power Systems (EPSs) in order to protect the power bus from failed systems, in order to isolate failed systems, and in order to protect the bus voltage from collapsing. A dedicated process is available for the design of fault protection systems. In the beginning of the design process, the load profile of each load connected to the spacecraft EPS needs to be determined and analyzed. The load profile and duty cycle of all subsystems connected to the EPS need to be determined for all operational modes and phases in the first step of the process. The transient behavior of connected loads also needs to be taken into account. Fault protection methods can be differentiated between centralized approaches and decentralized approaches. After the load profile has been determined, a decision needs to be made if a centralized or decentralized approach shall be utilized in the second step. The individual requirements of all connected loads need to be taken into account. In the third step, the fault protection system needs to be designed for the spacecraft EPS. Active and passive systems for electrical fault protection systems are available. In the design of fault protection systems, it needs to be ensured that failed systems are isolated from the power bus. Some types of systems disconnect failed loads permanently, but there are also systems available that are commendable and do not disconnect failed systems permanently. Electrical fault conditions have implications on the EPS design because high currents are flowing in the case of an electrical failure. The cable harness needs to be designed in a way that it allows to conduct the currents occurring during electrical failure situations. Energy storages need to be designed in such a way that after the occurrence of failure currents, enough energy is still available to transition the spacecraft into a safe state operation mode.

6.9 Electrical Power System Design

Electrical Power Systems (EPSs) are integrated into spacecraft in order to supply electrical loads with energy. They consist of the Power Management and Distribution (PMAD) system, sources of primary energy, energy storages, as well as power electronics. Electrical loads are connected to and supplied by the EPS. Important parameters in the design of spacecraft EPSs are the load power budget, the utilized generative and energy storage systems, as well as their degradation.

A dedicated process is available for the design of spacecraft EPS. The first step of the EPS design process includes the identification and collection of the EPS requirements, as well as their analysis. EPS requirements can result from mission-specific parameters, such as the mission duration, the spacecraft configuration, the mission type, or the requirements of a special payload. Input parameters, such as the power demand during nominal and peak periods or the environmental conditions in the spacecraft location, need to be precisely analyzed and characterized. Examples for the EPS requirements are the nominal and peak power requirements, as well as the power demand profile. The requirements of the loads connected to the EPS need to be analyzed, such as the desired supply voltages and the tolerances to voltage variations. A common requirement is that the variation of the bus voltage needs to be limited to ±1% [147] [228]. EPSs need to be designed in a way that even in the case of double failures, they can be in operation without interruptions. Optimal redundancies can be achieved by implementing redundant systems. All components integrated into an EPS need to be certified for the application in space and, if possible, already flight-proven. Preliminary designs are developed utilizing different systematic options and bus topologies in the second step. For spacecraft employing power buses in the kilowatt class, fully regulated bus topologies are commonly utilized. Power source and energy storage designs are developed with reference to the EPS requirements in the third step. The primary power source is selected with reference to the mission type, the power budget, and the duration of sunlit and eclipse periods. Usually, the systematic primary power source options are reduced to dynamic or static solar generators or nuclear generators. Energy storages for solar powered EPSs are designed with reference to the duration of sunlit and eclipse periods and with reference to the load demand during nominal and peak periods. Important energy storage-specific parameters are the storage technology, the storage capacity, as well as the storage mass and volume. When solar power is employed, generative systems need to be designed in a way that they can supply the spacecraft loads and the energy storages during power generation periods, and power storages need to be designed in a way that continuous and uninterrupted supply of the EPS loads is guaranteed during eclipse periods. In the fourth step, the method utilized for the power regulation and control has to be identified. Depending on the load requirements and the capabilities of the EPS systems, either Maximum Power Point Tracking or Direct Energy Transfer regulators can be utilized for the regulation of solar power. The choice of specific control algorithms can offer advantages. This step also includes the decision on the thermal control system for the regulators integrated into the EPS. Individual designs are discussed and evaluated in the fourth step. Based on the results of the evaluation, the system design process is iterated multiple times until an optimal solution has been found.

For the power transfer through wires, losses in the magnitude of between 1% and 6% can be accounted in a good approximation, so that the efficiency of a certain wire can be estimated between 94% and 99% during the initial development phase. For the power conversion in regulators, up to 10% of the input power can be estimated for internal power dissipation. During initial development phases, efficiencies between 95.0% and 98.5% can be estimated for power converters. A decrease in the solar generator output power resulting from internal power dissipation and mismatching of a Maximum Power Point Tracking algorithm also needs to be taken into account. During the initial design phase, a possible decrease in the output power in the magnitude of 10% can be estimated, so that an efficiency value in the magnitude of 90% can be utilized for the solar regulator. An energy storage efficiency in the magnitude of 85% during sunlit conditions and of 65% during eclipse conditions can be estimated when a Direct Energy Transfer (DET) regulator has been utilized. In case that a Maximum Power Point Tracking regulator has been utilized, the preliminary efficiencies need to be reduced by 5%. The power consumption of the Power Distribution Unit (PDU) can be estimated in a magnitude of $30\frac{\text{W}}{\text{kW}_{\text{Load}}}$.

The design of a certain power bus is specific to the bus topology and will be illustrated for the example of a solar-powered fully regulated bus in the following sections. Usually, sunlit and eclipse periods are differentiated in the design of solar powered EPSs. In the beginning of the design, the load power demand needs to be defined for sunlit and eclipse periods. Efficiencies need to be assigned for the regulators and converters utilized in the power bus. Power dissipation occurring within the cable harness can be neglected in a simplified analysis. However, in a more precise analysis, efficiencies need to be assigned for power transmission cables in the same way as this has been done here for power converters and regulators. The topology of the analyzed EPS, as well as the employed efficiencies, are visualized in Figure 6.2.

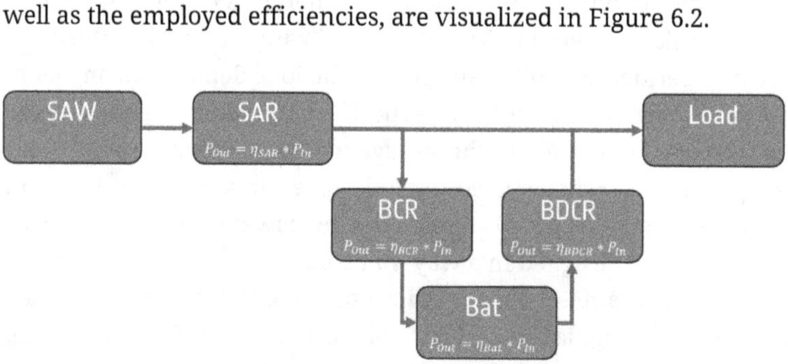

Figure 6.2 Topology of a fully regulated bus with Direct Energy Transfer solar regulation including assigned efficiencies for regulators

The EPS design starts with the load power demand as the initial parameter [228]. The EPS loads are supplied by the Solar Array Wings (SAW) through the Solar Array Regulator (SAR) during sunlit periods and by the battery through the Battery Discharge Regulator (BDCR) during eclipse periods. Subsequently, the output power of a BDCR needs to be equal to the power demand during eclipse periods, and during sunlit periods, the BDCR is out of operation. The input power of the BDCR is equal to the sum of the BDCR output power and the internally dissipated power. Subsequently, the BDCR input power has a greater value than the BDCR output power. The BDCR input power can be calculated as the division of the BDCR output power by the BDCR efficiency.

$$P_{BDCR,Out} = \begin{cases} 0 & \textit{Sunlit} \\ P_{Load,Nom,E} & \textit{Eclipse} \end{cases}$$

$$P_{BDCR,In} = \begin{cases} 0 & \textit{Sunlit} \\ \frac{P_{BDCR,Out}(Eclipse)}{\eta_{BDCR}} & \textit{Eclipse} \end{cases}$$

In the design of the energy storage, it is differentiated between the power during the discharged period and during the charge period. Energy storages are usually implemented employing batteries (BAT) in space applications. The power at the battery is usually counted positively during discharge operation. The battery power is equal to the BDCR input power during eclipse periods. The energy discharged from the energy storage during this period can be calculated as the product of discharge power and the duration of the discharge period. The power required for recharging the energy storage during the sunlit period can be calculated as the division of the discharged energy by the duration of the charge period. Power dissipation can also be observed inside energy storages. Because of this, the required charge power needs to be scaled up employing the energy storage efficiency.

$$P_{Bat} = \begin{cases} -\frac{P_{BDCR,In}(Eclipse) \cdot T_{Bat,Discharge}}{T_{Bat,Charge}} \cdot \frac{1}{\eta_{Bat}} & \textit{Sunlit} \\ P_{BDCR,In}(Eclipse) & \textit{Eclipse} \end{cases}$$

The Battery Charge Regulator (BCR) needs to be in operation during sunlit periods and will be out of operation during eclipse periods. During sunlit periods, the output power of the BCR needs to be equal to the power required to recharge the energy storage. In order to account for the internal power dissipation inside the BCR, the input power of the BCR needs to feature a value equal to the sum of the BCR output power and the internally dissipated power. The BCR input power can be calculated as the division of the BCR output power by the BCR efficiency.

$$P_{BCR,Out} = \begin{cases} P_{Bat}(Sunlit) & \textit{Sunlit} \\ 0 & \textit{Eclipse} \end{cases}$$

$$P_{BCR,In} = \begin{cases} \frac{P_{BCR,Out}(Sunlit)}{\eta_{BCR}} & \textit{Sunlit} \\ 0 & \textit{Eclipse} \end{cases}$$

The solar generators need to supply the bus loads and the BCR during sunlit periods. Subsequently, the output power of the Solar Array Regulator (SAR) needs to be equal to the sum of the load power demand and the BCR input power demand during sunlit periods. It features a value of zero during eclipse periods. Power dissipation also occurs inside the SAR. In order to account for this, the SAR input power needs to be equal to the sum of the SAR output power and the internally dissipated power. The SAR input power can be calculated as the division of the SAR output power by the SAR efficiency.

$$P_{SAR,Out} = \begin{cases} P_{BCR,In}(Sunlit) + P_{Load,Nom,S} & Sunlit \\ 0 & Eclipse \end{cases}$$

$$P_{SAR,In} = \begin{cases} \frac{P_{SAR,Out}(Sunlit)}{\eta_{SAR}} & Sunlit \\ 0 & Eclipse \end{cases}$$

The Solar Array Wing (SAW) needs to be designed in a way that its output power is sufficient to supply the required input power of the SAR during sunlit periods. Following this, the SAW output power needs to be equal to the SAR input power during sunlit periods.

$$P_{SAW} = \begin{cases} P_{SAR,In}(Sunlit) & Sunlit \\ 0 & Eclipse \end{cases}$$

6.10 Thermal Control System Design

Thermal control systems are usually designed employing a dedicated process. In the first step, the spacecraft is analyzed in order to identify internal sources of thermal energy and in order to estimate the amount of thermal energy dissipated from the spacecraft environment. Surfaces facing the Sun usually dissipate a greater amount of thermal energy than surfaces facing away from it. Possible locations for radiators and internal heat sources have to be identified in the second step. A good approach for the placement of radiators is to integrate them in locations on the spacecraft structure that are deflected from receiving thermal radiation. The dissipation of thermal radiation by spacecraft subsystems that has formally been radiated by the radiator systems also needs to be taken into account in the design and minimized. In the third step, different thermal control system approaches need to be designed. The implications of different thermal control system designs and radiator locations need to be analyzed and discussed in the fourth step. Following this, the thermal control system design process will be iterated multiple times until an optimal solution has been identified.

6.11 Space Based Solar Power

Space-Based Solar Power (SBSP) systems can be designed employing a dedicated design process. It is highly topology-specific and similar to the design process of power buses. The initial parameter for the design of a SBSP system is the power budget of the supplied systems. The amount of power that needs to be received by the rectenna to supply the surface systems can be calculated by scaling the power budget with the conversion efficiency of the rectenna from wirelessly received power to electrical DC power. When continuous contact between solar satellite and ground receiver cannot be guaranteed, the value of the power budget needs to be determined in a way that the amount of energy received during a contact period is sufficient to power the supplied systems for the duration of one contact and one Loss of Contact (LOC) period.

$$P_{Rec} = \frac{P_{Load,Nom}}{\eta_{WPT2DC}}$$

The area density of the rectenna power can be employed in order to calculate the required area size of the rectenna. The area size can be calculated as the fraction of the rectenna power requirement by the area density of the rectenna power.

$$A_{Rec} = \frac{P_{Rec}}{S_{Rec}}$$

At the rectenna, the radius of the power beam needs to be equal to or greater than the radius of the rectenna. Pointing errors also need to be taken into account for the SBSP application. Because of this, the beam radius at the rectenna needs to be equal to or greater than the sum of the rectenna radius and the offset distance error at the rectenna. The offset distance error at the rectenna can be calculated as the product of the transmission distance with the pointing angle error at the satellite.

$$r_{Beam} \geq \Delta z_{Distance} + r_{Rec} = z_{Distance} \cdot \sigma_{Point} + r_{Rec}$$

The link efficiency increases with increasing receiver area and decreases with increasing distance between transmitter and receiver. It can be calculated as the division of the received power by the transmitted power or as the division of the receiver area size by the beam area size.

$$\eta_{WPT} = \frac{P_{Rec}}{P_{Trans}} = \frac{A_{Rec}}{A_{Beam}} = \frac{r_{Rec}^2}{(z_{Distance} \cdot \sigma_{Point} + r_{Rec})^2}$$

The beam collection efficiency is a measure for the share of the power emitted at the transmission antenna of the solar satellite that is received by the rectenna. Decreasing beam collection efficiencies can be observed with increasing transmission wavelength and increasing transmission distance. It can be calculated as the division of the square root of the product of the area sizes of the receiving and transmitting antenna by the product of the WPT wavelength and the transmission distance. It can also be defined with respect to the WPT efficiency. Subsequently, it allows to calculate the WPT efficiency with respect to the area sizes of the receiving and transmitting antennas, the transmission wavelength, and the transmission distance.

$$\eta_{Beam,Collection} = \sqrt{-\ln\left(1 - \eta_{WPT}\right)} = \frac{\sqrt{A_{Rec} \cdot A_{Trans}}}{\lambda \cdot z_{Transmission}}$$

The power transmission requirement can be defined as the amount of power that needs to be transmitted at the solar satellite in order to satisfy the power receiving requirement at the rectenna. It can be calculated by scaling the rectenna power requirement with the WPT efficiency and the DC to WPT conversion efficiency at the solar satellite.

$$P_{Trans} = \frac{P_{Rec}}{\eta_{DC2WPT} \cdot \eta_{WPT}}$$

The power generation requirement at the solar satellite can be calculated by scaling the power transmission requirement at the solar satellite with the product of the photovoltaic conversion efficiency and the solar array regulator efficiency. The power generation requirement is equal to the product of the solar array front size and the value of the solar constant in the location of the solar satellite in orbit. In this way, the required solar array front side area size can be determined. The area size determined in this way is the projected area to the received solar irradiance. When perfect incidence conditions cannot be assumed with the desired system, the area size needs to be scaled up employing the solar incidence angle.

$$P_{SAW} = \frac{P_{Trans}}{\eta_{PV} \cdot \eta_{Regulator}} = A_{SAW} \cdot E_0$$

$$A_{SAW} = \frac{P_{Trans}}{\eta_{PV} \cdot \eta_{SAR}} \cdot \frac{1}{E_0}$$

Only in stationary orbits, solar satellites are always in contact conditions to a receiving rectenna. When other orbits are utilized, there will be a Loss of Contact (LOC) period, during which no power transfer from solar satellite to the rectenna is possible. During these periods, the supplied systems need to be supplied by another source of energy. A common solution for this problem is the utilization of energy storages.

$$E_{Bat} = P_{Load,LOC} \cdot T_{LOC}$$

Energy storages can be designed employing the duration of LOC periods and the nominal power demand of the supplied systems during LOC periods. The size of an energy storage can be calculated as the product of both parameters.

6.12 Mass and Cost Estimations of Electrical Power Systems

System masses and costs are important parameters in the design of space systems. Between 15% and 40% of the total spacecraft mass is usually accounted for the Electrical Power System (EPS). Generative systems, including additional systems and structures, like mounting or tracking systems, the energy storage system and the Power Management and Distribution (PMAD) system usually feature equal shares of the total EPS mass. The mass of photovoltaic-based generative systems is to a large extend influenced by the employed mounting and tracking method. Photovoltaic generators can either be implemented body-mounted on the outer surface of the spacecraft, or they can be implemented as Solar Array Wings (SAWs), which consist of external structures mounted to a spacecraft and can include additional tracking systems. Body-mounted generators contribute to the spacecraft mass with between 5.5 g/W and 6 g/W and Solar Array Wings (SAWs) contribute to the spacecraft mass with between 17 g/W and 40 g/W [97] [98]. Different energy storage technologies can also be differentiated by the storage mass. Batteries employing nickel-hydrogen technology contribute to the spacecraft mass with around 22 g/W, while nickel-cadmium batteries contribute with 29 g/W [98]. Increases in the converter and regulator mass can be observed with increasing voltage variability, and decreases in the mass can be observed with increasing voltage. The mass of DC/DC converters at a bus voltage level of 28 V can be estimated with 2.5 g/W, at a bus voltage level of 50 V, it can be estimated with 2.2 g/W, and at a bus voltage level of 100 V, it can be estimated with 2.0 g/W [97]. Power Management and Distribution (PMAD) systems usually feature masses up to 1 g/W, but this can increase to 20 g/W in the case that power is transferred over long distances [74]. The cable harness contributes to the system mass with between 10 g/W and 40 g/W [98]. The Power Control Unit contributes to the mass with 20 g/W [98]. The different parameters for system mass estimation are summarized in Table 6.5.

Table 6.5 Mass estimation of a power system [74] [97] [98]

	System Mass
photovoltaic generator	body-mounted: $5.5\frac{g}{W}$ to $6\frac{g}{W}$ solar wings: $17\frac{g}{W}$ to $40\frac{g}{W}$
power electronics	DET regulators: $42\,g \cdot \frac{P_{SAW}}{V_{bus}}$ DCDC converters: $2.5\frac{g}{W}$ for a 28 V bus $2.2\frac{g}{W}$ for a 50 V bus $2.0\frac{g}{W}$ for a 100 V bus
Power Management and Distribution (PMAD) system	$1\frac{g}{W}$ to $20\frac{g}{W}$
battery storage	NiCd: $29\frac{g}{W}$ NiH_2: $22\frac{g}{W}$
Power Control Unit (PCU)	$20\frac{g}{W}$
cable harness	$10\frac{g}{W}$ to $40\frac{g}{W}$
Thermal Control System (TCS)	2–5% of the spacecraft dry mass

The spacecraft EPS usually accounts for 20% of the spacecraft costs [148]. The total costs for uncrewed spacecraft vary between 15 €/g and 150 €/g [97]. For photovoltaic generator power at the Beginning of Life (BOL), between 300 €/W and 1200 €/W need to be accounted for [97]. The costs for battery storage capacity at the BOL vary between 50 €/Wh and 400 €/Wh [97]. The costs of power electronics can be estimated in the magnitude from 70,000–90,000 €/kg, and the costs of PMAD units can be estimated in the magnitude from 30,000 €/unit to 160,000 €/unit [97]. The different parameters for system cost estimation are summarized in Table 6.6.

Table 6.6 Cost estimation of a power system [97]

	System Costs
uncrewed spacecrafts	$15\frac{€}{g}$ to $150\frac{€}{g}$
solar generator	$300\frac{€}{W}$ to $1200\frac{€}{W}$
battery capacity	$50\frac{€}{Wh}$ to $400\frac{€}{Wh}$
Power Management and Distribution Unit (PMDU)	$30{,}000\frac{€}{Module}$ to $160{,}000\frac{€}{Module}$
power electronics	$70{,}000\frac{€}{kg}$ to $90{,}000\frac{€}{kg}$

The launch costs decreased drastically in the recent years. In 2002, the launch cost to Low Earth Orbit (LEO) had a magnitude of 10,000 $/kg, and the launch cost to Geostationary Orbit (GEO) had a magnitude of up to 50,000 $/kg [228]. The launch cost for missions to LEO decreased to 5000 $/kg utilizing a Falcon 9 Launch Vehicle and to 13,000 $/kg utilizing an Atlas V Launch Vehicle in 2018 [228]. The different parameters of the launch costs are summarized in Table 6.7.

Table 6.7 Estimation of the launch costs [228] [331] [337]

	2002	2011	2020
launch to LEO	$10,000 \frac{\$}{kg}$	Falcon 9: $5700 \frac{\$}{kg}$ STS: $54,500 \frac{\$}{kg}$	Falcon 9: $5000 \frac{\$}{kg}$ Atlas V: $13,000 \frac{\$}{kg}$
launch to GEO	$50,000 \frac{\$}{kg}$	–	–

6.13 Technology Evaluation

During the design process of space systems, a number of different system designs is usually developed (Table 6.8). At the end of the process, an evaluation has to be carried out in order to compare the different designs against each other. System designs are evaluated with reference to a set of important mission specific criteria. Examples of suitable criteria for lunar surface EPSs are constructability, maintainability, reliability, adaptability, system safety, or technology risks. Different designs are compared against each other utilizing a System Design Score (SDS). Each of the mission specific criteria have to be evaluated for each system design with an individual scoring value. An example could be a range from 1 to 3. In this case, a score of 1 represents complicated system implementations and those systems that do not yet feature a proof of concept. Systems that already feature a proof of concept in applications on the Earth or systems featuring an intermediate complexity level are rated with a score of 2. Systems featuring a simple complexity level or that are already classified as flight-proven have a score rating of 3. The final system design score can be calculated in the following as the sum of the individual criteria scores for each system design. After the evaluation, the system concept is chosen featuring the highest value of the system design score.

Table 6.8 Technology assessment

Evaluation Criteria	System Design 1	...	System Design M
criteria 1	$C_{1,1}$	...	$C_{M,1}$
criteria 2	$C_{1,2}$	...	$C_{M,2}$
...	...	...	...
criteria N	$C_{1,N}$	...	$C_{M,N}$
system design score	$SDS_1 = \sum_{n=1}^{N} C_{1,n}$	...	$SDS_M = \sum_{n=1}^{N} C_{M,n}$

7 Formula Symbols

Formula symbol	Explanation
absorptance factor	
α	absorptance
α_{Vis}	absorptance factor for the visible spectrum
$\alpha_{LunarSurface}$	mean absorptance of regolith simulants, 0.6645
acceleration	
$\frac{d\omega}{dt}$	angular acceleration (in $\frac{1}{s^2}$)
albedo factor	
a_{Albedo}	albedo factor
$a_{Albedo,mean}$	mean albedo factor, 0.306 for the Earth, 0.11 for the Moon [11] [97]
$A_{Albedo,Moon}$	albedo factor for the lunar equator, 0.08 [134]
$A_{Aledo,Moon,0}$	albedo factor for the lunar equator, 0.08 [134]
$a_{Albedo,Moon,0}$	albedo factor for the lunar latitude variation, 0.045 [134]
$b_{Albedo,Moon,0}$	albedo factor for the lunar latitude variation, 0.15 [134]
altitude	
H_{Sat}	altitude of a spacecraft over the surface of a celestial body (in m)
amplitude	
$A_{Shepherd}$	amplitude of the exponential domain of Shepherd's battery model (in V)

Formula symbol	Explanation
angle	
$\theta_{Incidence}$	incidence angle (in degree)
$\theta_{Incidence,C}$	critical incidence angle from which on proton radiation penetrates both cover glass and solar cell
$\theta_{Surface}$	angle between the surface normal of the lunar surface and the surface normal of the receiving surface (in degree)
Ω	solid angle, $d\Omega = 2\pi \cdot \sin(\theta)d\theta$
anomaly	
$v_{Anomaly}$	true anomaly (in degree)
area	
A	area (in m^2)
$A_{Surface}$	total surface area of the Sun (in m^2)
A_{Sphere}	surface area of a sphere (in m^2)
$A_{Observer}$	total surface area of the receiving surface (in m^2)
$A_{projected}$	surface area of the receiver projected on a theoretic sphere (in m^2)
A_{Earth}	visible area of the Earth
$A_{SAW,Collector}$	total solar generator collector area
A_{SAW}	total solar generator area
$A_{SolarCell,Collector}$	collector area of a single solar cell
A_{Beam}	beam area (in m^2)
A_{Rec}	rectenna area requirement (in m^2)
A_{Trans}	transmitter area requirement (in m^2)
$A_{Crystal}$	semiconductor cross-sectional area (in m^2)
A_{Cross}	cross-sectional area of the solar cell perpendicular to irradiance (in m^2)
A_{Wire}	cross-sectional area of a wire (usually in mm^2)
σ_{DCS}	displacement of the cross-section area (in cm^2)
bandgap	
ΔE_{Gap}	bandgap (in eV)

Formula symbol	Explanation
calendar life	
$\alpha_{SoH,Calender}$	calendar life (in %)
$k_{SoH,a}$	calendar life degradation parameter
caloric value	
H_U	caloric value
$I_{FC,theo}$	theoretical maximum Fuel Cell current (in A)
$I_{FC,Terminal}$	real Fuel Cell terminal current (in A)
capacity	
Q_{Bat}	battery capacity at a certain point in time t (in Ah)
$Q_{Bat,0}$	battery capacity (in Ah)
$Q_{Bat,var}$	time variant battery capacity (in Ah)
$Q_{Bat,np}$	nameplate battery capacity (in Ah)
$Q_{Bat,Discharge}$	charge drawn from the battery during the discharge operation (in Ah)
$Q_{Bat,Charge}$	charge supplied to the battery during the charge operation (in Ah)
$Q_{BAT,BOL}$	battery storage capacity at the Beginning of Life (in Ah)
$Q_{BAT,EOL}$	battery storage capacity at End of Life (in Ah)
C	capacitor capacity (in F)
capacity fading	
$\alpha_{CF}(t)$	capacity fading (in %)
$Q_{Bat,FC}$	discharge capacity in the characteristic point full charge (in Ah)
$Q_{Bat,Exp}$	discharge capacity in the characteristic point end of the exponential domain (in Ah)
$Q_{Bat,Nom}$	discharge capacity in the characteristic point end of the nominal domain (in Ah)
charge	
qe	elemental charge of an electron, $1.60217733 \cdot 10^{-19}$ As [109]
Q_C	capacitor charge (in As)
$Q_{El.,Semiconductor}$	electron charge passing through a semiconductor

Formula symbol	Explanation
charge carrier concentration	
n_i	intrinsic charge carrier concentration
n_{i0}	intrinsic charge carrier concentration at $T = 0$ K, $4.62 \cdot 10^{15} \frac{1}{\mathrm{cm}^3 \cdot \mathrm{K}^{\frac{3}{2}}}$ for Si [109] [160]
charge carrier life duration	
τ_n	charge carrier life duration of electrons
τ_p	charge carrier life duration of holes
chemical	
A_{Chem}	product or reactant chemical
$B_{Chem.}$	product or reactant chemical
$AB_{Chem.}$	product or reactant chemical
coefficient	
D_{Uni}	damage coefficient for the unidirectional radiation
$D_{Uni,e}$	damage coefficient for the unidirectional electron radiation
$D_{Uni,p}$	damage coefficient for the unidirectional proton radiation
$D_{Uni,p,HM}$	damage coefficient for the unidirectional proton radiation featuring high and medium particle energies
$D_{Uni,p,L}$	damage coefficient for the unidirectional proton radiation featuring low particle energies
D_{omni}	damage coefficient for the omnidirectional radiation
$D_{omni,e}$	damage coefficient for the omnidirectional electron radiation
$D_{omni,p,L}$	damage coefficient for the omnidirectional radiation featuring low energies
$D_{omni,p,HM}$	damage coefficient for the omnidirectional radiation featuring high or medium energies
$D_{omni,I_{SC}}$	omnidirectional damage coefficients for the proton radiation for I_{SC}
$D_{Omni,p,V_{OC}}$	omnidirectional damage coefficients for the proton radiation for V_{OC}
$\alpha_{pElectric}$	temperature coefficient of the specific resistances (in $\frac{1}{\mathrm{K}}$)
$\beta_{pElectric}$	temperature coefficient of the specific resistances (in $\frac{1}{\mathrm{K}}$)

Formula symbol	Explanation
$K_{Shepherd}$	polarization resistance coefficient of Shepherd's battery model
$\alpha_{Shepherd}$	curve fitting coefficient, usually equal to 3 [276]
$p_{Bat,l}$	coefficient of the temperature-dependent formulation of the model coefficients
$q_{Bat.l}$	coefficient of the temperature-dependent formulation of the model coefficients
$\frac{dV_{OC}}{dT}(\phi)$	temperature coefficient of the open circuit voltage (in $\frac{1}{K}$ or $\frac{V}{K}$)
$\frac{dV_{MPP}}{dT}(\phi)$	temperature coefficient of the Maximum Power Point voltage (in $\frac{1}{K}$ or $\frac{V}{K}$)
$\frac{dI_{SC}}{dT}(\phi)$	temperature coefficient of the short circuit current (in $\frac{1}{K}$ or $\frac{A}{K}$)
$\frac{dI_{MPP}}{dT}(\phi)$	temperature coefficient of the Maximum Power Point current (in $\frac{1}{K}$ or $\frac{A}{K}$)
conduction density	
n	intrinsic conduction density of electrons
p	intrinsic conduction density of holes
conductivity	
κ	electrical conductivity
constant	
k_B	Boltzmann constant, $1.38 \cdot 10^{-23} \frac{Ws}{K}$ or $8.62 \cdot 10^{-5} \frac{eV}{K}$ [109]
σ	Boltzmann constant (in $\frac{W}{m^2 \cdot K^4}$)
ε_0	electric field constant, $8.85418781762 \cdot 10^{-12} \frac{As}{Vm}$ [109]
h	Planck's constant, $6.6260755 \cdot 10^{-34}$ Js [109] [151]
N_A	Avogadro constant, $6.022 \cdot 10^{26} \frac{1}{kMol}$
F	Faraday constant, $96{,}485 \frac{As}{mol}$
D_N	diffusion constant of electrons
D_P	diffusion constant of holes
$B_{Shepherd}$	inverse time constant of the exponential domain of Shepherd's battery model (in $\frac{1}{Ah}$)
c-rate	
C_{Rate}	battery c-rate

Formula symbol	Explanation
current	
I	current (in A)
I_{Wire}	current through a transmission line (in A)
I_{Diode}	diode current (in A)
I_0	saturation current (in A)
I_{Ph}	photocurrent (in A)
$I_{SolarCell}$	solar cell terminal current (in A)
$I_{SolarCell,BOL}$	terminal current of a single generator cell at the BOL (in A)
I_{SC}	resulting short circuit current (in A)
I_{MPP}	Maximum Power Point current (in A)
$I_{SC}^0(\phi)$	short circuit current under test conditions (in A)
$I_{MPP}^0(\phi)$	Maximum Power Point current under test conditions (in A)
I_{R_P}	parallel resistance current (in A)
$I_{String\ n_{SolarArray},\ Cell\ m_{SolarArray}}$	solar cell current at the cell *m* in string *n* (in A)
$I_{String\ n_{SolarArray}}$	string current for the solar cell string *n* (in A)
I_{SAW}	solar array terminal current (in A)
$I_{SAW,BOL}$	total solar generator terminal current at the BOL (in A)
$I_{Segment\ j_{Segment}}$	current through the cell segment connected anti-parallel to a bypass diode (in V)
$I_{sc}(\phi)$	short circuit current after the exposure to radiation (in A)
$I_{sc}(BOL)$	short circuit current at the Beginning of Life (in A)
I_{Diode}	diode current (in A)
I_T	transistor current (in A)
I_{Ref}	reference output current (in A)
I_{Bat}	battery current (in A)
$I_{Bat,Discharge}$	battery discharge current (in A)
$I_{Bat,Charge}$	battery charge current (in A)
$I_{Bat,1h}$	battery discharge current that would discharge the battery after one hour (in A)

Formula symbol	Explanation
$I_{Bat,Cell}$	current of a battery cell (in A)
$I_{Bat,System}$	current of a battery system (in A)
I_P	parallel resistance current (in A)
I_L	inductor current (in A)
I_C	capacitor current (in A)
I_F	field current (in A)
j_F	field current density (in $\frac{\mathrm{A}}{\mathrm{m}^2}$)
I_R	resistor current (in A)
cycle life	
$\alpha_{CF,Cycle}$	cycle life (in %)
$k_{CF,c}$	cycle life degradation parameter
damage equivalent ratio	
D_{PE}	proton to electron damage equivalent ratio, conversion factor between the proton and electron damage equivalent fluence $D_{PE} = \begin{cases} 3000, & Si: V_{OC}, I_{SC}, P_{MPP} \\ 1400, & GaAs: V_{OC} \\ 400, & GaAs: I_{SC} \\ 1000, & GaAs: P_{MPP} \end{cases}$ [162] [171] [218]
day	
$d_{Year,Earth}$	day in a year on Earth, [0,365]
days of autonomy	
DoA	days of autonomy (in Earth days)
degradation	
$n_{Critical}$	parameter degradation at critical fluence (usually 20% or 25%)
density	
N_C	effective density of states in the conduction band (in $\frac{1}{\mathrm{cm}^3}$) [109]
N_V	effective density of states in the valence band (in $\frac{1}{\mathrm{cm}^3}$) [109]
n_A	density of acceptor atoms
n_{Don}	density of donator atoms
N_d	density of atomic displacements (in $\frac{1}{\mathrm{cm}^3}$)

Formula symbol	Explanation
$\rho_{Cover,l}$	density of a certain material layer of the front or back side laminate
$\rho_{Cover,TC}$	cover glass density that was utilized in laboratory experiments during damage
$\rho_{Silica} = 2.2 \frac{\mathrm{g}}{\mathrm{cm}^3}$	coefficient determination when fused silica cover glass has been utilized
S_{Rec}	rectenna power area density (in $\frac{\mathrm{W}}{\mathrm{m}^2}$)
depth	
$d_{Material}$	depth in material
$d_{depletion}$	total depth of the depletion zone
$d_{depletion,n}$	depth of the depletion zone in the *n*-semiconductor
$d_{depletion,p}$	depth of the depletion zone in the *p*-semiconductor
$d_{Penetration}$	penetration depth in the solar cell
$d_{SolarCell}$	depth of the solar cell (in mm)
depth of discharge	
DoD	depth of discharge, [0, ..., 1]
difference spectrum	
$\Delta\phi_e(E)$	difference spectrum of electron radiation (in $\frac{\mathrm{particles}}{\mathrm{cm}^2 \cdot \mathrm{s}}$)
$\Delta\phi_p(E)$	difference spectrum of proton radiation (in $\frac{\mathrm{particles}}{\mathrm{cm}^2 \cdot \mathrm{s}}$)
displacements	
$\overline{\nu}$	average displacements per primary displacement
distance	
$r_{observer}$	distance between the center of the Sun and the receiving surface (in m)
r_{Source}	distance between the center of the source and its surface (in m)
r_{Orbit}	distance between the orbiting electron and the atomic nucleus (in m)
l	distance/length (in m)
$z_{Transmission}$	transmission distance (in m)
$\Delta z_{Distance}$	offset distance error at the rectenna

Formula symbol	Explanation
dose	
dose	absorbed dose (in rad)
$D_{Absorbed}$	absorbed dose at a material depth of *d*
$dD_{Absorbed}$	absorbed dose per unit fluence for a certain energy and cover material depth
duration	
T_s	switching interval duration (in s)
T_{On}	conducting subinterval duration (in s)
T_{Off}	blocking subinterval duration (in s)
T_{EoDC}	duration of a discharge operation from full charge to full discharge state
T_{EoC}	duration of a charge operation from full discharge to full charge state
T_{Sunlit}	duration of sunlit periods (in h)
$T_{Eclipse}$	duration of eclipse periods (in h)
$T_{Mission,years}$	scheduled mission duration from the BOL until the EOL (in years)
$T_{Discharge}$	duration of the discharge operation (in h)
$T_{Bat,Charge}$	duration of the battery charge operation (in h)
$T_{Bat,discharge}$	duration of the battery discharge operation (in h)
T_{LOC}	duration of LOC periods (in h)
duty cycle	
D_{PE}	duty cycle
efficiency	
η	conversion efficiency of the photovoltaic generator
$\eta_{Internal}$	internal quantum efficiency
$\eta_{External}$	external quantum efficiency
$\eta_{SolarCell,Theo}$	theoretically maximum possible achievable efficiency with a real solar cell
$\eta_{Converter}$	efficiency of a power converter
$\eta_{Bat,Ah}$	ampere hour efficiency

Formula symbol	Explanation
$\eta_{Bat,Wh}$	watt hour efficiency
$\eta_{Bat,Discharge}$	efficiency in discharge operation, [0, 1]
$\eta_{FC,rev}$	thermodynamic Fuel Cell efficiency
$\eta_{FC,V}$	Fuel Cell voltage efficiency
$\eta_{FC,I}$	Fuel Cell current efficiency
$\eta_{FC,FG}$	Fuel Cell fuel gas efficiency
$\eta_{FC,Sys}$	Fuel Cell system efficiency
η_{FC}	total Fuel Cell efficiency
$\eta_{BUS,Sunlit}$	power bus efficiency in sunlit operation, $\begin{cases} 0.85 & \text{DET} \\ 0.80 & \text{MPPT} \end{cases}$ [101]
$\eta_{BUS,Eclipse}$	power bus efficiency in eclipse operation, $\begin{cases} 0.65 & \text{DET} \\ 0.60 & \text{MPPT} \end{cases}$ [101]
η_{BUS}	power bus efficiency
η_{BDCR}	battery discharge regulator efficiency, [0, 1]
η_{Bat}	battery efficiency, [0, 1]
η_{BCR}	battery charge regulator efficiency, [0, 1]
η_{SAR}	solar array regulator efficiency, [0, 1]
η_{PV}	photovoltaic conversion efficiency, in [0, 1]
η_{WPT2DC}	WPT to DC conversion efficiency, typically 0.4 or 40% [144]
η_{WPT}	WPT efficiency, typically 0.54–0.93 or 54–93% [144] [332]
$\eta_{Beam,Collection}$	beam collection efficiency, in [0, 1]
η_{DC2WPT}	DC to WPT conversion efficiency
emittance	
$\varepsilon_{Lunar\ Surface}$	emittance factor of lunar regolith, 0.95 [134]
ε_{IR}	emittance factor for the infrared spectrum
energy	
E_C	energy state of the lower conduction band edge (in Ws or eV)
E_V	energy state of the upper valence band edge (in Ws or eV)
$\Delta E_{Reaction}$	energy released by the reaction (in MeV)

Formula symbol	Explanation
E_{Total}	total energy released by all reactions (in MeV)
$E_{n_{Orbit}}$	energy of an electron in the orbit n_{orbit} (in $\frac{kg \cdot m^2}{s^2}$)
$\Delta E_{n_{Orbit}}$	energy difference between two electron shells
$E_{n_{Orbit},old}$	energy state of the old shell
$E_{n_{Orbit},new}$	energy state of the new shell
$E_{n_{Orbit}}$	energy of an electron in the orbit n_{orbit} (in eV)
$E_{m_{Orbit}}$	energy of an electron in the orbit m_{orbit} (in eV)
E_{ACC}	ionization energy of acceptors, energy to lift the surplus hole into the conduction band
E_{Don}	ionization energy of donators
E_{Photon}	energy of a photon (in Ws)
E_d	displacement energy (in MeV)
E_{th}	threshold energy (in MeV)
$\Delta E_{Particle}$	particle energy increment (in MeV)
$E_{Particle}$	particle energy
$E_{Particle,0}$	particle energy at the front side transition from space to cover glass
$E_{Particle,n}$	particle energy after traveling through the cover material, at the transition from cover material to solar cell
E_L	inductor energy (in Ws)
E_C	capacitor energy (in Ws)
$\Delta W_{Bat,El}$	electrical energy supplied to the battery(in Wh)
$E_{Bat,nom}$	nominal battery energy (in Wh)
$E_{Bat,Discharge}$	energy drawn from the battery during discharge operation (in Wh)
$E_{Bat,Charge}$	energy supplied to the battery during charge operation (in Wh)
$E_{Bat,Specific}$	specific energy (in $\frac{Wh}{kg}$)
$E_{Bat,Nom}$	nominal battery energy (in Wh)
E_{El}	electrical energy (in Ws)
$E_{Kin,Rot}$	rotational kinetic energy (in J = Ws)

Formula symbol	Explanation
$E_{SAW,S,E}$	energy generation during a sunlit/eclipse cycle (in Wh)
E_{BackUp}	backup energy to supply the load demand in emergency situations (in Wh)
E_F	Fermi energy (in Ws or eV)
enthalpy	
$\Delta H_{Chem.}$	reaction enthalpy
$H_{Chem.,A}$	reaction enthalpy of chemical A
$H_{Chem.,B}$	reaction enthalpy of chemical B
$H_{Chem.,AB}$	reaction enthalpy of chemical AB
$\Delta G_{Chem.}$	free enthalpy
entropy	
$\Delta S_{Chem.}$	entropy change
$S_{Chem.,A}$	entropy of chemical A
$S_{Chem.,B}$	entropy of chemical B
$S_{Chem.,AB}$	entropy of chemical AB
error	
σ_{Point}	pointing angle error at satellite
factor	
Ω_0	1, factor for the geometry of the radiation emissions (in sr)
K	factor for the reflexive properties of the Earth's surface
n_i0	factor, $4.62 \cdot 10^{15} \frac{1}{\mathrm{cm}^3 \cdot \mathrm{K}^{\frac{3}{2}}}$ for Silicon [109] [160]
n_{Diode}	diode ideality factor, equal to1 for perfect diodes, between 1 and 2 for solar cells
C_{Ph}	photocurrent factor, relating solar irradiance to photocurrent (in $\frac{\mathrm{m}^2}{\mathrm{V}}$)
α_{Optic}	absorptivity factor
$\alpha_{Optic,d}$	depth-dependent absorptance factor, $\begin{cases} 7.2\frac{1}{\mathrm{mm}} & Si \\ 630.0\frac{1}{\mathrm{mm}} & GaAs \end{cases}$ [109]
τ_{Optic}	transmittance factor
ρ_{Optic}	reflectance factor

Formula symbol	Explanation
$\rho_{Optic,ARC}$	reflectance factor after the application of anti-reflex coating
$\alpha_{Optic}(\lambda)$	absorptance factor with reference to a certain wavelength
$\tau_{Optic}(\lambda)$	transmittance factor with reference to a certain wavelength
$\rho_{Optic}(\lambda)$	reflectance factor with reference to a certain wavelength
$\alpha_{SolarCell,V}$	irradiance factor for voltages
$\alpha_{SolarCell,I}$	irradiance factor for currents
$R_{V_{OC}}(\phi_{Ref})$	remaining factor of the open circuit voltage
$R_{I_{SC}}(\phi_{Ref})$	remaining factor of the short circuit current
$R_{V_{MPP}}(\phi_{Ref})$	remaining factor of the Maximum Power Point voltage
$R_{I_{MPP}}(\phi_{Ref})$	remaining factor of the Maximum Power Point current
$R_{\frac{dV_{OC}}{dT}}(\phi_{Ref})$	remaining factor of the open circuit voltage temperature coefficient
$R_{\frac{dI_{SC}}{dT}}(\phi_{Ref})$	remaining factor of the short circuit current temperature coefficient
$R_{\frac{dV_{MPP}}{dT}}(\phi_{Ref})$	remaining factor of the Maximum Power Point current temperature coefficient
$R_{\frac{dI_{MPP}}{dT}}(\phi_{Ref})$	remaining factor of the Maximum Power Point voltage temperature coefficient
FF	fill factor
K_F	feed-forward gain factor, usually equal to 1
K_P	proportional gain factor
K_I	integral gain factor
K_D	derivative gain factor
$\alpha_{SAW,PF}$	packaging factor, typically 0.9 [359]
$\alpha_{SAW,W2M}$	factor relating solar array power to mass (in $\frac{\mathrm{W}}{\mathrm{kg}}$)
$\alpha_{SAW,W2}$	factor relating solar array power to volume (in $\frac{\mathrm{W}}{\mathrm{m}^3}$)
field strength	
E_{El}	electrical field strength (in $\frac{\mathrm{V}}{\mathrm{m}}$)
filter	
$f_{Bat,LPF}$	low pass filter function for the battery current

Formula symbol	Explanation
fluence	
ϕ_{Ref}	equivalent fluence of a certain type of reference radiation that can be reproduced in a laboratory environment, for example 1 MeV electrons or 10 MeV protons (in $\frac{\text{particles}}{\text{cm}^2}$)
ϕ	radiation fluence (in $\frac{\text{e}}{\text{cm}^2}$)
$\phi(\geq E_d)$	radiation fluence for particles with energies greater or equal to E_d (in $\frac{1}{\text{cm}^2}$)
$\phi_{Critical}$	critical fluence
$\phi_{Critical,e}$	critical fluence for electron radiation of a certain energy
$\phi_{Critical,p}$	critical fluence for proton radiation of a certain energy
ϕ_{1MeVe}	equivalent fluence for 1 MeV electron radiation (in $\frac{\text{particles}}{\text{cm}^2\cdot\text{s}}$)
$\phi_{1MeV p,I_{SC}}$	equivalent fluence for 1 MeV proton radiation for I_{SC} (in $\frac{\text{particles}}{\text{cm}^2\cdot\text{s}}$)
$\phi_{1MeV p,V_{OC}}$	equivalent fluence for 1 MeV proton radiation for V_{OC} (in $\frac{\text{particles}}{\text{cm}^2\cdot\text{s}}$)
ϕ_{1MeVe}	fluence of 1 MeV electron radiation
ϕ_{10MeVp}	fluence of 10 MeV proton radiation
flux	
$\phi_{Sun,total}$	radiant flux of the Sun (in W)
$\phi(E_{th})$	differential flux (in $\frac{\text{particles}}{\text{cm}^2\cdot\text{s}\cdot\text{MeV}}$)
$\phi(> E_{th})$	integral flux (in $\frac{\text{particles}}{\text{cm}^2\cdot\text{s}}$)
$\Delta\phi(E_{th})$	difference flux (in $\frac{\text{particles}}{\text{cm}^2\cdot\text{s}}$)
ϕ_n	omnidirectional flux normal to a surface
ϕ_0	total omnidirectional flux
$\phi_e(> E)$	integral flux of electron radiation (in $\frac{\text{particles}}{\text{cm}^2\cdot\text{s}}$)
$\phi_p(> E)$	integral flux of proton radiation (in $\frac{\text{particles}}{\text{cm}^2\cdot\text{s}}$)
ψ_L	magnetic flux linkage (in Vs)
ϕ_L	magnetic flux (in Vs)
ϕ_{total}	radiant flux (in W)

Formula symbol	Explanation
force	
$F_{Coulomb}$	Coulomb force between nucleus and orbiting electron (in $\frac{kg \cdot m}{s^2}$) [109]
F_{CF}	centrifugal force (in $\frac{kg \cdot m}{s^2}$) [109]
frequency	
f	frequency (in $\frac{1}{s}$)
f_s	switching frequency (in $\frac{1}{s}$)
half axis	
s	minor half axis of the visible Earth disc
incidence angle	
$\theta_{Incidence}$	incidence angle (in degree)
index	
ms	index for solar or battery cells connected in series
n_P	index for solar or battery cell strings connected in parallel
$j_{Segment}$	index of a bypass diode in a solar array
$k_{Cells/Diode}$	index of solar cells in a string segment parallel to a bypass diode
inductance	
L	inductance (in H or $\frac{Vs}{A}$)
irradiance	
E_{Albedo}	albedo irradiance in a certain lunar surface location (n $\frac{W}{m^2}$)
$E_{Moon,IR}$	lunar infrared irradiance (in $\frac{W}{m^2}$)
E_0	solar irradiance (in $\frac{W}{m^2}$)
$E_{0,VQ}$	mean irradiance calculated by the equation of Volker Quaschning (in $\frac{W}{m^2}$)
$E_{0,Ley}$	mean irradiance calculated by the equation of Wilfried Ley (in $\frac{W}{m^2}$)
$E_{0,Perigee}$	mean irradiance received during the perigee on Earth (in $\frac{W}{m^2}$)
$E_{Moon,IR}$	lunar infrared irradiance (in $\frac{W}{m^2}$)

Formula symbol	Explanation
$E_{SolarCell,R}$	irradiance reflected at the surface (in $\frac{W}{m^2}$)
$E_{SolarCell,P}$	received irradiance penetrating the solar cell (in $\frac{W}{m^2}$)
$E_{SolarCell,T}$	received irradiance transmitted through the solar cell (in $\frac{W}{m^2}$)
$E_{SolarCell,A}$	received irradiance absorbed in the solar cell (in $\frac{W}{m^2}$)
E_{TC}	irradiance during standardized laboratory tests (in $\frac{W}{m^2}$)
$E_{String\ n_{SolarArray},\ Cell\ m_{SolarArray}}$	solar irradiance at the cell $m_{SolarArray}$ in string $n_{SolarArray}$ (in $\frac{W}{m^2}$)
horizontal irradiance	
$E_{Hor.,Albedo}$	horizontal albedo irradiance received by a certain surface in a certain lunar surface location (in $\frac{W}{m^2}$)
$E_{Hor.}$	horizontal irradiance (in $\frac{W}{m^2}$)
$E_{Hor.,Visible}$	horizontal irradiance in the visible spectrum (in $\frac{W}{m^2}$)
$E_{Hor.,IR}$	horizontal irradiance in the infrared spectrum (in $\frac{W}{m^2}$spectral)
$E_{Hor.,Moon,IR}$	horizontal lunar infrared irradiance received by a certain surface in a certain lunar surface location (in $\frac{W}{m^2}$)
spectral irradiance	
E_{λ}	spectral irradiance (in $\frac{W}{m^2}$)
$E_{\lambda,Albedo}$	spectral albedo irradiance in a certain lunar surface location (in $\frac{W}{m^2}$)
$E_{\lambda,Hor.}$	spectral horizontal irradiance (in $\frac{W}{m^2}$)
$E(\lambda)$	monochromatic irradiance of a certain wavelength (in $\frac{W}{m^2}$)
latitude	
θ	latitude (in degree)
length	
l	length (in m)
l_{Wire}	length of a wire (in m)
$l_{Crystal}$	length of the semiconductor crystal (in m)
l_{pen}	penetration path length of a particle
l_{range}	length of the particle range

Formula symbol	Explanation
diffusion length	
L_N	diffusion length of electrons
L_P	diffusion length of holes
losses	
$\Delta V_{FC,P}$	penetration losses in a Fuel Cell (in V)
$\Delta V_{FC,R}$	voltage loss at the parasitic inner resistance in a Fuel Cell (in V)
$\Delta V_{FC,Diff}$	diffusion loss voltage in a Fuel Cell (in V)
mass	
m_e	electron mass, $9.1093897 \cdot 10^{-31}$ kg [109]
m_{Bat}	battery mass (in kg)
M	atomic mass, M(Si) = 28 Da [171]
mass defect	
$\Delta m_{reaction}$	mass defect of the reaction (in u)
Δm_{Total}	total mass defect of all reactions (in u)
M	molar mass (in $\frac{\mathrm{g}}{\mathrm{Mol}}$, $M(\mathrm{H_2}) = 2.02\frac{\mathrm{g}}{\mathrm{Mol}}$)
M_{SAW}	mass of a solar generator (in kg)
mass-energy equivalence	
$m_e c^2$	electron mass-energy equivalence, 0.511 MeV [171]
mass flow	
$\frac{dm}{dt}$	mass flow rate (in $\frac{\mathrm{kg}}{\mathrm{s}}$)
mobility	
$\mu_n(T)$	electron mobility at a temperature T (in $\frac{\mathrm{m^2}}{\mathrm{Vs}}$)
$\mu_p(T)$	proton mobility at a temperature T (in $\frac{\mathrm{m^2}}{\mathrm{Vs}}$)
μ_{0n}	electron mobility at a temperature T_0 (in $\frac{\mathrm{m^2}}{\mathrm{Vs}}$)
μ_{0p}	proton mobility at a temperature T_0 (in $\frac{\mathrm{m^2}}{\mathrm{Vs}}$)
moment	
J	moment of inertia (in $\mathrm{kg} \cdot \mathrm{m^2}$)

Formula symbol	Explanation
momentum	
L_e	orbital momentum (in $\frac{\mathrm{kg \cdot m}}{\mathrm{s}}$)
number	
n_Z	number of positively charged protons in the atomic core [109]
n_{Orbit}	integer number of the orbital track
m_{Orbit}	integer number of the orbital track m
N_E	total number of acceptors/electrons
$m_{Refractive}$	positive integer number or zero, $m \in 0, 1, 2, \ldots$
$N_{Photons}$	total number of photons reaching the solar cell
$N_{Photons}(\lambda)$	total number of photons featuring a certain wavelength
$N_{EHPairs}$	total number of generated electron-hole pairs
M_S	total number of solar or battery cells connected in series to a string
N_P	total number of solar or battery cell strings connected in parallel to a solar array
M_{E_i}	total number of solar cells receiving irradiance in the magnitude of E_i in the string $n_{SolarArray}$
$K_{Cells/Diode}$	total number of solar cells connected antiparallel to a bypass diode
n_a	number of atoms per unit volume of the absorber (in $\frac{\mathrm{Atoms}}{\mathrm{cm^3}}$, $n_a(Si) = 5 \cdot 10^{22} \frac{\mathrm{Atoms}}{\mathrm{cm^3}}$)
N_{td}	total number of displacements, which result from proton radiation featuring a certain particle energy
N_L	total number of inductor turns
$N_{Channel}$	total number of S3R channels
n_e	number of the employed free electrons in the reaction
$n_{Bat,days}$	total number of days in operation (in Earth days)
$n_{Bat,cycles}$	total number of charge/discharge cycles

Formula symbol	Explanation
parameter	
γ_{Diode}	curve fitting parameter for the diode exponential
$\alpha_{SAW,Deg}$	degradation parameter for the solar generator degradation during a mission
$\alpha_{SAW,\frac{Deg}{Year}}$	annual degradation parameter, $\begin{cases} 3.75\frac{\%}{\text{year}} & \text{Si} \\ 2.75\frac{\%}{\text{year}} & \text{SJ}-\text{GaAs} \\ 0.50\frac{\%}{\text{year}} & \text{MJ}-\text{GaAs} \end{cases}$ [101]
particle	
${}^{4}_{2}\alpha$	helium core
e^{+}	positron
${}^{1}_{1}p$	proton core
permittivity	
ε_r	relative permittivity of the medium
phase	
ξ_{Phase}	Earth phase parameter
power	
electrical power	
$P_{SolarCell}$	solar cell terminal power, (in W)
P_{MPP}	Maximum Power Point power (in W)
P_{El}	electrical power (in W)
$P_{Dissipated}$	dissipated power (in W)
$P^{+}_{Dissipated,Wire}$	dissipated power in the supply wire (in W)
$P^{-}_{Dissipated,Wire}$	dissipated power in the back wire (in W)
$P_{Dissipated,Wire}$	dissipated power in the transmission line (in W)
$P_{Dissipated,Harness}$	dissipated power in the complete cable harness (in W)
P_L	inductor power (in W)
P_C	capacitor power (in W)
P_R	resistor power (in W)
P_D	diode power (in W)
P_T	transistor power (in W)

Formula symbol	Explanation
$P_{Converter,Input}$	input power of a power converter (in W)
$P_{Converter,Output}$	output power of a power converter (in W)
$P_{Converter,Dissipation}$	dissipated power of a power converter (in W)
P_{Bat}	battery power (in W)
$P_{Bat,Nom}$	nominal battery power (in W)
$P_{Bat,Discharge,Nom}$	nominal discharge power (in W)
$P_{Bat,Specific}$	specific power (in $\frac{\mathrm{Wh}}{\mathrm{kg}}$)
$P_{Bat,Density}$	power density (in $\frac{\mathrm{W}}{\mathrm{m}^3}$)
P_{El}	electrical power
$P_{Load,Nom,S/E}$	nominal electrical load power during a sunlit/eclipse cycle (in W)
$P_{Load,Nom,S}$	nominal electrical load power during sunlit periods (in W)
$P_{Load,Nom,E}$	nominal electrical load power during eclipse periods (in W)
$P_{SAW,EOL}$	power of a solar generator at the End of Life (in W)
P_{SAW}	solar array generator output power (in W)
$P_{SAW,BOL}$	power of a solar generator at the Beginning of Life (in W)
$P_{SAW,EOL}$	power of a solar generator at the End of Life (in W)
$P_{SolarCell,BOL}$	power of a solar cell at the Beginning of life (in W)
$P_{SolarCell,EOL}$	power of a solar cell at the End of Life (in W)
$P_{Load,Nom}$	nominal electrical load power (in W)
$P_{BDCR,Out}$	battery discharge regulator output power (in W)
$P_{BDCR,In}$	battery discharge regulator input power (in W)
$P_{Bat,Charge}$	battery charge power (in W)
$P_{Bat,Discharge}$	battery discharge power (in W)
$P_{BCR,In}$	battery charge regulator input power (in W)
$P_{BCR,Out}$	battery charge regulator output power (in W)
$P_{SAR,In}$	solar array regulator input power (in W)
$P_{SAR,Out}$	solar array regulator output power (in W)
$P_{BCR,In}$	battery charge regulator input power (in W)
P_{Rec}	rectenna power (in W)

Formula symbol	Explanation
P_{Trans}	transmitter power (in W)
$P_{Load,LOC}$	power demand during LOC periods (in W)
kinetic power	
P_{Kin}	kinetic power (in W)
optical power	
$P_{Radiant}$	optical power (in W)
thermal power	
P_{TM}	thermal power (in W)
radiant exitance	
M	radiant exitance (in $\frac{W}{m^2}$)
M_{Sun}	radiant exitance (in $\frac{W}{m^2}$)
spectral radiance	
L_{Planck}	spectral radiance (in $\frac{W}{m^3 \cdot sr}$)
spectral radiant exitance	
$M(\lambda, T_{Radiation})$	spectral radiant exitance (in $\frac{W}{m^2}$)
radiant power	
P	radiant power (W)
$P_{Emission}$	emitted radiant power (in W)
$P_{Absorbed}$	absorbed radiant power (in W)
$P_{Absorbed,Visible}$	absorbed radiant power in the visible spectrum (in W)
$P_{Absorbed,IR}$	absorbed radiant power in the infrared spectrum (in W)
$P_{Reflexion}$	reflected radiant power (in W)
$P_{Transmission}$	transmitted radiant power (in W)
$P_{Incoming}$	received radiant power (in W)
horizontal radiant power	
$P_{Hor.,Visible}$	horizontal radiant power in the visible spectrum (in W)
$P_{Hor.,IR}$	horizontal radiant power in the infrared spectrum (in W)
$P_{Converted}$	converted radiant power (in W)

Formula symbol	Explanation
radius	
ρ	angular radius of a celestial body (in degree)
r_{Earth}	radius of the Earth (in m)
r_{Beam}	beam radius (in m)
r_{Rec}	receiver radius (in m)
range function	
$R(E_{Particle,0})$	range function
$R^{-1}(l_{Range})$	inverse range function
reflectance	
β	reflectance
refractive index	
$n_{Refractive,I}$	refractive index of the medium *i*
resistance	
ρ	electrical resistance
R_S	resistance value of the series resistance (in Ω)
R_P	resistance value of the series resistance (in Ω)
$\rho_{Electric}$	specific resistances (in $\frac{\Omega \cdot mm^2}{m}$)
$\rho^0_{Electric}$	specific resistances at test conditions, usually 20 °C (in $\frac{\Omega \cdot mm^2}{m}$)
R_{Wire}	internal resistance of a wire (in Ω)
R^+_{Wire}	internal resistance of a supply wire (in Ω)
R^-_{Wire}	internal resistance of a back wire (in Ω)
R_{Line}	internal resistance of a power transmission line (in Ω)
R	resistance (in Ω)
R_{Int}	internal resistance (in Ω)
sensitivity	
$S(\lambda)$	spectral sensitivity (in $\frac{A}{W}$)
speed	
c_0	speed of light in vacuum, $299{,}792 \frac{km}{s} = 2.99792458 \cdot 10^8 \frac{m}{s}$ [109] [151]
c_i	speed of light in the medium *i*

Formula symbol	Explanation
state of charge	
SoC	state of charge, [0, ..., 1]
stopping power	
$\frac{dE}{dx}$	electron stopping power of the material under investigation (in $\frac{\text{MeV}\cdot\text{cm}^2}{\text{g}\cdot\text{m}}$)
temperature	
T	temperature (in K)
T_0	reference temperature (in K)
T_{TC}	temperature in standardized laboratory tests (in °C or in K)
ΔT	temperature deviation (in °C or in K)
$T_{Surface}$	surface temperature (in K)
$T_{Radiation}$	radiation temperature of the source of radiation (in K)
$T_{Surface}(\theta)$	lunar surface temperature at a certain latitude during noon (in K)
$T_{Bat,Cell}$	battery cell temperature (in °C)
thermal energy transfer	
$\Delta Q_{Therm.}$	thermal energy transfer to the reaction material, positive for reactions that heat up during discharging and negative for reactions that heat up during charging
thermal model	
$f_{Thermal\ Model}$	thermal model integrated into the simulation system
thickness	
d_{ARC}	required thickness of anti-reflex coating in order to achieve best case results
dx	material thickness
$t_{Cover,eq}$	equivalent shielding thickness
$t_{Cover,l}$	thickness of a certain material layer of the front or back side laminate
time	
t_0	time at the beginning of an interval
Δt	time delta (in s)

Formula symbol	Explanation
torque	
$M = J \cdot \frac{d\omega}{dt}$	torque (in $\text{Nm} = \frac{\text{kg}\cdot\text{m}^2}{\text{s}^2}$)
transfer function	
$f_{Transfer,V}$	voltage transfer function
$f_{Transfer,I}$	current transfer function
velocity	
ω	angular velocity (in $\frac{1}{\text{s}}$)
$\omega_{n_{Orbit}}$	angular velocity of the orbiting electron (in $\frac{1}{\text{s}}$)
$v_{n_{Orbit}}$	velocity of the orbiting electron (in $\frac{\text{m}}{\text{s}}$)
v_{Drift}	drift velocity (in $\frac{\text{m}}{\text{s}}$)
voltage	
V	voltage (in V)
$V_{Diffusion}$	diffusion voltage (in V)
V_{Diode}	diode voltage (in V)
$V_{Diode,j}$	diode voltage of diode $j_{Segment}$ (in V)
V_{Temp}	temperature voltage (in V)
$V_{SolarCell}$	solar cell terminal voltage (in V)
$V_{SolarCell,BOL}$	terminal voltage of a generator cell at the BOL (in V)
V_{OC}	open circuit voltage (in V)
V_{MPP}	Maximum Power Point voltage (in V)
$V_{OC}^0(\phi)$	open circuit voltage under test conditions (in V)
$V_{MPP}^0(\phi)$	Maximum Power Point voltage under test conditions (in V)
V_{R_S}	voltage at the series resistance (in V)
V_{R_P}	voltage at the parallel resistance (in V)
$V_{String\ n_{SolarArray},\ Cell\ m_{SolarArray}}$	solar cell voltage at the cell m in string n (in V)
$V_{String\ n_{SolarArray}}$	string voltage for the solar cell string n (in V)
V_{SAW}	solar array terminal voltage (in V)

Formula symbol	Explanation
$V_{SAW,BOL}$	total solar generator terminal voltage at the BOL (in V)
$V_{Segment\ j_{Segment}}$	voltage at the string segment connected antiparallel to a bypass diode (in V)
$V_{Segment\ j_{Segment},\ Cell\ k_{Cells/Diode}}$	cell voltage of a solar cell in a string segment (in V)
V_{Wire}	voltage occurring at a wire (in V)
V_{Wire}^{+}	voltage occurring at a supply wire (in V)
V_{Wire}^{-}	voltage occurring at a back wire (in V)
V_{Line}	voltage occurring at a transmission line (in V)
V_{In}	input voltage (in V)
V_{Out}	output voltage (in V)
V_L	inductor voltage (in V)
V_C	capacitor voltage (in V)
V_R	resistor voltage (in V)
V_D	diode voltage (in V)
$V_{D,F}$	diode forward voltage (in V)
V_T	transistor voltage (in V)
$V_{T,F}$	transistor forward voltage (in V)
V_{Error}	error voltage between measured and reference voltage (in V)
$V_{Meas.}$	measured voltage (in V)
$V_{Ref.}$	reference voltage (in V)
$V_{In,Converter}$	input voltage at the input of the converter schematic (in V)
$V_{Out,Converter}$	output voltage at the output of the converter schematic (in V)
$V_{\ Out}$	converter output voltage (in V)
$V_{Ref.,Out}$	output reference voltage (in V)
$V_{Out,F}$	output value of the feed-forward control block (in V)
$V_{Out,P}$	output value of the proportional control block (in V)
$V_{Out,I}$	output value of the integral control block (in V)
$V_{Out,D}$	output value of the derivative control module (in V)
$V_{Out,FPID}$	output value of the FPID control system (in V)

Formula symbol	Explanation
$V_{MEAS.,BUS}$	measured bus voltage (in V)
$V_{Ref.,BUS}$	reference bus voltage (in V)
$V_{Ref.,Channel}$	channel reference voltage (in V)
$V_{Error,Channel}$	channel error voltage (in V)
ΔV_{Min}	S3R threshold voltage (in V)
ΔV_{Max}	maximum voltage error (in V)
V_{Bat}	battery voltage (in V)
$V_{Geo,Bat}$	battery volume (in m^3)
$V_{Bat,Therm.}$	thermal equilibrium battery voltage (in V)
$V_{Shepherd,OCV}$	voltage at the variable voltage source of Shepherd's battery model (in V)
$V_{Shepherd,0}$	constant battery voltage of Shepherd's battery model (in V)
$V_{Bat.FC}$	voltage in the characteristic point full charge (in V)
$V_{Bat,Exp}$	voltage in the characteristic point end of the exponential domain (in V)
$V_{Bat.Nom}$	voltage in the characteristic point end of the nominal domain (in V)
$V_{Bat.FDC}$	voltage in the characteristic point full discharge (in V)
$V_{Bat,Cell}$	voltage of a battery cell (in V)
$V_{Bat,System}$	voltage of a battery system (in V)
$V_{FC,Rev}$	reversible Fuel Cell voltage
$V_{FC,Terminal}$	Fuel Cell terminal voltage (in V)
$V_{FC,OC}$	Fuel Cell open circuit voltage (in V)
$V_{SAW,Geo}$	volume of a solar generator (in m^3)
$V_{Battery}$	voltage at the battery terminals (in V)
$V_{Cathode}$	voltage between the battery's cathode and a reference electrode (in V)
V_{Anode}	voltage between the battery's anode and a reference electrode (in V)

Formula symbol	Explanation
volume	
$V_{Crystal}$	semiconductor crystal volume (in m^3)
wavelength	
$\lambda_{Emission,Ma}$	maximum wavelength of the radiation spectrum (in µm)
λ	wavelength (in m)
$\lambda_{Solar,max}$	maximum wavelength for the conversion into electrical energy (in m)

8 Bibliography

[1] *ESA/NASA/Pesquet, T.:* Picture of the International Space Station. *https://www.esa.int/Science_Exploration/Human_and_Robotic_Exploration/International_Space_Station/Where_is_the_International_Space_Station*

[2] *ESA:* Gateway to the Moon. 11/03/2019. *https://www.esa.int/Science_Exploration/Human_and_Robotic_Exploration/Gateway_to_the_Moon*

[3] *McGuinness, J./Russell, J.:* NASA Provides Update to Astronaut Moon Lander Plans Under Artemis. March 23, 2022. *https://www.nasa.gov/press-release/nasa-provides-update-to-astronaut-moon-lander-plans-under-artemis*

[4] *NASA:* NASA Artemis. Presentation. European Astronaut Centre (EAC) 2019

[5] *NASA:* Render of the lunar gateway with Orion and the Moon. *https://www.esa.int/Science_Exploration/Human_and_Robotic_Exploration/Exploration/Gateway*

[6] *Royal Museums Greenwich:* Follow the NASA Artemis missions. *https://www.rmg.co.uk/stories/topics/nasa-moon-mission-artemis-program-launch-date*

[7] *Airbus:* Fast forward to Artemis IV with ESM-4 arrival in Bremen. June 23, 2022. *https://www.airbus.com/en/newsroom/stories/2022-06-fast-forward-to-artemis-iv-with-esm-4-arrival-in-bremen*

[8] *NASA:* Artemis Mission Maps 2023. 2023. *https://nasa-external-ocomm.app.box.com/s/pczky3724zgcnmlth1zb2853qhtnss1t*

[9] *Lloyd, V./Johnson, A./Hambleton, K.:* NASA Identifies Candidate Regions for Landing Next Americans on Moon. August 19, 2022. *https://www.nasa.gov/news-release/nasa-identifies-candidate-regions-for-landing-next-americans-on-moon*

[10] *ESA:* NASA und ESA einigen sich auf die Lunar Gateway-Umlaufbahn. 25/07/2019. *https://www.esa.int/Space_in_Member_States/Germany/NASA_und_ESA_einigen_sich_auf_die_Lunar_Gateway-Umlaufbahn*

[11] *Williams, D. R.:* NASA Planetary Factsheets. NASA Goddard Space Flight Center 2019. *https://nssdc.gsfc.nasa.gov/planetary/factsheet*

[12] *NASA:* NASA Artemis. Presentation. European Astronaut Centre (EAC) 2019

[13] *ESA:* 3D rendering of the moon village. *https://www.esa.int/ESA_Multimedia/Images/2019/10/The_Moon_as_a_spaceport_to_the_Universe*

[14] *Seedhouse, E.:* Lunar Outpost. The Challenges of Establishing a Human Settlement on the Moon. Springer & Praxis Publishing Ltd., Chichester, UK 2009 (ISBN: 978-0-387-09746-6)

[15] NASA's Exploration Systems Architecture Study. NASA 2005

[16] *NASA:* NASA's Lunar Exploration Program Overview – Artemis Plan. September 2020. *https://www.nasa.gov/sites/default/files/atoms/files/artemis_plan-20200921.pdf*

[17] *ESA:* ESA Strategy for Science at the Moon

[18] *Jia, Y./Lin, Z. W.:* The Radiation Environment on the Moon from Galactic Cosmic Rays in a Lunar Habitat. In: Radiation Research, Vol. 173, 2010, pp. 238–244. *https://www.jstor.org/stable/40539837*

[19] *Cucinotta, F. A./Hu, S./Schwadron, N. A./Kozarev, K./Townsend, L. W./Kim, M.-H. Y.:* Space radiation risk limits and Earth-Moon-Mars environmental models. In: Space Weather, Vol. 8, No. 12, Dec. 2010 (DOI: 10.1029/2010SW000572)

[20] *Friedl, S.:* Modelling and Simulation of the Primary Power Distribution of a Lunar Habitat

[21] *NASA/JSC/Schmitt, H. H.:* Astronaut Harrison Schmitt standing next to boulder during third EVA. *https://science.nasa.gov/image-detail/amf-as17-146-22294*

[22] *NASA/JSC/Aldrin, E. B.:* Buzz Aldrin Stands Beside the U.S. Flag on the Moon. *https://science.nasa.gov/image-detail/amf-as11-40-5875*

[23] *NASA:* Artemis 1 Mission Page. *https://www.nasa.gov/specials/artemis-i*

[24] *Wörner, J.:* Space19+. Presentation. European Astronaut Centre (EAC) 2019

[25] *Watson-Morgan, L./Chavers, G./Connolly, J./Crowe, K./Krupp, D./Means, L./Percy, T./Polsgrove, T./Turpin, J.:* NASA's Initial and Sustained Artemis Human Landing Systems. In: 2021 IEEE Aerospace Conference (50100), Big Sky, MT/USA, Mar. 2021, pp. 1–11 (DOI: 10.1109/AERO50100.2021.9438179)

[26] *NASA:* Orion Reference Guide. *https://www.nasa.gov/wp-content/uploads/2023/02/orion-reference-guide-111022.pdf*

[27] *ESA:* Artemis I. *https://www.esa.int/Science_Exploration/Human_and_Robotic_Exploration/Orion/Artemis_I*

[28] *ESA:* Artemis II. *https://www.esa.int/Science_Exploration/Human_and_Robotic_Exploration/Orion/Artemis_II*

[29] *ESA:* Artemis III. *https://www.esa.int/Science_Exploration/Human_and_Robotic_Exploration/Orion/Artemis_III*

[30] *NASA:* Rendering of 13 candidate landing regions for Artemis III. *https://www.nasa.gov/news-release/nasa-identifies-candidate-regions-for-landing-next-americans-on-moon*

[31] *ESA:* Artemis IV. *https://www.esa.int/Science_Exploration/Human_and_Robotic_Exploration/Orion/Artemis_IV*

[32] *ESA:* Orion and I-HAB. *https://www.esa.int/ESA_Multimedia/Videos/2023/06/Orion_and_I-Hab*

[33] *ESA:* What is ESPRIT? *https://www.esa.int/ESA_Multimedia/Images/2020/10/What_is_ESPRIT*

[34] *ESA:* Gateway: Lunar View. *https://www.esa.int/Science_Exploration/Human_and_Robotic_Exploration/Gateway_Lunar_View*

[35] *ESA:* Gateway refuelling module gets a boost. *https://www.esa.int/Science_Exploration/Human_and_Robotic_Exploration/Gateway_refuelling_module_gets_a_boost*

[36] *ESA:* Artemis V. *https://www.esa.int/Science_Exploration/Human_and_Robotic_Exploration/Orion/Artemis_V*

[37] *ESA:* Delivering Esprit to Gateway. *https://www.esa.int/ESA_Multimedia/Videos/2023/06/Delivering_Esprit_to_Gateway*

[38] *Jakupca, I.:* NASA Fuel Cell and Hydrogen Activities. 2024. p. 12. *https://ntrs.nasa.gov/api/citations/20240000908/downloads/Fuel%20Cell%20Project%20Overview%20for%20Robinson%20Res%20Rev2.pdf*

[39] *NASA:* Space Launch System Flip Book

[40] *NASA:* Space Launch System Factsheet. *https://www.nasa.gov/sites/default/files/atoms/files/sls_fact_sheet.pdf*

[41] *NASA/Shiflett, K.:* Artemis I Rocket with Orion and European Service Module on the Launchpad. *https://www.esa.int/var/esa/storage/images/esa_multimedia/images/2022/11/artemis_i_rocket_with_orion_and_european_service_module_on_the_launchpad/24572464-1-eng-GB/Artemis_I_rocket_with_Orion_and_European_Service_Module_on_the_launchpad_pillars.jpg*

[42] *NASA:* SLS Lift Capabilities for Different Vehicle Configurations. *https://www.nasa.gov/wp-content/uploads/2021/12/sls_lift_capabilities_configurations_12022021_woleo.jpg*

[43] *NASA:* Render of the Orion Spacecraft. *https://www.esa.int/Science_Exploration/Human_and_Robotic_Exploration/Orion/The_spacecraft*

[44] *Moskowitz, C.:* NASA to Launch New Spacecraft. December 1, 2014. *https://www.scientificamerican.com/article/nasa-to-launch-new-spacecraft*

[45] *ESA-K. Oldenburg:* Orion: dimensions. *https://www.esa.int/ESA_Multimedia/Images/2019/03/Orion_dimensions*

[46] *ESA-K. Oldenburg:* Orion power. *https://www.esa.int/ESA_Multimedia/Images/2018/11/Orion_power*

[47] *ESA:* Power. *https://www.esa.int/Science_Exploration/Human_and_Robotic_Exploration/Orion/Power*

[48] *ESA Academy:* Ladybird Guide to Spacecraft Operations

[49] *NASA OCHMO:* Vehicle Hatches. Technical Brief

[50] *Airbus:* Orion European Service Module. *https://www.airbus.com/en/OrionESM*

[51] *ESA:* ESA workhorse to power NASA's Orion spacecraft. *https://www.esa.int/Science_Exploration/Human_and_Robotic_Exploration/Orion/ESA_workhorse_to_power_NASA_s_Orion_spacecraft*

[52] *NASA:* Astronauts inside the Orion Crew Exploration Module. *https://images-assets.nasa.gov/image/jsc2022e044165/jsc2022e044165~large.jpg?w=1920&h=1247&fit=clip&crop=faces%2Cfocalpoint*

[53] *NASA:* NASA's Readiness for the Artemis II Crewed Mission to Lunar Orbit. May 1, 2024

[54] *ESA/Lochtenberg, K.:* Render of Lunar Gateway Modules. *https://www.esa.int/Space_in_Member_States/Germany/NASA_und_ESA_einigen_sich_auf_die_Lunar_Gateway-Umlaufbahn*

[55] *EoPortal:* ESA. Lunar Gateway. 2019. *https://www.eoportal.org/other-space-activities/lunar-gateway#esa-participation-in-nasas-lunar-gateway-venture*

[56] *Gargioni, G./Peterson, M./Alexandre, D./Schroeder, K.:* Multiple Asteroid Retrieval Mission from Lunar Orbital Platform-Gateway Using Reusable Spacecraft. In: IEEE Aerospace Conference, March 1, 2019, p. 13

[57] *ESA:* New names for the Gateway. 03/04/2024. *https://www.esa.int/Science_Exploration/Human_and_Robotic_Exploration/New_names_for_the_Gateway*

[58] *ESA:* Gateway: Lunar I-Hab. *https://www.esa.int/Science_Exploration/Human_and_Robotic_Exploration/Gateway_Lunar_I-Hab*

[59] *Catchpole, J. E.:* The International Space Station Building for the Future. 2008

[60] *Thales:* Thales Alenia Space Chosen by Northrop Grumman to Provide the Pressurized Module For Halo. *https://www.thalesgroup.com/en/worldwide/space/press-release/thales-alenia-space-chosen-northrop-grumman-provide-pressurized*

[61] *Thales Alenia Space:* Lunar I-Hab mock-up interior. [Online]. Available: *https://www.esa.int/var/esa/storage/images/esa_multimedia/images/2024/04/lunar_i-hab_mock-up_interior/26024185-2-eng-GB/Lunar_I-Hab_mock-up_interior_pillars.jpg*

[62] *ESA-S. Corvaja:* Gateway's Lunar I-Hab module under construction in Turin. *https://www.esa.int/Science_Exploration/Human_and_Robotic_Exploration/Gateway_Lunar_I-Hab*

[63] *NASA/Donaldson, A. A.:* NASA, United Arab Emirates Announce Artemis Lunar Gateway Airlock. January 7, 2024.*https://www.nasa.gov/news-release/nasa-united-arab-emirates-announce-artemis-lunar-gateway-airlock*

[64] *Jensen, J.:* Starship Users Guide

[65] *Witt, M.:* As Artemis Moves Forward, NASA Picks SpaceX to Land Next Americans on Moon. April 16, 2021. *https://www.nasa.gov/press-release/as-artemis-moves-forward-nasa-picks-spacex-to-land-next-americans-on-moon*

[66] *SpaceX:* Starship on the Launchpad in Boca Chica. *https://t3n.de/news/spacex-starship-startversuche-1559864*

[67] *Wattles, J.:* Bezos-founded Blue Origin lands major NASA moon lander project. May 19, 2023. *https://edition.cnn.com/2023/05/19/world/nasa-lunar-lander-spacex-blue-origin-scn/index.html*

[68] *Berger, E.:* Once again, NASA leans into the future by picking an innovative lunar lander. May 19, 2023. *https://arstechnica.com/space/2023/05/blue-origin-wins-pivotal-nasa-contract-to-develop-a-second-lunar-lander*

[69] *O'Shea, C. A.:* NASA Selects Blue Origin as Second Artemis Lunar Lander Provider. May 19, 2023. *https://www.nasa.gov/news-release/nasa-selects-blue-origin-as-second-artemis-lunar-lander-provider*

[70] Render of the Blue Moon Spacecraft on the Lunar Surface. *https://www.nasa.gov/wp-content/uploads/2023/05/bluemoon_nasa_option_2023-05-19_01.29.31_0.jpg*

[71] *Blue Origin:* Render of the Blue Moon MK1 Spacecraft. *https://d3ezn0y6hdgq62.cloudfront.net/assets/blue-moon/mk-1/mk1-full_crop.png*

[72] *Blue Origin:* Blue Moon Mark 1 Lunar Lander. *https://www.blueorigin.com/blue-moon/mark-1*

[73] *Dueñas, M.:* Dynetics to develop NASA's Artemis Human Lunar Landing System. April 30, 2020. *https://www.leidos.com/insights/dynetics-develop-nasas-artemis-human-lunar-landing-system*

[74] *Hickman, J. M./Curtis, H. B./Landis, G. A.:* Design Considerations For Lunar Base Photovoltaic Power Systems. 21st Photovoltaic Specialists Conference, sponsored by the Institute of Electrical and Electronics Engineers, Kissimmee, Florida, May 21–25, 1990, p. 10

[75] *Khan, Z.:* Power System Concepts for the Lunar Outpost. A Review of the Power Generation, Energy Storage, Power Management and Distribution (PMAD) System Requirements and Potential Technologies for Development of the Lunar Outpost. In: AIP Conference Proceedings, Albuquerque, New Mexico (USA), 2006, pp. 1083–1092 (DOI: 10.1063/1.2169289)

[76] *Frazier, S.:* NASA Announces Artemis Concept Awards for Nuclear Power on Moon. June 21, 2022. *https://www.nasa.gov/press-release/nasa-announces-artemis-concept-awards-for-nuclear-power-on-moon*

[77] *NASA/Skelly, C.:* NASA, Industry to Mature Vertical Solar Array Technologies for Lunar Surface. March 23, 2021. *https://www.nasa.gov/feature/nasa-industry-to-mature-vertical-solar-array-technologies-for-lunar-surface*

[78] *Heinicke, C./Orzechowski, L./Arnhof, M.:* Updated design concepts of the Moon and Mars Base Analog (MaMBA). 2018

[79] *Thiébaut, L./Cowley, A.:* Microwave processing of regolith A 1D-printing cavity for enabling lunar construction technology. 8th European Conference for Aeronautics And Space Sciences (EUCASS 2019 (DOI: 10.13009/EUCASS2019-917)

[80] *ESA/Carril, P.:* Artist impression of a Moon Base concept. *https://www.esa.int/var/esa/storage/images/esa_multimedia/images/2019/07/artist_impression_of_a_moon_base_concept/19471139-2-eng-GB/Artist_impression_of_a_Moon_Base_concept_pillars.jpg*

[81] *Heinicke, C.:* Mamba – Moon And Mars Base Analog. *https://www.zarm.uni-bremen.de/de/forschung/unabhaengige-gruppen/extraterrestrial-habitation.html*

[82] *ESA:* LUNA infographics. *https://www.esa.int/ESA_Multimedia/Images/2024/09/LUNA_infographics*

[83] *ESA/DLR:* LUNA – Europe's Moon on Earth. *https://www.esa.int/ESA_Multimedia/Images/2024/09/LUNA_-_Europe_s_Moon_on_Earth*

[84] *ESA/Romeo, A.:* Moonwalk in Lanzarote

[85] *ESA/Romeo, A.:* Caves for space exploration. *https://www.esa.int/Science_Exploration/Human_and_Robotic_Exploration/CAVES_and_Pangaea/A_new_journey_into_Earth_for_space_exploration*

[86] *Reynolds, W. T.:* Architecture analysis of wireless power transmission for lunar outposts. 2015

[87] *Mintz, T./Maslowski, E. A./Colozza, A./McFarland, W./Prokopius, K. P./George, P. J./Hussey, S. W.:* Simulation of a Lunar Surface Base Power Distribution Network for the Constellation Lunar Surface Systems. 2010. p. 39

[88] *NASA/Goddard Space Flight Center/Arizona State University:* Lunar Far Side. *https://science.nasa.gov/resource/lunar-far-side-2*

[89] *NASA/Goddard Space Flight Center/Arizona State University:* Lunar Near Side. *https://science.nasa.gov/resource/lunar-near-side-2*

[90] *NASA:* Image of the Sun. 2015. *https://solarsystem.nasa.gov/system/resources/detail_files/768_oct_1_2015_flare.jpg*

[91] *NASA:* Image of the Earth. *https://www.nasa.gov/centers/goddard/images/content/638831main_globe_east_2048.jpg*

[92] *NASA:* Image of the Moon. *https://www.nasa.gov/sites/default/files/thumbnails/image/as11-44-6667.jpg*

[93] *Guzik, M. C./Jakupca, I./Gilligan, R./Bennett, W. R./Smith, P. J./Fincannon, J.:* Regenerative Fuel Cell Power Systems for Lunar and Martian Surface Exploration. In: AIAA SPACE and Astronautics Forum and Exposition, American Institute of Aeronautics and Astronautics, Orlando, Florida, September 2017 (DOI: 10.2514/6.2017-5368)

[94] *Mason, S./Bloomfield, S.:* SP-100 Power System Conceptual Design for Lunar Base Applications. Sixth Symposium on Space Nuclear Power Systems, sponsored by the Institute for Space Nuclear Power Studies, Albuquerque, New Mexico, January 8–12, 1989, p. 24

[95] *Balint, T. S.:* Comparison of Power System Options Between Future Lunar and Mars Missions. 2005, p. 8

[96] *Scheuring, R. A./Jones, J. A./Novak, J. D./Polk, J. D./Gillis, D. B./Schmid, J./Duncan, J. M./Davis, J. R.:* The Apollo Medical Operations Project: Recommendations to improve crew health and performance for future exploration missions and lunar surface operations. In: Acta Astronaut., Vol. 63, No. 7–10, October 2008, pp. 980–987 (DOI: 10.1016/j.actaastro.2007.12.065)

[97] *Ley, W./Wittmann, K./Hallmann, W. (eds.):* Handbuch der Raumfahrttechnik. 5., aktualisierte und erweiterte Auflage. Carl Hanser Verlag, München 2019

[98] *Wertz, J. R./Larson, W. J.:* Space Mission Analysis and Design. 7th Edition. Space Technology Library. Microcosm Press/Kluwer Academic Publishers 2005

[99] *Bentley, A. N./Jonckheere, C. G. de/Miller, D. E.:* Geometric albedo of the Moon in the ultraviolet. In: Astron. J., Vol. 79, March 1974, p. 401 (DOI: 10.1086/111558)

[100] *Bhanderi, D./Bak, T.:* Modeling Earth Albedo for Satellites in Earth Orbit. In: AIAA Guidance, Navigation, and Control Conference and Exhibit, American Institute of Aeronautics and Astronautics, San Francisco, California, August 2005 (DOI: 10.2514/6.2005-6465)

[101] *Larson, W. J./Wertz, J. R.:* Space Mission Analysis and Design. 2nd Edition. Microcosm Inc. 1992

[102] *Kieffer, H. H./Stone, T. C.:* The Spectral Irradiance of the Moon. In: Astron. J., Vol. 129, No. 6, June 2005, pp. 2887–2901 (DOI: 10.1086/430185)

[103] *Stephens, G. L./O'Brien, D./Webster, P. J./Pilewski, P./Kato, S./Li, J.:* The Albedo of Earth. In: Rev. Geophys., Vol. 53, No. 1, March 2015, pp. 141–163 (DOI: 10.1002/2014RG000449)

[104] *Kaczmarzyk, M./Gawronski, M./Piatkowski, G.:* Global database of direct solar radiation at the Moon's surface for lunar engineering purposes. E3S Web Conf., Vol. 49, 2018. *https://doi.org/10.1051/e3sconf/20184900053*

[105] *Li, X./Yu, W./Wang, S./Li, S./Tang, H./Li, Y./Zheng, Y./Tsang, K. T./Ouyang, Z.:* Condition of Solar Radiation on the Moon. In: *Badescu, V. (ed.):* Moon. Prospective Energy and Material Resources. Springer, Berlin/Heidelberg 2012

[106] *Hu, G.-P./Zheng, Y.-C./Xu, A.-A./Tang, Z.-S.:* Lunar Surface Temperature of Global Moon: Preparation of Database With Topographic and Albedo Effects. In: IEEE Geosci. Remote Sens. Lett., Vol. 13, No. 1, January 2016, pp. 110–114 (DOI: 10.1109/LGRS.2015.2499305)

[107] *Kieffer H. H./Stone, T. C.:* The Spectral Irradiance of the Moon. In: US Geological Survey. Flagstaff 2005. *http://modresvetlo.cz/PDF/The_Spectral_Irradiance_of_the_Moon.pdf*

[108] *USGS:* Robotic Lunar Observatory *https://astrogeology.usgs.gov/moon-cal*

[109] *Quaschning, V.:* Regenerative Energiesysteme. Technologie – Berechnung – Klimaschutz. 10th Edition. Carl Hanser Verlag, München 2019 (DOI: 10.3139/9783446461147)

[110] *NASA:* NASA SP-8105. Spacecraft Thermal Control. May 1973. *https://ntrs.nasa.gov/api/citations/19730018163/downloads/19730018163.pdf*

[111] *Zünkler, J.-H.:* Development of a simulator for the operational conditions of photovoltaic power generators on the lunar surface and implications of environmental conditions to photovoltaic generator design and optimization. Master's Thesis. University of Paderborn, Paderborn 2023

[112] *Wertz, J. R./Larson, W. J. (eds.):* Space Mission Analysis and Design. 3rd Edition. Space Technology Library. Microcosm/Kluwer, El Segundo, California/USA/Boston/Dordrecht 1999

[113] *NASA:* Image of Orion Spacecraft with Earth and Moon. *https://blogs.nasa.gov/artemis/2022/11/28/artemis-i-flight-day-13-orion-goes-the-max-distance*

[114] *rustimation.eu:* Mondphase und -beleuchtung ausrechnen. *https://www.rustimation.eu/index.php/mondphase-und-beleuchtung-ausrechnen*

[115] *Gaier, J. R./Street, K. W./Gustafson, R. J.:* Measurement of the Solar Absorptance and Thermal Emittance of Lunar Simulants. 2010. p. 20

[116] *Thelen, P.:* Introduction to the Space Radiation Environment. 2018. p. 23. *https://www.osti.gov/biblio/1524958*

[117] *Dachev, T. P./Tomov, B. T./Matviichuk, Y. N./Dimitrov, P. S./Vadawale, S. V./Goswami, J. N./Angelis, G. de/Girish, V.:* An overview of RADOM results for earth and moon radiation environment on Chandrayaan-1 satellite. In: Adv. Space Res., Vol. 48, No. 5, September 2011, pp. 779–791 (DOI: 10.1016/j.asr.2011.05.009)

[118] *Andrews, D./Wetzel, E.:* Solar Electric SpaceTug to Support Moon and Mars Exploration Missions. In: Space 2005, American Institute of Aeronautics and Astronautics, Long Beach, California, August 2005 (DOI: 10.2514/6.2005-6739)

[119] *Ley, W./Wittmann, K./Hallmann, W. (eds.):* Handbook of Space Technology. Wiley, Chichester, UK 2009

[120] *Royal Observatory of Belgium:* International sunspot number Sn: monthly mean and 13-month smoothed number. *https://www.sidc.be/SILSO/IMAGES/GRAPHICS/wolfmms.png*

[121] *Goldstein, B. E.:* Observations of electrons at the lunar surface. In: J. Geophys. Res., Vol. 79, No. 1, January 1974, pp. 23–35 (DOI: 10.1029/JA079i001p00023)

[122] *Halekas, J. S./Poppe, A./Delory, G. T./Farrell, W. M./Horányi, M.:* Solar wind electron interaction with the dayside lunar surface and crustal magnetic fields: Evidence for precursor effects. In: Earth Planets Space, Vol. 64, No. 2, Feb. 2012, pp. 73–82 (DOI: 10.5047/eps.2011.03.008)

[123] *Goswami, J. N.:* An Overview of the Chandrayaan-1 Mission. Presented at the 41st Lunar and Planetary Science Conference. 2010

[124] *Vadawale, S. V./Goswami, J. N./Dachev, T. P./Tomov, B. T./Girish, V.:* Radiation Environment in Earth–Moon Space: Results from Radom Experiment Onboard Chandrayaan-1. In: Advances in Geosciences, July 2011, pp. 121–132 (DOI: 10.1142/9789814355377_0009)

[125] *Spence, H. E. et al.:* CRaTER: The Cosmic Ray Telescope for the Effects of Radiation Experiment on the Lunar Reconnaissance Orbiter Mission. In: Space Sci. Rev., Vol. 150, No. 1–4, January 2010, pp. 243–284 (DOI: 10.1007/s11214-009-9584-8)

[126] *Roam, D.:* Render of Chandrayaan-1 Spacecraft. *https://www.space.com/40114-chandrayaan-1.html*

[127] *Reitz, G./Berger, T./Matthiae, D.:* Radiation exposure in the moon environment. In: Planet. Space Sci., Vol. 74, No. 1, December 2012, pp. 78–83 (DOI: 10.1016/j.pss.2012.07.014)

[128] *LaBel, K. A.:* Space Radiation Effects on Electronics: Simple Concepts and New Challenges. 2004. *https://nepp.nasa.gov/files/25295/MRS04_LaBel.pdf*

[129] *Suparta, W./Zulkeple, S. K.:* Investigating Space Radiation Environment Effects on Communication of Razaksat-1. In: J. Aerosp. Technol. Manag., Vol. 10, May 2018 (DOI: 10.5028/jatm.v10.815)

[130] *Jaeger, J. C.:* The Surface Temperature of The Moon. 1952

[131] *Jaeger, J. C.:* Conduction of heat in a solid with periodic boundary conditions, with an application to the surface temperature of the moon. In: Math. Proc. Camb. Philos. Soc., Vol. 49, No. 2, April 1953, pp. 355–359 (DOI: 10.1017/S0305004100028450)

[132] *Jaeger, J. C./Harper, A. F. A.:* Nature of the Surface of the Moon. In: Nature, Vol. 166, No. 4233, December 1950, pp. 1026–1026 (DOI: 10.1038/1661026a0)

[133] *Kuskov O. L./Kronrod, V. A.:* Constitution of the Moon 5. Constraints on composition, density, temperature, and radius of a core. 1998. p. 22

[134] *Williams, J.-P./Paige, D. A./Greenhagen, B. T./Sefton-Nash, E.:* The global surface temperatures of the Moon as measured by the Diviner Lunar Radiometer Experiment. In: Icarus, Vol. 283, Feb. 2017, pp. 300–325 (DOI: 10.1016/j.icarus.2016.08.012)

[135] *Sharp, T./Urrutia, D. E.:* What is the Temperature on the Moon? space.com, last updated May 5, 2023. *https://www.space.com/18175-moon-temperature.html*

[136] *Racca, G. D.:* Moon surface thermal characteristics for moon orbiting spacecraft thermal analysis. In: Planet. Space Sci., Vol. 43, No. 6, Jun. 1995, pp. 835–842 (DOI: 10.1016/0032-0633(95)00067-F)

[137] *NASA:* LRO Deviner Dataset. *https://pds-geosciences.wustl.edu/missions/lro/diviner.htm*

[138] *Colozza, A. J.:* Analysis of Lunar Regolith Thermal Energy Storage

[139] *Fleith, P./Cowley, A./Canals Pou, A./Valle Lozano, A./Frank, R./López Córdoba, P./González-Cinca, R.:* In-situ approach for thermal energy storage and thermoelectricity generation on the Moon. Modelling and simulation. In: Planet. Space Sci., Vol. 181, Feb. 2020, p. 104789 (DOI: 10.1016/j.pss.2019.104789)

[140] *Krauter, S.:* Solar Electric Power Generation

[141] *NASA:* Image of the ISS. *https://www.nasa.gov/sites/default/files/styles/full_width_feature/public/thumbnails/image/iss056e201352.jpg*

[142] *NASA Analytical Mechanics Associates-Johnson Space Center Advanced Concepts Lab (JSC ACL):* Render of Orion. *https://www.researchgate.net/profile/Jacob-Williams-3/publication/269049229/figure/fig5/AS:668862076956683@1536480554942/Orion-spacecraft-Image-credit-Analytical-Mechanics-Associates-Johnson-Space-Center.png*

[143] *Schmiegel, A. U.:* Energiespeicher für die Energiewende. Auslegung und Betrieb von Speichersystemen. Carl Hanser Verlag, München 2019

[144] *Gillespie, D./Wilson, A. R./Martin, D./Mitchell, G./Filippi, G./Vasile, M.:* Comparative Analysis of Solar Power Satellite Systems to Support a Moon

[145] *Ramasamy, R./Paridon, D. van/Summerer, L.:* Solar Power Satellites for Lunar Rover Exploration. IAC-18-C3410, Oct. 2018

[146] *Kenny, B. H./Cull, R. C./Kankam, M. D.:* An Analysis of Space Power System Masses. In: Proceedings of the 25th Intersociety Energy Conversion Engineering Conference, IEEE, Reno, Nevada 1990, pp. 484–489 (DOI: 10.1109/IECEC.1990.716942)

[147] *Kamel, A. M./ElWakeel, A. S./AbouZayed, U. R.:* Sequential Switching Shunt Regulator for Satellite Power Control System. In: IOSR J. Electr. Electron. Eng., Vol. 12, No. 03, Jun. 2017, pp. 78–83 (DOI: 10.9790/1676-12030278839

[148] *Surampudi, S.:* Overview of the Space Power Conversion and Energy Storage Technologies. p. 45

[149] *Holm-Nielsen, J. B./Sanjeevikumar, P.:* Power Electronic Converter Configuration and Control for DC Microgrid Systems. MDPI 2020 (DOI: 10.3390/books978-3-03936-432-9)

[150] *Baez, A.:* Design Considerations for High Power Spacecraft Electrical Systems. 2012. *www.nasa.gov*

[151] *Mertens, K.:* Photovoltaik. Lehrbuch zu Grundlagen, Technologie und Praxis. 4., aktualisierte Auflage. Fachbuchverlag Leipzig im Carl Hanser Verlag, München 2018

[152] *Bennett, W. R./Smith, P. J./Jakupca, I. J.:* Analysis of 100-W Regenerative Fuel Cell Demonstration. 2020. p. 20

[153] *Böcker, J.:* Antriebe für umweltfreundliche Fahrzeuge. 2018

[154] *Tulay, G./Iskender, I.:* Design of a Novel Converter Between Li-Ion Battery and Supercapacitor to Feed Synthetic Aperture Radar Loads for Satellite Applications. International Journal of Renewable Energy Research (IJRER), Vol. 10, No. 4, 2020. p. 10

[155] *De, A.:* Architectural Design Criteria for Spacecraft Solar Arrays. In: *Kosyachenko, L. A. (ed.):* Solar Cells. Thin-Film Technologies. InTech 2011 (DOI: 10.5772/21312)

[156] *NASA:* Orion Solar Array Wing. *https://www.esa.int/var/esa/storage/images/esa_multimedia/videos/2019/04/orion_spreads_its_wings/19331727-1-eng-GB/Orion_spreads_its_wings_pillars.png*

[157] *Iles, P. A.:* Evolution of space solar cells. In: Sol. Energy Mater, 2001, p. 13

[158] *NREL:* Best Research-Cell Efficiencies. *https://www.nrel.gov/pv/cell-efficiency.html*

[159] Das Tafelwerk interaktiv. Formelsammlung für die Sekundarstufe I. Cornelsen

[160] *Rückert, U.:* Skript zur Vorlesung Halbleiterbauelemente. Heinz Nixdorf Institut, 2003

[161] *Voigt, T.:* Leistungselektronik für die Energiewende. Chapter "Photovoltaik"

[162] *Anspaugh, B. E.:* GaAs Solar Cell Radiation Handbook. p. 328

[163] *Thiede, A.:* Werkstoffe der Elektrotechnik. 2013.

[164] *Quaschning, V.:* Simulation der Abschattungsverluste bei solarelektrischen Systemen. In: Beiträge zur Solarforschung, No. 6, 1996, Köster, Berlin

[165] *Förstner, J.:* Vorlesung EM Wellen. Universität Paderborn, 2018

[166] *King, R./Fetzer, C. M./Law, D. C./Edmondson, K. M./Yoon, H./Kinsey, G. S./Krut, D. D./Ermer, J. H./Hebert, P./Cavicchi, B. T./Karam, N. H.:* Advanced III-V Multijunction Cells for Space. In: 2006 IEEE 4th World Conference on Photovoltaic Energy Conference, IEEE, Waikoloa, Hawaii, 2006, pp. 1757–1762 (DOI: 10.1109/WCPEC.2006.279831)

[167] *CESI:* Datasheet CTJ-30. Solar Cell. *https://www.cesi.it/space-solar-cells*

[168] *Vergura, S.:* A Complete and Simplified Datasheet-Based Model of PV Cells in Variable Environmental Conditions for Circuit Simulation. In: Energies, Vol. 9, No. 5, Apr. 2016, p. 326 (DOI: 10.3390/en9050326)

[169] *Krauter, S.:* Lecture Energy Transition

[170] *Krauter, S.:* Lecture Solar Electric Energy Systems

[171] *Tada, H. Y./Carter, J. R. Jr./Anspaugh, B. E./Downing, R. G.:* Solar Cell Radiation Handbook, 3rd Edition. NASA JPL 1982

[172] *AZUR Space:* Datasheet HNR 0005979-01-01. Solar Cell. *http://www.azurspace.com/index.php/en/products/products-space/space-solar-cells*

[173] *AZUR Space:* Datasheet HNR 0006050-01-00. Solar Cell. *http://www.azurspace.com/index.php/en/products/products-space/space-solar-cells*

[174] *AZUR Space:* Datasheet HNR 0003429-01-01. Solar Cell. *http://www.azurspace.com/index.php/en/products/products-space/space-solar-cells*

[175] *AZUR Space:* Datasheet HNR 0004148-00-01. Solar Cell. *http://www.azurspace.com/index.php/en/products/products-space/space-solar-cells*

[176] *AZUR Space:* Datasheet HNR 0003384-01-02. Solar Cell. *http://www.azurspace.com/index.php/en/products/products-space/space-solar-cells*

[177] *AZUR Space:* Datasheet HNR 0003421-01-02. Solar Cell. *http://www.azurspace.com/index.php/en/products/products-space/space-solar-cells*

[178] *AZUR Space:* Datasheet HNR 0003422-02-02. Solar Cell. *http://www.azurspace.com/index.php/en/products/products-space/space-solar-cells*

[179] *AZUR Space:* Datasheet HNR 0002490-00-04. Solar Cell. *http://www.azurspace.com/index.php/en/products/products-space/space-solar-cells*

[180] *AZUR Space:* Datasheet HNR 0002162-00-03. Solar Cell. *http://www.azurspace.com/index.php/en/products/products-space/space-solar-cells*

[181] *AZUR Space:* Datasheet Silicon Bypass Diode

[182] *AZUR Space:* Datasheet Handling of semiconductor products

[183] *CESI:* Datasheet CTJ-30 –Thin. Solar Cell. *https://www.cesi.it/space-solar-cells*

[184] *CESI:* Datasheet CTJ-LC. Solar Cell. *https://www.cesi.it/space-solar-cells*

[185] *SolarAero:* Datasheet IMM_Alpha Solar Cell. *https://solaerotech.com/space-solar-cells-cics*

[186] *SolarAero:* Datasheet ZTJ+. Solar Cell. *https://solaerotech.com/space-solar-cells-cics*

[187] *SolarAero:* Datasheet ZTJ. Solar Cell. *https://solaerotech.com/space-solar-cells-cics*

[188] *SolarAero:* Datasheet ZTJ-Omega. Solar Cell. *https://solaerotech.com/space-solar-cells-cics*

[189] *SolarAero:* Datasheet Z4J. Solar Cell. *https://solaerotech.com/space-solar-cells-cics*

[190] *Spectrolab:* Datasheet Spectrolab C3P5J. Concentrator Cell. *https://www.spectrolab.com/photovol taics.html*

[191] *Spectrolab:* Datasheet Spectrolab C4MJ. Concentrator Cell. *https://www.spectrolab.com/photovoltaics.html*

[192] *Spectrolab:* Datasheet Spectrolab XTJ-Prime. Solar Cell. *https://www.spectrolab.com/photovoltaics.html*

[193] *Spectrolab:* Datasheet Spectrolab XTJ-CIC. Solar Cell. *https://www.spectrolab.com/photovoltaics.html*

[194] *Spectrolab:* Datasheet Spectrolab XTE-HF. Solar Cell. *https://www.spectrolab.com/photovoltaics.html*

[195] *Spectrolab:* Datasheet Spectrolab XTE-SF. Solar Cell. *https://www.spectrolab.com/photovoltaics.html*

[196] *Spectrolab:* Datasheet Spectrolab XTE-LILT. Solar Cell. *https://www.spectrolab.com/photovoltaics.html*

[197] *Spectrolab:* Datasheet Spectrolab UTJ-CIC. Solar Cell. *https://www.spectrolab.com/photovoltaics.html*

[198] NASA High Efficiency 30–40% Solar Cell. Patent No. 9,117,948; 9,418,844

[199] *Mikhaylik, Y./Kovalev, I./Schock, R./Kumaresan, K./Xu, J./Affinito, J.:* High Energy Rechargeable Li-S Battery Development at Sion Power and BASF. In: ECS Transactions 25(35), April 2010. p. 26.

[200] *Yamaguchi, M./Lee, K./Araki, K./Kojima, N./Okuno, Y./Imaizumi, M.:* Analysis for nonradiative recombination loss and radiation degradation of Si space solar cells. In: Prog. Photovolt. Res. Appl., Vol. 29, No. 1, Jan. 2021, pp. 98–108 (DOI:10.1002/pip.3346)

[201] *Lamb, D. A./Irvine, S. J. C./Baker, M. A./Underwood, C. I./Mardhani, S.:* Thin film cadmium telluride solar cells on ultra-thin glass in low earth orbit—3 years of performance data on the AlSat-1N CubeSat mission. In: Prog. Photovolt. Res. Appl., Vol. 29, No. 9, Sep. 2021, pp. 1000–1007, (DOI: 10.1002/pip.3423)

[202] *Li, J./Aierken, A./Liu, Y./Zhuang, Y./Yang, X./Mo, J./Fan, R./ Chen, Q. Y./Zhang, S./Huang, Y./Zhang, Q.:* A Brief Review of High Efficiency III-V Solar Cells for Space Application. In: Front. Phys., Vol. 8, Feb. 2021, p. 631925 (DOI: 10.3389/fphy.2020.631925)

[203] *Marvin D. C./Nocerino, J. C.:* Evaluation of multijunction solar cell performance in radiation environments. In: Conference Record of the Twenty-Eighth IEEE Photovoltaic Specialists Conference 2000 (Cat. No.00CH37036), IEEE, Anchorage, AK/USA 2000, pp. 1102–1105 (DOI: 10.1109/PVSC.2000.916079)

[204] *Blanes, J. M./Carrasco, J. A./Garrigós, A./Marroquí, D./Torres, C.:* Sequential Switching Shunt Regulator Parallel Power Processing Control for High Capacitance Solar Arrays. In: Energies, Vol. 14, No. 2, Jan. 2021, p. 429 (DOI: 10.3390/en14020429)

[205] *Takamoto, T./Waschio, H./Juso, H.:* Application of InGaP/GaAs/InGaAs Triple Junction Solar Cells to Space Use and Concentrator Photovoltaics. In: 2014 IEEE 40th Photovoltaic Specialist Conference (PVSC), Denver, CO/USA 2014, pp. 0001–0005

[206] *ESA/SJM Photography:* Bendy, ultra-thin solar cell. *https://www.esa.int/esearch?q=flexible+solar+cell*

[207] *Liu, Y./Li, J./Aierken, A./Zhuang, Y./Yang, X./Mo, J. H./Fan, R. K./Chen, Q. Y./Zhang, S. Y./Huang, Y. M./Zhang, Q.:* Large Area Multijunction III-V Space Solar Cells Over 31% Efficiency. In: 2017 IEEE 44th Photovoltaic Specialist Conference (PVSC), Washington, D. C./USA, June 2017, pp. 2094–2098 (DOI: 10.1109/PVSC.2017.8366024)

[208] *Guter, W./Dunzer, F./Ebel, L./Hillerich, K./Köstler, W./Kubera, T./Meusel, M./Postels, B./Wächter, C.:* Space Solar Cells – 3G30 and Next Generation Radiation Hard Products. In: E3S Web Conference, Vol. 16, 2017, p. 03005 (DOI: 10.1051/e3sconf/20171603005)

[209] *Li, X./Li, G./Lu, H./Zhang, W.:* > 35% 5-junction space solar cells based on the direct bonding technique. In: Journal of Semiconductors, Volume 42, Number 12, 2021, p. 7

[210] *Bett, A. W. et al.:* Highest Efficiency Multi-Junction Solar Cell for Terrestrial and Space Applications. 2009. p. 7

[211] *Meusel, M. et al.:* Development Status of European Multi-Junction Space Solar Cells with High Radiation Hardness. 2005. p. 7

[212] *Llorens, J. M./Vasileiou, T./Raya, A./Buencuerpo, J./Ripalda, J. M./Hildebrant, K./Izzo, D./Summerer, L.:* Light Absorption Enhancement with Bio-inspired Nanostructures. 2020

[213] *Vagidov, N. Z./Montgomery, K. H./Bradshaw, G. K./Wilt, D. A.:* Light trapping structures for radiation hardness enhancement of space solar cells. In: Sol. Energy Mater. Sol. Cells, Vol. 182, Aug. 2018, pp. 136–141 (DOI: 10.1016/j.solmat.2018.03.036)

[214] *Sharps, P. et al.:* Next Generation Radiation Hard IMM Space Solar Cells. In: E3S Web Conference, Vol. 16, 2017, p. 03002 (DOI: 10.1051/e3sconf/20171603002)

[215] *King, R. R./Sherif, R. A./Kinsey, G. S./Kurtz, S./Fetzer, C. M./Edmondson, K. M./Law, D. C./Cotal, H. L./ Krut, D. D./Ermer, J. H./Karam, N. H.:* Bandgap Engineering in High-Efficiency Multijunction Concentrator Cells. In: International Conference on Solar Concentrators for the Generation of Electricity or Hydrogen, Scottsdale, Arizona/USA, 1–5 May 2005, p. 7

[216] *Cappelluti, F. et al.:* Novel Concepts for High-Efficiency Lightweight Space Solar Cells. In: E3S Web Conference, Vol. 16, 2017, p. 03007 (DOI: 10.1051/e3sconf/20171603007)

[217] *Mourra, O./Fernandez, A./Tonicello, F.:* Buck Boost Regulator (B2R) for Spacecraft Solar Array Power Conversion. In: 2010 Twenty-Fifth Annual IEEE Applied Power Electronics Conference and Exposition (APEC), pp. 1313–1319. *https://www.semanticscholar.org/paper/Buck-Boost-Regulator-(B2R)-for-spacecraft-Solar-Mourra-Fern%C3%A1ndez/9390c4992392479f7da5dfa088c21077406ce73c*

[218] *Messenger, S. R./Summers, G. P./Burke, E. A./Walters, R. J./Xapsos, M. A.:* Modeling solar cell degradation in space. A comparison of the NRL displacement damage dose and the JPL equivalent fluence approaches. In: Prog. Photovolt. Res. Appl., Vol. 9, No. 2, Mar. 2001, pp. 103–121 (DOI: 10.1002/pip.357)

[219] *Gonzales, M./Walters, R. J./Messenger, S./Cress, C./Maximenko, S.:* Modeling radiation degradation in solar cells extends satellite lifetime. *https://spie.org/news/3417-modeling-radiation-degradation-in-solar-cells-extends-satellite-lifetime?SSO=1*

[220] *SPENVIS – SPace ENVironment Information System:* Background: Radiation damage to solar cells. *https://www.spenvis.oma.be/help/background/soldam/soldam.html*

[221] Personal Correspondence with Bruno Sousa.

[222] *Sgorlon Gaiatto, C./Antonello, F./Segneri, D./Sousa, Bruno, S./Abascal Palacios, B./Schiavo, A./Bartesaghi, M./Maass, J./Inojosa, A.:* A novel physics-based computational framework to model spacecraft solar array power under degradation. Application to European Space Agency (ESA) Cluster mission. In: Acta Astronaut., Vol. 226, Jan. 2025, pp. 341–348, pp. 341–348 (DOI: 10.1016/j.actaastro.2024.10.052)

[223] *Sousa, B.:* Plot of the time variant Solar Array Power of the Cluster Spacecrafts during the mission

[224] *ESA:* Sentinel-1 impact. *https://www.esa.int/ESA_Multimedia/Images/2016/08/Sentinel-1_impact*

[225] *Yamaguchi, M.:* Radiation-resistant solar cells for space use. In: Sol. Energy Mater., 2001, p. 23

[226] SPENVIS – SPace ENVironment Information System. *https://www.spenvis.oma.be*

[227] *Messenger, S. R./Summers, G. P./Burke, E. A./Walters, R. J./Xapsos, M. A.:* Modeling solar cell degradation in space: A comparison of the NRL displacement damage dose and the JPL equivalent fluence approaches. In: Prog. Photovolt. Res. Appl., Vol. 9, No. 2, Mar. 2001, pp. 103–121 (DOI: 10.1002/pip.357)

[228] *Magalhães, R. O. de/Moreira, H. J. P.:* Space Power Topology Selection and its System Level Modeling and Control. In: J. Aerosp. Technol. Manag., No. 12, Jul. 2020, p. e2720 (DOI: 10.5028/jatm.v12.1158)

[229] *Jiang, Z./Liu, S./Dougal, R. A.:* Design and testing of spacecraft power systems using VTB. In: IEEE Trans. Aerosp. Electron. Syst., Vol. 39, No. 3, Jul. 2003, pp. 976–989 (DOI: 10.1109/TAES.2003.1238750)

[230] *Linow, S.:* Energie – Klima – Ressourcen. Quantitative Methoden zur Lösungsbewertung von Energiesystemen. Carl Hanser Verlag, München 2020

[231] *Summerer, L.:* ENSaF – Steps towards a European Space Nuclear Safety Framework. European Space Agency – ESTEC, Advanced Concepts Team, Noordwijk, Netherlands. p. 6

[232] *U.S. Department of Energy (DOE):* Space and Defense Power Systems | Department of Energy. p. 7

[233] *Fathy, M.:* Plutonium-238 oxide pellet glowing from its internal generated heat. *https://www.researchgate.net/figure/Plutonium-238-oxide-pellet-glowing-from-its-internal-generated-heat_fig1_360719626*

[234] *Maset, E./Ejea, J. B./Ferreres, A./Sanchis-Kilders, E./Jordan, J./Esteve, V.:* High-Efficiency Weinberg Converter for Battery Discharging in Aerospace Applications. In: Twenty-First Annual IEEE Applied Power Electronics Conference and Exposition, APEC '06. IEEE 2006, pp. 1510–1516 (DOI: 10.1109/APEC.2006.1620740)

[235] *Albach, M.:* Elektrotechnik. Pearson

[236] *Shinyakov, Y. A./Semyonov, V. D./Sukhorukov, M. P./Li, D./Borodin, D. B./Kabirov, V. A.:* Modernization of the Weinberg's Converter for the Implementation of a Charge-Discharge Device in the Power Supply System of a Spacecraft. In: 2018 19th International Conference of Young Specialists on Micro/Nanotechnologies and Electron Devices (EDM), IEEE, Erlagol, Russia, June 2018, pp. 533–539 (DOI: 10.1109/EDM.2018.8434944)

[237] *Kornetzky, P./Moussaoui, Z./Batarseh, I./Hawasly, S./Kennedy, C.:* Modeling technique for DC to DC converters using Weinberg topology. In: Proceedings of SOUTHEASTCON '96, IEEE, Tampa, Florida/USA 1996, pp. 551–556 (DOI: 10.1109/SECON.1996.510133)

[238] *Shtein, D. A./Klassen, S. V./Klassen, T. S./Konyakhin, V. S.:* Multiport Push-Pull/S3R/S4R DC-DC Converter for Spacecraft Power Systems. In: 2020 21st International Conference of Young Specialists on Micro/Nanotechnologies and Electron Devices (EDM), IEEE, Chemal, Russia, June 2020, pp. 400–406 (DOI: 10.1109/EDM49804.2020.9153547)

[239] *Böcker, J.:* Power Electronics. Lecture. University of Paderborn 2017

[240] *Böcker, J.:* Grundlagen Electrotechnik. Teil B

[241] *Mertsching, B.:* Basics in electrical Engineering. 2012

[242] *Kahveci, H./Çorapsız, M. R.:* Voltage Control Strategy for DC – DC Buck – Boost Converter. In: 4th Int. Conf. Adv. Nat. Appl. Sci., July 2019

[243] *Zhang, H./Quan, L./ Gao, Y.:* Dynamic modelling and small signal analysis of push-pull bidirectional DC-DC converter. In: Math. Comput. Model. 2014, p. 5

[244] *Yadav, G./Gowrishankara, C. K/Hariharan, V. K.:* Analysis and development of Weinberg converter with fault tolerant feature. In: 2016 Biennial International Conference on Power and Energy Systems. Towards Sustainable Energy (PESTSE), IEEE, Bangalore, January 2016, pp. 1–6 (DOI: 10.1109/PESTSE.2016.7516476)

[245] *Zenk, H.:* In Push-Pull Converter Output Voltage Stability Comparison with Using Fuzzy Logic, PI and PID Controllers. In: Fuzzy Log., Vol. 03, No. 12, p. 7

[246] *Terma Space:* S3R Shunt Regulation Module Datasheet. *www.terma.com*

[247] *Nelson, C.:* Theoretical considerations for buck mode switching regulators. In: Analog Circuit Design, Elsevier, 2013, pp. 89–134 (DOI: 10.1016/B978-0-12-397888-2.00005-5)

[248] *Zehendner, M./Ulmann, M.:* Power Topologies Handbook. Texas Instruments, Inc. 2016. p. 199

[249] *Green, M.:* Design Calculations for Buck-Boost Converters. 2012. p. 12

[250] *Hung, C.-Y./Lee, C. Q./Lee, H. T.:* Analysis and modeling of Weinberg converter system with output current limiter. In: IAS '95. Conference Record of the 1995 IEEE Industry Applications Conference Thirtieth IAS Annual Meeting, IEEE, Orlando, Florida/USA 1995, pp. 2430–2437 (DOI: 10.1109/IAS.1995.530612)

[251] *Hung, C.-Y./Lee, C. Q./Lee, H. T.:* Modeling of high power DC/DC converter system based on Weinberg topology. In: Proceedings of PESC '95 – Power Electronics Specialist Conference, IEEE, Atlanta, Georgia/USA 1995, pp. 1222–1228 (DOI: 10.1109/PESC.1995.474970)

[252] *Gianninoto, G./Scorzafava, E./Carlani, D.:* A New, Single Diode, Sequential Switching Shunt Regulator. In: E3S Web Conference, Vol. 16, 2017, p. 14013 (DOI: 10.1051/e3sconf/20171614013)

[253] *Ibrahim, O./Yahaya, N. Z./Saad, N.:* Comparative studies of PID controller tuning methods on a DC-DC boost converter. In: 2016 6th International Conference on Intelligent and Advanced Systems (ICIAS), IEEE, Kuala Lumpur, Malaysia, August 2016, pp. 1–5 (DOI: 10.1109/ICIAS.2016.7824044)

[254] *Silva, J. F./Pinto, S. F.:* Control Methods for Switching Power Converters. In: Power Electronics Handbook, Elsevier, 2007, pp. 935–998 (DOI: 10.1016/B978-012088479-7/50052-3)

[255] *Al-Ogaili, A. S./bin Aris, I./Othman, M. L. B./Hoon, Y./Isa, D./Azis, N. B./Abbas, H. F.:* A Review of Different Control Methods in Power Electronic. In: Journal of Engineering and Applied Sciences 13 (Special Issue 1) 2018, p. 13

[256] *Böcker, J.:* Power Electronics for Energy Transition. Lecture

[257] *Li, Z.-Q./Zhang, X.-L./Wen, X.:* Simulation Research on Control Strategy for Push-pull Forward DC-DC Converter. In: Proceedings of the 2016 International Conference on Computer Engineering and Information Systems. Atlantis Press, Shanghai 2016 (DOI: 10.2991/ceis-16.2016.42)

[258] *Vidal-Idiarte, E./Restrepo, C./El Aroudi, A./Calvente, J./Giral, R.:* Digital Control of a Buck Converter Based on Input-Output Linearization. An Interpretation Using Discrete-Time Sliding Control Theory. In: Energies, Vol. 12, No. 14, July 2019, p. 2738 (DOI: 10.3390/en12142738)

[259] *El-Sharif, I./Zerek, A. R./Hareb, F. O.:* Design of discrete-time PID controller. In: Int. Conf. Control. p. 6.

[260] *Toochinda, D.:* Discrete-time PID Controller Implementation. Rungis, France, March 15, 2022. *https://www.scilab.org/discrete-time-pid-controller-implementation*

[261] *Hlaili, M./Mechergui, H.:* Comparison of Different MPPT Algorithms with a Proposed One Using a Power Estimator for Grid Connected PV Systems. In: Int. J. Photoenergy, January 2016, pp. 1–10 (DOI: 10.1155/2016/1728398)

[262] *Quevedo, D. E.:* Advanced Control. 2018

[263] *Quevedo, D. E.:* Advanced Systems Theory

[264] *Mannes, Q.:* A Hybrid PWM Control Method for the Sequential Switching Shunt Regulator (HPWM-S3R). 2018 (DOI: 10.13140/RG.2.2.11156.76162)

[265] *Barreras, J. V./Schaltz, E./Andreasen, S. J./Minko, T.:* Datasheet-based modeling of Li-Ion batteries. In: 2012 IEEE Vehicle Power and Propulsion Conference, IEEE, Seoul, South Korea, October 2012, pp. 830–835 (DOI: 10.1109/VPPC.2012.6422730)

[266] *MIT Electric Vehicle Team:* A Guide to Understanding Battery Specifications. 2008. p. 3. *http://web.mit.edu/evt/summary_battery_specifications.pdf*

[267] *NASA/Howell, T.:* In Their Own Words. Preserving International Space Station Knowledge. *https://appel.nasa.gov/2009/09/01/in-their-own-words-preserving-international-space-station-knowledge*

[268] *Bugga, R./Smart, M./Whitacre, J./West, W.:* Lithium Ion Batteries for Space Applications. In: 2007 IEEE Aerospace Conference, Big Sky, Montana/USA 2007, pp. 1–7 (DOI: 10.1109/AERO.2007.352728)

[269] *McKissock, B./Loyselle, P./Vogel, E.:* Guidelines on Lithium-ion Battery Use in Space Applications. 2009. p. 54

[270] *Dalton, P. J./Balcer, S.:* International Space Station Lithium-Ion Battery. 2016. *https://ntrs.nasa.gov/citations/20160012048*

[271] *Hinz, H.:* Comparison of Lithium-Ion Battery Models for Simulating Storage Systems in Distributed Power Generation. In: Inventions, Vol. 4, No. 3, August 2019, p. 41 (DOI: 10.3390/inventions4030041)

[272] *Johnson, V. H.:* Battery performance models in ADVISOR. In: J. Power Sources, Vol. 110, No. 2, August 2002, pp. 321–329 (DOI: 10.1016/S0378-7753(02)00194-5)

[273] *Sousa, B.:* Plot of the time variant Battery Capacity of the Cluster Spacecrafts during the mission

[274] *He, H./Xiong, R./Fan, J.:* Evaluation of Lithium-Ion Battery Equivalent Circuit Models for State of Charge Estimation by an Experimental Approach. In: Energies, Vol. 4, No. 4, March 2011, pp. 582–598 (DOI: 10.3390/en4040582)

[275] *Tremblay, O./Dessaint, L.-A.:* Experimental Validation of a Battery Dynamic Model for EV Applications. In: World Electr. Veh. J., Vol. 3, No. 2, June 2009, pp. 289–298 (DOI: 10.3390/wevj3020289)

[276] *Tremblay, O./Dessaint, L.-A./Dekkiche, A.-I.:* A Generic Battery Model for the Dynamic Simulation of Hybrid Electric Vehicles. In: 2007 IEEE Vehicle Power and Propulsion Conference, IEEE, Arlington, Texas/USA, September 2007, pp. 284–289 (DOI: 10.1109/VPPC.2007.4544139)

[277] *ABSL:* Datasheet 3s4p

[278] *ABSL:* Datasheet 8s 104p

[279] *ABSL:* Datasheet 8s3p

[280] *ABSL:* Datasheet 8s10p

[281] *ABSL:* Datasheet 8s16p

[282] *ABSL:* Datasheet 8s20p

[283] *ABSL:* Datasheet 8s52p

[284] *ABSL:* Datasheet 72 Ah Li-Ion Cell

[285] *ABSL:* Datasheet 8s16p

[286] *ABSL:* Datasheet 8s3p

[287] *EaglePicher:* Datasheet SZHR25

[288] *EaglePicher:* Datasheet SCL-21060-003

[289] *EaglePicher:* Datasheet SLC-201

[290] *EaglePicher:* Datasheet SLB-101

[291] *EaglePicher:* Datasheet SAR-10215

[292] *EaglePicher:* Datasheet SAR-10197

[293] *EaglePicher:* Datasheet SAR-10199

[294] *EaglePicher:* Datasheet SAR-10207

[295] *EaglePicher:* Datasheet LP-32975

[296] *EaglePicher:* Datasheet LP-33081

[297] *EaglePicher:* Datasheet LP-33165

[298] *EaglePicher:* Datasheet LP-33330

[299] *EaglePicher:* Datasheet LP-33450

[300] *EaglePicher:* Datasheet LP-33333

[301] *EaglePicher:* Datasheet LP-34100

[302] *EaglePicher:* Datasheet LiBG18EV-1

[303] *EaglePicher:* Datasheet SLC-16050

[304] *EaglePicher:* Datasheet SZHR50

[305] *EaglePicher:* Datasheet LP-33037

[306] *EAS:* Datasheet HE341440NCA10

[307] *EAS:* Datasheet HE602030NCA55

[308] *EAS:* Datasheet HP601300LFP22

[309] *EAS:* Datasheet HP602030LFP40

[310] *EAS:* Datasheet MP602030LFP50

[311] *EAS:* Datasheet UHP341440NCA

[312] *GS Yuasa:* Datasheet LSE Gen 3 Energy

[313] *GS Yuasa:* Datasheet LSE Gen 3 Power

[314] *Nakagawa, Y./Fuse, H./Honda, G./Segawa, M./Shizuki, T./Watanabe, K./Naito, H.:* Development of generation IV large format Li-ion cells for space applications. 2019. p. 6

[315] *Saft:* Datasheet 4S1P

[316] *Saft:* Datasheet Overview

[317] *Saft:* Datasheet BA685D

[318] *Saft:* Datasheet BA685C

[319] *Saft:* Datasheet VES16

[320] *Chen, M./Rincon-Mora, G. A.:* Accurate electrical battery model capable of predicting runtime and I-V performance. In: IEEE Trans. Energy Convers., Vol. 21, No. 2, June 2006, pp. 504–511 (DOI: 10.1109/TEC.2006.874229)

[321] *Shepherd, C. M.:* Design of Primary and Secondary Cells – Part 2. An Equation Describing Battery Discharge. In: J. Electrochem. Soc., July 1965. *http://citeseerx.ist.psu.edu/viewdoc/download?doi=10.1.1.1026.7710&rep=rep1&type=pdf*

[322] *Iannicelli, F./Luisi, D. F.:* Modelling, simulation and experimentation of a regenerative energetic system for space applications. Politecnico di Torino 2016

[323] *Mutschke, G./Eckert, K.:* Grüner Wasserstoff. Forschung für mehr Effizienz. 9. Dezember 2019. *https://www.hzdr.de/db/Cms?pOid=59948&pNid=0&pLang=de*

[324] *Brinkert, K./Richter, M. H./Akay, Ö./Liedtke, J./Giersig, M./Fountaine, K. T./Lewerenz, H.-J.:* Efficient solar hydrogen generation in microgravity environment. In: Nat. Commun., Vol. 9, No. 1, December 2018, p. 2527 (DOI: 10.1038/s41467-018-04844-y)

[325] *Bashkatov, A./Hossain, S./Yang, X./Mutschke, G./Eckert, K.:* Oscillating Hydrogen Bubbles at Pt Microelectrodes. In: Phys. Rev. Lett., Vol. 123, No. 21, November 2019, p. 214503 (DOI: 10.1103/PhysRevLett.123.214503)

[326] *Walter, U.:* Höllenritt durch Raum und Zeit. Astronaut Ulrich Walter erklärt die Raumfahrt. 2017

[327] *NASA:* Advanced Space Power Systems (ASPS). Regenerative Fuel Cells (RFC). 2015. *https://techport.nasa.gov/view/13652*

[328] *Gietl, E. B.:* The Electric Power System of the International for Power. In: Technol. Dev. 2000, p. 12

[329] *ESA/Azimut Space:* Artificial regolith brick in a vacuum chamber. *https://www.esa.int/ESA_Multimedia/Images/2019/07/Artificial_regolith_brick_in_a_vacuum_chamber*

[330] *Malaviya, P./Sarvaiya, V./Shah, A./Thakkar, D./Shah, M.:* A comprehensive review on space solar power satellite: an idiosyncratic approach. In: Environ. Sci. Pollut. Res., Vol. 29, No. 28, Jun. 2022, pp. 42476–42492 (DOI: 10.1007/s11356-022-19560-w)

[331] *Summerer, L./Jacques, L.:* IAC-11-C3.1.3 Prospects for Space Solar Power in Europe. p. 15

[332] *Fay, E. H./Cull, R. C.:* Lunar Orbiting Microwave Beam Power System. NASA Tech. Memorandum 103211. 1990

[333] *Rodenbeck, C. T./Jaffe, P./Strassner, B. H./Hausgen, P. E./Mcspadden, J./Kazemi, H./Shinohara, N./Tierney, B. B./DePuma, C. B./Self, A. P.:* Microwave and Millimeter Wave Power Beaming. In: IEEE J. Microw., Vol. 1, No. 1, January 2021, pp. 229–259 (DOI: 10.1109/JMW.2020.3033992)

[334] *Chattopadhyay, G./Manohara, H./Mojarradi, M./Vo, T./Mojarradi, H./Bae, S./Marzwell, N.:* Millimeter-wave wireless power transfer technology for space applications. In: 2008 Asia-Pacific Microwave Conference, IEEE, Macau, December 2008, pp. 1–4 (DOI: 10.1109/APMC.2008.4957899)

[335] *Grandidier, J./Wright, M. W./Krut, D./Nesmith, B. J./Brophy, J. R.:* Power Beaming for Deep Space and Permanently Shadowed Regions. 2020

[336] *Hartbrich, I.:* Raus aus dem Erdschatten. In: VDI Nachrichten, No. 22, November 2022

[337] *Baraskar, A./Yoshimura, Y./Nagasaki, S./Hanada, T.:* Space solar power satellite for the Moon and Mars mission. In: J. Space Saf. Eng., Vol. 9, No. 1, Mar. 2022, pp. 96–105 (DOI: 10.1016/j.jsse.2021.10.008)

[338] *ESA:* Stages of space-based solar power. *https://www.esa.int/Space_in_Member_States/United_Kingdom/ESA_accelerates_the_race_towards_clean_energy_from_space*

[339] *ESA:* Concept for an in-orbit demonstrator of a Space-Based Solar Power beaming satellite. *https://www.esa.int/Enabling_Support/Preparing_for_the_Future/Discovery_and_Preparation/ESA_reignites_space-based_solar_power_research*

[340] *NASA:* Reference Guide to the International Space Station. Utilization Edition.

[341] *European Space Agency:* Columbus. Europe's Laboratory on the International Space Station

[342] *NASA:* Spacewalkers Thomas Pesquet of ESA (European Space Agency) and Akihiko Hoshide of JAXA (Japan Aerospace Exploration Agency) set up the 4A channel on the International Space Station's P4 (Port) truss segment for the installation of an roll-out solar array. *https://www.nasa.gov/international-space-station/international-space-station-assembly-elements*

[343] *Richardson, D.:* Problems prevent astronauts from deploying new solar array. June 16, 2021. *https://www.orbital-velocity.com/news/problems-prevent-spacewalkers-from-deploying-new-solar-array*

[344] *Garcia, M.:* NASA Spacewalkers Finish Installing Roll-Out Solar Array. June 15, 2023. *https://blogs.nasa.gov/spacestation/2023/06/15/nasa-spacewalkers-finish-installing-roll-out-solar-array*

[345] *Dalton, P. J./Bowens, E./North, T./Balcer, S.:* International Space Station Lithium-Ion Battery Status. 2019. *https://ntrs.nasa.gov/citations/20200000004*

[346] *Malik, T.:* NASA picks 3 companies to build vertical solar arrays for the moon. August 31, 2022. *https://www.space.com/moon-solar-arrays-companies-nasa-picks*

[347] *Schwab, A. J.:* Elektroenergiesysteme. Smarte Stromversorgung im Zeitalter der Energiewende. Springer, Berlin/Heidelberg, 2020 (DOI: 10.1007/978-3-662-60374-1)

[348] ESA – Spaceship EAC: Render of Flexhab

[349] *NASA:* Render of a Nuclear Fisson Reactor on the Moon. *https://www.nasa.gov/wp-content/uploads/2023/03/kilopower_moon_render_1.jpg*

[350] *E. S. R. and T. C. Thermal Control and Life Support Division:* ESA PSS-03-40, Vol. 2, Issue 1, Environmental Control and Life Support

[351] *E. S. R. and T. C. Power and Energy Conversion Division:* ESA PSS-02-10, Vol. 1, Issue 1, Power Standard

[352] *E. S. R. and T. C. Power and Energy Conversion Division:* ESA PSS-02-10, Vol. 2, Issue 1, Rationale for the Power Standard

[353] *Tonicello, F./Mauret, B./Jensen, H./Croci, L./Deplus, N./Neugnot, N./Rumler, P./Kocma, T.:* ECSS-E-ST&HB-20-20C – A Power Interface Standard & Handbook for Products Development. In: E3S Web Conference, Vol. 16, 2017, p. 13001 (DOI: 10.1051/e3sconf/20171613001)

[354] *NASA:* Constellation Program Human-Systems Integration Requirements. March 2009

[355] *Tonicello, F.:* ECSS on Power Interface Standardisation. Introduction

[356] *ESA/Foster + Partners:* 3D-printed lunar base design. *https://www.esa.int/Enabling_Support/Space_Engineering_Technology/Building_a_lunar_base_with_3D_printing*

[357] *Bastien, L.:* Novel materials and approaches for exploration spacesuits

[358] *Jensen, N. E./Battrick, B.:* ESA Satellite Power Systems. 2003

[359] *Asif, S.:* Evolutionary Computation Based Multi-Objective Design Search and Optimization of Spacecraft Electrical Power Subsystems. 2008. p. 151

Index

A

B

C

R

S

www.ingramcontent.com/pod-product-compliance
Lightning Source LLC
LaVergne TN
LVHW081928200926
848484LV00009B/2393